Editor
Michel Étienne
Institut de la Recherche Agronomique
Avignon
France

ISBN 978-94-017-8556-3 ISBN 978-94-017-8557-0 (eBook)
DOI 10.1007/978-94-017-8557-0
Springer Dordrecht Heidelberg New York London

Library of Congress Control Number: 2013955561

Éditions Quæ, R10, 78026 Versailles cedex, France www.quae.com

Printed on acid-free paper

Springer is part of Springer Science+Business Media (www.springer.com)

Michel Étienne
Editor

Companion Modelling

A Participatory Approach to Support Sustainable Development

Springer

Companion Modelling

Foreword

The goal of this book is to get the reader familiar with the companion modelling approach by reporting the scientific stance that founds it, the tools and methods used, the specific role that models play in it, and the ways of integrating several levels of decision, taking into account power games and evaluating the impact of the process on the stakeholders involved. The expression 'companion modelling' is used when referring to the global approach; it is replaced by the 'ComMod process' when dealing with a specific application. The ComMod acronym comes from the English 'Companion Modelling' and identifies the team of researchers that elicited and formalized the approach on the basis of a set of case studies dealing with the interactions between nature and society (i.e. socio-ecological systems). The book is complementary to the handbook (Daré et al. 2009), which provides methodological skills for those who wish to apply a companion modelling approach and agree to respect the key ethical principles stated in the 'ComMod Charter'.

This person is called the 'commodian' in all the chapters of the book. The approach systematically involves a group of people who directly participate in its development (i.e. partners), and among these some share the central issue with the commodians (i.e. stakeholders). From time to time, specific events permit the team to share viewpoints (i.e. collective key moments). As the leader of the ComMod process, the commodian can be considered either as the 'designer' when he decides how to run the approach or as the 'facilitator' when he is acting in a collective key moment. A ComMod process can go through several iterations tackling different issues in the same study case; each iteration is called a 'loop'. Inside a loop, the process goes through several steps called 'phases'.

Reference

Daré W, Ducrot R, Botta A, Étienne M (2009) Repères méthodologiques pour la mise en oeuvre d'une démarche de modélisation d'accompagnement. Cardère Éditions, Laudun, p 127

Acknowledgments

This book is one of the achievements of the Commod project "Companion Modelling: A Research Practice to Support Sustainable Development", realized with the financial support of The French National Research Agency, in the framework of the programme 'Agriculture and Sustainable Development' (ANR-05-PADD-007).

It is the output of a reflection on the many issues discussed since 2000 among the members of the ComMod group, and of the workshops and seminars during which companion modelling was discussed and criticized by French and foreign researchers.

Finally, this book could not have been published without the intense involvement of project managers, institutional partners and local stakeholders who accepted to lead and share with us, many times during several years, a ComMod process in each of the 27 study cases that make up the primary material of this book.

Contents

Authors

Géraldine Abrami UMR G-EAU, IRSTEA, Montpellier, France, e-mail: geraldine.abrami@irstea.fr

Martine Antona UR GREEN, CIRAD, Montpellier, France, e-mail: martine.antona@cirad.fr

Patrick d'Aquino UR GREEN, CIRAD, Dakar, Senegal, e-mail: patrick.daquino@cirad.fr

Sigrid Aubert UR GREEN, CIRAD, Antananarivo, Madagascar, e-mail: sigrid.aubert@cirad.fr

Cécile Barnaud UMR DYNAFOR, INRA, Toulouse, France, e-mail: cecile.barnaud@toulouse.inra.fr

Olivier Barreteau UMR G-EAU, IRSTEA, Montpellier, France, e-mail: olivier.barreteau@irstea.fr

Nicolas Becu AGILE, Université de La Rochelle, La Rochelle, France, e-mail: nicolas.becu@univ-lr.fr

Pierre Bommel UR GREEN, CIRAD, Brasilia, Brazil, e-mail: pierre.bommel@cirad.fr

Aurélie Botta UR GREEN, CIRAD, Montpellier, France, e-mail: aurelie.botta@cirad.fr

François Bousquet UR GREEN, CIRAD, Montpellier, France, e-mail: francois.bousquet@cirad.fr

William's Daré UR GREEN, CIRAD, Montpellier, France, e-mail: william's.dare@cirad.fr

Anne Dray Swiss Federal Institute of Technology, Zurich, Switzerland, e-mail: anne.dray@usys.uthz.ch

Raphaèle Ducrot UMR G-EAU, CIRAD, Maputo, Mozambique, e-mail: raphaele.ducrot@cirad.fr

Michel Étienne Ecodevelopment Unit, INRA, Avignon, France, e-mail: etienne@avignon.inra.fr

Stefano Farolfi UMR G-EAU, CIRAD, Maputo, Mozambique, e-mail: stefano.farolfi@cirad.fr

Christine Fourage UMR ESO-CARTA, Université Catholique, Angers, France, e-mail: cfourage@uco.fr

Natalie Jones University of Queensland, Townsville, Australia, e-mail: n.jones3@uq.edu.au

Erwann Lagabrielle University of Port-Elizabeth, Port Elizabeth, South Africa, e-mail: erwann.lagabrielle@gmail.com

Christophe Le Page UR GREEN, CIRAD, Montpellier, France, e-mail: christophe.le_page@cirad.fr

Raphaël Mathevet CEFE-CNRS, Montpellier, France, e-mail: raphael.mathevet@cefe.cnrs.fr

Claude Monteil UMR DYNAFOR, INP-ENSA, Toulouse, France, e-mail: monteil@ensat.fr

Jean-Pierre Müller UR GREEN, CIRAD, Montpellier, France, e-mail: jean-pierre.muller@cirad.fr

AnneMarie Van Paassen University of Wageningen, Wageningen, The Netherlands, e-mail: Annemarie.vanPaassen@wur.nl

Pascal Perez SMART, University of Wollongong, Wollongong, Australia, e-mail: pascal.perez@uow.edu.au

Jérôme Queste UR GREEN, CIRAD, La Réunion, France, e-mail: jerome.queste@cirad.fr

Véronique Souchère UMR SADAPT, INRA, Grignon, France, e-mail: souchere@grignon.inra.fr

Guy Trébuil UR GREEN, CIRAD, Montpellier, France, e-mail: guy.trebuil@cirad.fr

Chapter 1
Introduction

François Bousquet, Michel Étienne and Patrick d'Aquino

In 1996, an interdisciplinary group of researchers working the field of renewable resources management set out the first components, of an approach named 'companion modelling' (Bousquet et al. 1996; Barreteau et al. 1997). For many years these researchers had been involved in environmental research, such as the environment programme of the National Centre for Scientific Research (CNRS) or the Institut de recherche pour le développement (IRD) research programme 'Dynamics and Use of Renewable Resources' (Gillon et al. 2000), which was based on research carried out in the 1980s, such as the CNRS action 'Ecosystems and Social Systems' (Jollivet 1992). Among the numerous conclusions produced by this research, was the recommendation to go beyond the multi-disciplinary juxtaposition and to commit to a cross-disciplinary approach in addressing environmental issues, and the proposal to rely, whenever possible, on the modelling method as a catalyst for the interaction process between researchers from different disciplines. During the same period, groups of researchers made exploratory inroads into new modelling tools and their suitability for facilitating interdisciplinary dialogue by creating a common representation. Based on the theoretical foundations of the sciences of complexity, methods such as multi-agent systems (MAS) (Bousquet and Le Page 2004), individual-based modelling (Grimm 1999) and micro-simulation emerged within some disciplinary communities. The

F. Bousquet (✉)
UR GREEN, CIRAD, Montpellier, France
e-mail: francois.bousquet@cirad.fr

M. Étienne
Ecodevelopment Unit, INRA, Avignon, France
e-mail: etienne@avignon.inra.fr

P. d'Aquino
UR GREEN, CIRAD, Dakar, Senegal
e-mail: patrick.daquino@cirad.fr

M. Étienne (ed.), *Companion Modelling*, DOI: 10.1007/978-94-017-8557-0_1,

research presented here is based on these findings as well as on the sharing of a few premises.[1]

- The socio-ecological systems that we study are complex objects and, therefore, supporting the decision-making process does not involve attempting to predict the future state of the system. It is more akin to understanding the organization in which it is found, to envision the organizations sought, to encourage the system of interactions that govern change, to monitor constantly and render explicit the changes in the system, to be able to suggest adaptations and to learn continuously by observing their effects.
- Every stakeholder in a social system has his own view of the reality of the system, a point of view that he has built up on what he has experienced during his life trajectory in the physical and social space. These constructions come from, and constitute, the system of representations specific to the culture to which the stakeholder belongs (Friedberg 1992). Where key issues and high uncertainties lead to constructions that are not just conditioned by scientific facts but also by the values of stakeholders, Funtowicz and Ravetz (1993) suggested using a post-normal approach. Then, decision quality relies on the quality of the decision-making process itself, among other things, the existence of prior dialogue between stakeholders, not just to check that these decisions are acceptable but also to construct them together.

The first experiments proposed models incorporating various types of disciplinary knowledge and were based on multi-agent modelling (Barreteau 1998). From a basic principle, to recognize and formalize the diversity of viewpoints in a complex system, rose other experiments aiming at interaction between the various bearers of knowledge, be they researchers or local stakeholders, using different tools, such as role-playing games and simulation models. The approach set up, the subject of this book, is designed as an iterative and sustained interaction process between scientists and other stakeholders involved in renewable resource management. It aims at structuring, even integrating, this heterogeneous set of knowledge into a comprehensive synthesis that helps in settling a dispute. Let us emphasize that the creation of a common representation does not aim at substituting it for plural representations; it is more a question of developing an agreement whereby different viewpoints can be expressed. Knowledge can involve natural dynamics as well as social dynamics or their interactions. Highlighting the various representations produced forces the stakeholders into an awareness of the diversity of individual viewpoints, to share them, enrich them, raise doubts over them and discuss them during collaborative exchanges of views from which a shared representation can emerge. As the scientists in this perspective are considered stakeholders like anyone else, this type of approach implies that their initial analyses can also be questioned.

[1] For fuller discussions on the theoretical foundations of companion modelling, please refer to Collectif ComMod (2009).

The research developed during the last decade was carried out in parallel with other developments in the field of participatory modelling. Although we do not consider participatory mapping here as it does not include the simulation of ecological and social processes, we retain from the conclusions of Fox (1998) that the formalization of space is not consistent with the flexible and fuzziness properties of boundaries. It also violates the right to keep the information confidential. We also question the influence of these activities on power relationships among stakeholders (Abbot et al. 1998; Chambers 2006). A tentative typology of participatory modelling experiences can be based on the seminal work of various initiators such as the group model Building (Vennix 1996) and mediated modelling (van den Belt 2004), or can be based on tools such as Bayesian networks, system dynamics or MAS. These typologies are not very useful because they hide similarities and differences among the approaches. We have selected here some relevant experiences. Costanza and Ruth (1998) proposed a three-steps approach (from an abstract model to a contextualized model), each step involving the stakeholders. Other scholars proposed the reverse approach starting from contextualized models. There are different categories of stakeholders (Hare and Pahl-Wostl 2002) that are involved in different ways as noticed by Lynam et al. (2007) after Pretty (1995) and Arnstein (1969). Pahl-Wostl and Hare (2004) assessed the impact of participatory modelling through the concept of social learning, while other scholars focused on organizational or technical transformations. In 2001, Korfmacher (2001) proposed general rules for participatory modelling: a transparent process, continuous involvement, appropriately representative involvement, influence of stakeholders in modelling decisions, and assessment of the modelling role in management. These guidelines have been confirmed in recent publications (Reed 2008; Voinov and Brown 2008). These controversies and guidelines orientated our research and stimulated the comparisons and synthesis proposed in this book.

The Beginnings

Two applications put these conceptual constructions to the test and, through their complementarity, laid the foundations for future experiments. In 1998, Barreteau and colleagues, who had recently developed a computer simulation model on the dynamics of irrigated perimeters in the Podor region of Senegal, used a role-playing game to present this model to the stakeholders with whom he had worked (Barreteau and Bousquet 1999; Barreteau et al. 2001). The task was to simplify the model to make it playable whilst maintaining the complexity of decisions and interactions, and allowing processes comparable to those actually observed to be revealed in the dynamics of the game. The stakeholders played, discussed the roles assigned to them, the system of interactions they could activate, the simplified representation of the biophysical model, and the global observations at the scale of the irrigated perimeter. This revealed characteristics of the system such as, among others, the crop success rate, the conflicts in accessing water and the problems of credit management.

A role-play computer model was thus developed and used with local stakeholders to explore rapidly various scenarios. A few months later, d'Aquino and colleagues, who were working on land use and allocation plans in the Senegal river delta region as part of the decentralization of natural resource management to the rural councils, organized a companion modelling workshop with this fledgling organization (d'Aquino et al. 2002c, 2003). This time the idea was to develop with—different stakeholders (i.e. herdsmen, fishermen, farmers) a shared representation expressing their multiple viewpoints. A three-day workshop was organized. On the first day, the protagonists constructed a space-resource model, which they shared and used to create a list of rules specific to each user. On the second day, this knowledge was tested with a role-playing game mobilizing the stakeholders in a situated dynamic using the rules identified on the previous day. The actual problems encountered came to light and the discussion focused on the confrontations of rationalities and the scenarios that could potentially accommodate them. On the third day, a computer model, the numeric equivalent of the role-play, was used to initiate discussion on the consequences of these scenarios.

Identity of the ComMod Network: The Charter

Following these first attempts that served to test the application of the principles enacted, and which allowed the articulated organization of theoretical research phases, field experiments and the design of appropriate tools, it proved necessary early in the 2000s to formalize the method accurately in the first instance, and above all, to specify the particular stance of the researchers involved in this 'companion modelling'. The fundamental principles of this gradually co-constructed approach (i.e. formalization of a diversity of viewpoints, scientific knowledge considered as one point of view among others in the consultation, priority on the iterative consultation process rather than on its products) needed to be formalized clearly to justify better the group's methodological choices. Documents were produced on the participatory process (d'Aquino et al. 2002a) and its ethical rules with a charter produced and published in English (ComMod Group 2003) and French (Collectif ComMod 2005) in the journal *Natures Sciences Sociétés*, where it was commented on by reviewers from different disciplines.

The ComMod stance is based on a dynamic perception of the decision-making process, considered as 'the result of an interaction process between individuals and/or collective actors with different weights and representations in negotiation' (Weber 1995b). The aim of the ComMod process is either to produce knowledge (intended for researchers and local stakeholders) under an improved understanding of a system of inter-actions, or to support negotiation under a process explicitly targeting a transformation of practices or forms of social and economic interactions. The approach uses modelling and simulation tools to construct a shared representation (which does not mean unifying) of the system studied, account for its dynamics and provide support for analysing scenarios. Lastly, the ComMod approach assumes a researcher stance, which we believe must be stated imperatively.

Producing Knowledge About Complex Systems

Few collective decisions involving interactions between a social group and its environment are predictable in technical, economic or social terms. This unpredictability suggests a need for a different approach, one that accepts the incompleteness of analyses and the subjectivity of future choices, which justifies the existence of potentially contradictory viewpoints and allows them to be taken into account and reassessed. The objective here is learning about the existence of these different viewpoints and the—consequences of their diversity on the functioning of the system.

When one or more stakeholders in a resource management system expresses the wish to gain a better understanding of the functioning of the system, the ComMod approach is to construct a representation of this system in one or more diverse forms (e.g. diagrams, maps, simulation models, role-playing games, videos). This co-construction normally unites a certain number of stakeholders in constructing this representation before submitting it to other stakeholders for comments, challenges or modifications. The preliminary phases such as identifying the issue, wording the question and selecting the stakeholders concerned are integral parts of the ComMod approach.

Supporting the Collective Decision-Making Process

The objective can be to go beyond sharing viewpoints, to committing to a process with the explicit purpose of modifying the functioning of the system. This objective can be issued directly or after previous work on the sharing of knowledge by all stakeholders. The support is upstream of the technical decision, to boost the thinking of the various stakeholders involved to reach a shared representation and possible routes towards engaging in a social process of taking charge of identified problems. In this instance, this involves giving the community the means of taking over the uncertainties of the situation examined jointly in the best possible way. The ComMod approach thus aims at encouraging the quality of the process behind the decision and at establishing conditions for monitoring and possibly revising it.

Models as Support Tool

The researchers who develop and use this companion modelling approach propose various modelling tools, such as diagrams, maps, videos, etc. Note, however, that in practice most operations have used MAS models to conceptualize a representation, which has then been converted into real-life computer simulations and role-playing games. These two tools are frequently combined: role-playing games are a simulation where the stakeholders play a role whereas computer simulations use virtual agents. The aim is to clarify and share viewpoints on the situation studied. The use of models is reflexive: the stakeholders learn together by creating, modifying or observing the models. Stakeholders can use these tools to issue hypotheses, suggest scenarios and jointly observe the consequences. It is sometimes said that these models are 'disposable', representations shared between a group of stakeholders at a given moment. They are frequently an extremely simplified representation of the problem yet sufficient to reflect the complexity of the system by taking the main dynamics and interactions into account.

The Position of the Researcher

The researcher is found in several positions in the companion modelling process. He is firstly a researcher in the classical sense, inasmuch as he produces, with other stakeholders, knowledge on the management context and on the participatory process. His results lend themselves to rebuttal as he provides the elements which led to the conclusions issued. However, as already stated, he is also a stakeholder in the system and his role can be questioned. The ComMod Charter provides an ethical framework highlighting the following points:

- the transparency of hypotheses and underlying procedures; graphic and spatial modelling and role-playing games have been developed with this very much in mind.
- the clear display of domains of use in the models developed.
- the researcher's involvement in the process; the researcher who adopts this approach is a bearer of knowledge like anyone else, although he frequently plays a singular role in initiating and facilitating the process.
- the ongoing undermining of the proposed process, be it by local stakeholders or by scientists; this undermining takes concrete shape in many circumstances through failure to commit to a ComMod process as the analysis of the social context brought risks to light, or by halting an ongoing process due to the refusal by key stakeholders to take part in the process.

The publication of this charter in *Natures Sciences Sociétés* gave rise to miscellaneous comments, which are presented in Chap. 3, principally the status of the scientific knowledge in the dialogue engaged. Although debating this viewpoint is consistent with the foundations laid down, should this scientific knowledge be placed on the same level as the knowledge of other stakeholders or should it be debated differently?

Diversity

From the early 2000s, many new applications have been carried out in various countries worldwide concerning different resources and in various ecological and social contexts. New researchers joined the network, each of them performing their experiments using a method they considered suitable, by organizing *ad hoc* stages and mobilizing specific modelling tools.

Let us observe a significant trajectory to illustrate the diversity of companion modelling implementation. Based on over 20 years research into forestry-pasture development and forest-fire prevention, the regional grassland specialist departments in Provence and Languedoc have set up several operations combining livestock breeding and forest-fire prevention. Faced with the sheer size of the areas involved and the emerging potential conflict of usage between foresters, breeders, hunters and other users of the Mediterranean forests, the Ministry of Agriculture (via the Groupement d'intérêt scientifique Incendies de forêts) and the French Forestry Commission (Office national des forêts—ONF) (under the forestry-pasture programme of the Var Department), sought a tool to facilitate consultation between these various stakeholders. The Ecodevelopment Unit of the French National Institute for Agricultural Research (INRA) suggested, therefore, constructing a model for use didactically in the three main types of forest in the Var, that is, cork forest with maquis undergrowth, Pin d'Alep pine forest with Kermes oak undergrowth, and white oak forest with broom undergrowth. This experiment illustrated the first attempt at companion modelling applied to forest development (Étienne 2003). The approach was divided into four phases:

- integration of available scientific knowledge on forest dynamics, sensitivity to fires, brush-clearing techniques and forestry and breeding practices most commonly used in this type of environment in the form of a multi-agent computer simulation model
- simplification of this model from simple management entities (i.e. forest plots, grazing units, fuel breaks) and three virtual territories, each representing major characteristics of the three types of forest
- situation simulation exercise for stakeholders taking part in most of the silvo-pastoral management plans, in the context of a role-playing game in a fictitious forest close to their real-life situation, to make them react to the forest dynamics and effects of grazing on these dynamics
- reconversion and adaptation of the role-playing game for use in teaching students in forestry, agronomic or veterinary colleges.

The feedback on this case study carried out in the 2000 s led to several institutions requesting adaptations of this type of approach for similar issues. In December/2005, the Gard Departmental Agricultural and Forestry Service (Direction départementale de l'Agriculture et de la Forêt—DDAF) suggested tackling the forest-fire prevention problem at the peri-urban interface. The Environment Service of Nîmes-Métropole Urban Community, keen to raise the awareness of its elected representatives to this issue, offered its area as a test zone. The approach was divided into four phases:

- compilation of available mapping data on the forest, the dynamics of urbanization and practices of the main local stakeholders (i.e. farmers, urban developers and foresters)
- development of a virtual map representing three typical adjacent municipalities in the northern Nîmes area and validation of this map by a group of technicians covering the main activities of the area
- co-construction, with the same group, of a conceptual model representing the current functioning of this area and the likely dynamics over the next 15 years, then implementation of this model by INRA researchers as a multi-agent computer simulation model
- A situation simulation exercise for elected representatives from 14-municipalities involved in discussions on the issue of forest-fire prevention in conjunction with urbanization, in the context of five sessions of role-playing games involving an urban developer, three mayors, a DDAF technician and a representative of Nîmes-Métropole.

The initiative culminated in a collective awareness of the implications, in terms of fire, of expanding urbanization at the expense of agricultural wastelands and natural areas. The importance of reflecting collectively on setting up fire-prevention devices was identified clearly, but the elected representatives debated long and hard on the failure to integrate these systems within urban planning projects and the lack of financial resources to maintain them. The quality of the interactions and learning prompted the various stakeholders to agree to it being standardized at district scale and to provide financing for this purpose.

This account of two companion modelling operations raises the question of the essential facet of this approach. Here administrative bodies place an order with recognized research for its expertise on a specific theme, there other administrative bodies call on the researcher's methodological skills to lead a dialogue and raise awareness of certain stakeholders. In one case, the aim was to trigger dialogue between users of spaces and resources (i.e. foresters and breeders) with very different powers, in the other it was to raise the awareness of elected representatives, armed with their decision-making and management powers, with economic strategy stakeholders (urban developers). In the first operation, the researchers summarized the scientific knowledge, incorporated it into a computer model and used a simplified role-playing game to bring the stakeholders into confrontation. In the second operation, the various stakeholders collectively constructed a conceptual model, which highlighted the representation of their knowledge. The approach aimed at triggering dialogue between land users was then extended for educational purposes the one aimed at raising the awareness of elected representatives in an urban community generated a dialogue arena where it was decided to disseminate on a wider scale. What are the common points of these two operations so that the researcher, and also the partners who were inspired by the first to commence the second, thought them similar enough to talk about companion modelling in both cases?

Diversity Giving Way to Invariants?

Although each member of the ComMod network thought they were conducting their experiments in the companion modelling approach, the group rapidly had to face two questions.

- What are the invariants when applying a ComMod approach? Given the diversity of operating procedures and the increasing demand for training and new applications, a reflexive analysis was clearly needed to better describe the ComMod approach.
- What are the effects of the method? The various experiments have shown the feasibility of companion modelling. Wherever models have been developed, so dialogue arenas have been created and interactions taken place. What have they produced? Is it possible to measure the resulting learning? Have there been actual changes through technological innovations, through concerted management planning or through organizational changes? If so, what influence has companion modelling had on these changes?

This tension between a homogeneous stance and sought-for diversity, pragmatism, adaptation to contexts, questions and issues arising from case studies, requires a period of reflection to understand the potential divergences and enrich the experience of the group.

The ComMod Approach Invariants and Its Assessment

To respond to these questions, the majority of the ComMod network members replied to a call from the National French Research Agency (Agence nationale de la recherche française—ANR) on the theme of agriculture and sustainable development. Four types of result were expected.

- Understanding and comparisons of the effects of the companion modelling approach when implemented in a variety of ecological, institutional and socio-political contexts. The context is taken to be the combination of a geographical situation, stakeholders concerned, resource or area in play, social context of the use of the resource (e.g. conflict, routine situation, existence or otherwise of formal or informal dialogue arenas), and questions asked. The effect of the context should be assessed by its consequences on the collective decisions made or the knowledge produced when implementing companion modelling.
- Understanding of the effect of context on the implementation of the approach itself and especially on the effectiveness of the link between research activities and the practical applications. As the practical implementation was open, the framework in which it was applied, and especially the questions asked, can result in steering its implementation in one direction or another whilst respecting the founding principles laid down in the charter.

- Production of a methodological guide proposing flexible 'know-how' to help implement the ComMod approach successfully and to disseminate it.
- Production of a methodology to assess the companion modelling approach, stating the indicators to be monitored and highlighting the points for later methodological development. Precise collective and individual social and economic indicators have to be identified given the difficulties in assessing the approach; these account for changes in the stakeholders involved in terms of networks, social representations and management practices.

Research during the project has produced several tools and results. The first task was to perfect a common canvas to describe a successful ComMod process. It was developed through tests on a sample of seven cases, to account for the diversity of case studies, then discussed and amended by all members of the group. This document known as the Montfavet canvas was completed by each team initiating and running a ComMod process. It describes the initial context, the origin of the request, the questions asked, and presents a time chart of the various activities undertaken and describes the operations achieved to develop the various models (as role-playing games or computer simulations). Another document was produced to describe the operations in progress, a 'logbook', which was also filled in by the team running and initiating any new ComMod process. Just like a log, it reports chronologically on all the operations undertaken and especially on the sequence of meetings where stakeholders share ideas and representations. The second task was to perfect a protocol to assess the effects of such an operation. The resulting Canberra Protocol is divided into two parts. The first part covers the approach designer(s) who must indicate their objectives, the anticipated individual and collective learning, the modifications to interactions between the stakeholders and changes in their practices. The investigation also covers the various tools used. The second part of the protocol is intended for participants and aims to assess the same elements.

Once this descriptive and assessment material was finished, 27 case studies (a description sheet for each one is available in the Appendix) were described and some 18 were assessed. This material is now available and through comparison and synthesis provides elements of reply to questions raised in the ComMod approach. This work is based on the experience of the ComMod network. It reports on a collective reflexive approach to practices at the interface between renewable natural resource management, a stance of intervention in collective decision-making processes and sustainable development. It has the twofold intention of clarifying what companion modelling is exactly and to put these definitions to the test.

Organization of the Work

Chapter 2 presents the elements for implementing a companion modelling approach, as it is and as it is applied in the case studies. It introduces the protagonists and the dynamic of interactions between them. It emphasizes in particular the notions of iteration, interaction between companion modellers and participants in a collaborative action dynamic, and the key points in exploring collectively a virtual world during what we called collective key moments. Based on the original gathering of these elements, this chapter shows the diversity of implementations, adaptation to the context and the skills used.

In Chap. 3 we stand aside from the principles founding the origins of the ComMod stance to show that adopting the stance comes from practices, methods and techniques mobilized and developed by the commodian to facilitate, in the sense of giving life and/or making live, the approach and associated partners.

Chapter 4 explains how models are developed with the objective of sharing representations of an actual system and how these models are used in workshops based on the exploratory simulation of scenarios where the results are interpreted with reference to the actual system. Special attention is paid to multi-agent simulation models, using human agents (role-playing game) or virtual agents (computer simulation model). The advantage of combining the two types of agent in a simulation tool and/or the two types of simulation tools when implementing the approach is analysed in particular.

Chapter 5 attempts to understand the effects of taking the context into account when implementing the approach and its results. Based on an analysis framework of the social and environmental context and the intervention context, this chapter discusses the consequences of considering the context, or not as the case may be, in the various ComMod case studies, defining objectives, choosing participants, the dynamics of the process and the decisions or actions resulting from it.

Faced with the need for the ComMod network to improve its formalization of the positions it assumes when taking the social and political context into account, particularly in terms of asymmetries of power, Chap. 6 suggests a way of explaining these positions. When applied to ComMod researchers, this method reveals the existence of contrasting profiles within the network, both dialogical and critical. However, beyond this variability, two major points of agreement stand out within the group: a changing, adapting positioning based on the intervention context and changes in the power issues during the process, as well as the desire to explain this positioning to ensure that it is legiti-mate. This second point forms the basis for discussing one of the founding principles of the ComMod approach: the systematic explanation of all the hypotheses behind the modelling approach.

Chapter 7 addresses the question of assessing the effects of the ComMod approach. Having justified the principle and bolstered the theoretical foundations of such a procedure, we present the assessment protocol. Before describing the results of the meta-analysis based on the 18 case studies assessed, we felt it important to illustrate the protocol appropriation process suggested by assessors with different profiles of contrasting contexts. Lastly, we suggest a series of worthwhile improvements to the current protocol.

Chapter 8 dissects the technologies used in our approaches, especially the simulation tools. The analysis covers compatibility with the cognitive framework of stakeholders, their ability to be manipulated directly by participants and their different effects. It also addresses participant perception of the validity of the tool and its link to reality, as well as the ability of the tool to explore possible evolution trajectories.

Chapter 9 shows how the ComMod approach is totally committed as a contribution of science to sustainable development. It addresses how the approach comprehends sustainable development as a process and commits to implementing the principle of participation. It then lays down the companion modelling boundaries faced with an obligation of means and results and finally, evokes the prospect of introducing a quality approach, based on a precise monitoring/assessment method.

In Chapter 10 the relevance of the hypothesis is discussed whereby the participation of stakeholders in a companion modelling process works, during developed interactions, towards modifying their viewpoints, opinions and representations, thus providing an insight into their interactions, relations with the environment and the dynamics of the socio-ecological system. This chapter demonstrates the importance of collective key moments in the individual and collaborative learning processes observed.

Chapter 11 presents the issues and questions raised for companion modelling by integrating multiple, frequently relative and changing scales and, therefore, the evolution of the approach with stakeholders' mobilizing scales other than those considered initially. Taking multiple organizational levels into account does in fact affect the tools to be mobilized as much as the stakeholder interactions in the discussion arenas. Having revisited a few definitions and issues specific to this problem, this chapter presents and discusses the formal and participative processes used to consider these multiple dimensions and levels and changes in the approach.

Chapter 12 discusses the teaching and training of the companion modelling approach. This knowledge transfer is examined under three contexts: academic teaching, training sessions or observation of an actual companion modelling process.

Chapter 2
Companion Modelling: A Method of Adaptive and Participatory Research

Olivier Barreteau, François Bousquet, Michel Étienne, Véronique Souchère and Patrick d'Aquino

The principles laid down in the ComMod Charter and presented in the general introduction relate to a stance or attitude towards how a specific issue and specific field are addressed by taking into account the various types of knowledge and perceptions already present and the use of certain tools. These principles suggest a framing for the teams committed to them, but the adaptation capacity in organizing the implementation of companion modelling in a given case study is in practice left to the commodian. This chapter aims to detail the diversity involved in implementing a ComMod process and the common points that emerge from it. The objective is to describe in order to understand better, with no normative intention.

We relied essentially on *ex post* analysis of case studies and documents listed in the introduction. Our analysis compiled real-life cases and practices that claim to be companion modelling[1] and which will, therefore, be considered as such in our

[1] The ComMod Charter as it exists at the time of this work authorizes any person who has joined the group, and been accepted into it, to consider if his research work follows, or not, a companion modelling approach.

O. Barreteau (✉)
UMR G-EAU, IRSTEA, Montpellier, France
e-mail: olivier.barreteau@irstea.fr

F. Bousquet
UR GREEN, CIRAD, Montpellier, France
e-mail: francois.bousquet@cirad.fr

M. Étienne
Ecodevelopment Unit, INRA, Avignon, France
e-mail: etienne@avignon.inra.fr

V. Souchère
UMR SADAPT, INRA, Grignon, France
e-mail: souchere@grignon.inra.fr

P. d'Aquino
UR GREEN, CIRAD, Dakar, Senegal
e-mail: patrick.daquino@cirad.fr

M. Étienne (ed.), *Companion Modelling*, DOI: 10.1007/978-94-017-8557-0_2,

analysis. We shall then discuss in the overall conclusion to the work whether the nature of diversity observed remains within the framework of adherence to the initial principles of the charter, or dilutes them.

For this analysis we used the documents compiled in the project presented in the introduction to this work. Despite the care taken in the collective writing of frameworks for these documents, a certain amount of subjectivity in completing them must be considered in the analysis. Heterogeneity is added to this subjectivity, as every Montfavet canvas, Canberra Protocol and logbook is completed by one (or more) different authors. To reduce this methodological risk, all interpretations and comparative analyses have been discussed with the contributors to the original documents.

Based on this comparative analysis, we can suggest, therefore, a few key points describing companion modelling. The section 'Organising a companion modelling approach' below thus describes the components of companion modelling. These components unite individuals with specific roles for which we suggest a typology that is used in the remainder of the work. Alongside the individuals is a description of the tools, especially the models mobilized in all these approaches. We then address the sequences used to describe the stages of a ComMod process before comparing the dynamics as they are generated. We emphasize in particular the existence of collective 'high points', that is, key moments used to understand the dynamics at work. Finally, this section concludes by presenting a conceptual model of the approach summarizing all its constituent components. The next section 'An iterative approach' compares the different possible iterative faces of the approach, as this iterative organization is featured in its presentation—right up to the logo of the ComMod group. This comparison specifies the various types of loop or iteration encountered and discusses the central role of the iterative nature in any ComMod process. The following section 'Invariants noted during implementation' addresses a few invariants observed, that is, importance of local anchoring to ensure legitimacy and trust, stakeholder involvement in the entire process, use of models and organization of debriefings. Lastly, the 'Discussion' opens up a debate on this approach, its originality, effectiveness, partners' perceptions and its potential for adaptation in the face of a diversity of situations and stakeholders.

Organizing a Companion Modelling Approach

Given its challenge of intervening in real life (i.e. a diversity of protagonists, with assumed or even expected heterogeneity of viewpoints and objectives), companion modelling finds itself faced with a complex 'stakeholder-orientated' problem that induces reflexivity. Those responsible for a case study consider, in particular, their own intervention objectives as a challenge that is not necessarily shared and must be capable of validation or amendment. These objectives must, therefore, be explained or debated. We present in this section the main components of a

ComMod process, its human and non-human protagonists[2] (Callon 1986; Latour 1999), its temporal organization and especially, the time set aside for exchanges. These are the elements used to narrate a ComMod process on a specific case study, to confront the principles of the charter and its actual implementation. They especially allow the understanding and possibility of (in)validation by the associated stakeholders of mobilized representations to be addressed at every stage of the initiative.

The Main Protagonists

Any companion modelling involves creating interrelationships between a certain number of individuals in managing one or more natural resources. They have the ability to advise or regulate the relationship between other individuals or with the resource, or to use the resource. Companion modelling aims to generate collective reflection: it introduces into the system new individuals specialized in the companion modelling process, who we call commodians, and gives certain individuals already mentioned a new role within the system—the bearer of the approach. Companion modelling also relies on sharing knowledge as a favoured method of advancing relationships between individuals and between individuals and the resource. We have, therefore, classified the protagonists in our case studies into six categories, distinguished fundamentally by the forms of knowledge mobilized during a ComMod process.

Four categories are internal to the system. The 'lay' category relates to people whose knowledge of the system comes from their empirical experience of the world, which is not necessarily formalized or explained. It is borrowed from the world of hybrid forums (Callon 1986), which aims to provide equal consideration to the various types of knowledge. The 'researcher' category relates to academic knowledge, organized and validated under encoded formats, frequently based on experimentation, constructed in an external and formalized analysis and intended to be tested in the case study. The 'technician' category relates to formalized knowledge, but based fundamentally on the knowledge and typology of a large number of situations and on specialist technical data. These stakeholders are generally not directly concerned with the question being considered, but can at any time contribute their knowledge and expertise to the operating part of the system. The 'institutional' category covers more political or economic knowledge of the system. It groups individuals with a specific knowledge of development issues and activities of local stakeholders, who also have system steering objectives. By extension, we shall designate later the knowledge attached to each category.

---

2 Considering 'non-humans' as protagonists can be surprising. We are following here the sociology of translation that considers world objects as stakeholders in social and political networks used to understand social dynamics.

Two other categories lie outside the system and are linked directly to the implementation of a companion modelling approach. The 'commodian' category includes researchers who are familiar with the approach, have committed to respecting its ethics by signing the ComMod Charter and who are basically going to use methodological and organizational knowledge. The 'student' category corresponds to commodian apprentices who are going to test their scientific knowledge and construct a representation of the approach by participating in one or more of its stages. When they remain part of the entire process, such as thesis writers, this apprenticeship gives them a chance to acquire a certain familiarity with the approach and to open up a still uncertain aspect of its implementation.

Associating a Virtual World

In addition to these human stakeholder categories, companion modelling approaches also mobilize a whole network of non-human agents. These are artefacts, with the majority intended to represent or evoke issues of renewable natural resource management shared by at least one portion of the protagonists mentioned above. By artefact we mean any (temporarily or permanently) stabilized component that can be used as a reference by a group of stakeholders or otherwise support their interactions. More often than not these are actual objects, such as a map, a mock-up, a document, etc. Some ethnomethod-ological research has shown that these objects, like the prototypes in a design office, play a major role in an interaction network (Conein and Jacopin 1994; Suchman et al. 2002). We shall extend this notion to intangible objects in the remainder of the collaborative work (Bossen 2002). It can act, for example, as a reference in oral yet duly ratified agreements, such as after a ritualization stage in an assets' initiative (Ollagnon 1989; Weber 1998), to which reference can be made in an interaction. This can be the case especially in oral cultures.

Among these artefacts, models in the widest sense play a special role in a ComMod process, by offering a virtual world in support of reflection. The term 'companion modelling' originates with them. Almost all case studies, therefore, use at least one explicit model. Chapter 4 describes these models, the types and their construction in more details. However, because models are at the heart of the approach, ever present in the network of interactions between the various types of stakeholder described above, their initial presentation can prove useful.

These are virtual models designed to represent an issue of renewable natural resources management from the world of the stakeholders, that is, irrigated system model, wetland model, shrub progression dynamics model, etc. These models are in all cases based on the viewpoint of renewable resources management, focused on the interface between the resource dynamics and the use dynamics of these resources. The introduction presented this viewpoint in greater detail. They include, therefore, at least a representation of individual and/or collective interactions of stakeholders with the resource, such as samples, maintenance activities and checks on its dynamics.

As a virtual world they simulate the processes assumed to be representative of the dynamics of the real world. They result from hypotheses chosen by the modeller(s) on the dominant dynamics in the actual system. This is thus a first level of intervention of the model in the ComMod process: the model authors the discussion of the principal dynamics of a system. For complex systems like those involved in renewable natural resource management, models offer an extension to experimental approaches (Legay 1997): they test through simulation combinations of hypotheses for the system, without being subject to ethical or logistical constraints inherent in performing experiments in actual systems.

The dynamics of a ComMod process are, therefore, entirely a matter of moving from exploring these virtual worlds towards questioning the implementation of what they can bring to the real world. We have a dual translation/interpretation process: translating the real world to the virtual world to ensure a minimum degree of representativeness; interpreting what is going on in the virtual world for the real world and to interpret the results of simulations in action modalities in the real world.

This dual translation/interpretation process is an essential driving force in companion modelling dynamics: the evolution of the virtual world produces new simulations and a chance to discuss their significance for the real world; the changes induced in the real world, or at least in the viewpoints stakeholders have of it, revise its representation in the virtual world. The involvement of participants in this translation and interpretation process makes the insertion of models in the network of stakeholders taking part in the approach, a fundamental issue in the success of its implementation.

Key Sequences

The canvas painted at the beginning of the project, as presented in the introduction, suggested a seven-stage description framework for a companion modelling approach.

- Raising the awareness of the main stakeholders to the ComMod approach and its options for application to the local problem.
- Inventory of scientific, expert and lay knowledge available through surveys, diagnostics and analyses of published works as well as knowledge elicitation for the model.
- Design of the model.
- Choice of tool (computer or otherwise) and implementation of the model.
- Checking, validating and calibrating the model with local stakeholders.
- Exploratory simulations with local stakeholders.
- Diffusion to stakeholders not taking part in the approach.

Following several training sessions (Chap. 12), an 'education' stage has quickly been added to these seven stages. Analysing canvases has also been conducted to further precisely the unfolding of a ComMod process and to suggest a 12-stage format, described in the final chapter of this book. Although these stages are not systematically mobilized or progress in a different order, they form a typical sequence, a sort of full model for implementing a companion modelling approach. These phases are more or less interwoven over time and frequently have to be repeated, either to deal with certain uncertainties or to incorporate new knowledge produced during the process, or due to down times linked to social (e.g. the non-availability of partners or researchers at a given moment) or economic (seeking financial support) constraints. Figure 2.3 below presents the time charts, illustrating the linking of these stages in all the case studies analysed in this book.

Collective Key Moments

The sequences described above relate to the process conducted by the commodians of the companion modelling process in its strictest sense. This interacts with a dynamic of collective action specific to the system within which the intervention on a question of natural resource management takes place. This dynamic of collective action exists regardless of the companion modelling process implemented. Its centre of gravity is in the lay world, extending possibly to institutional stakeholders and experts/technicians. The ComMod process is punctuated by collective key moments. The key moments are where these two dynamics meet. They are the structuring components of a ComMod approach. They are where the principle of confrontation of viewpoints and exchange of representations is explicitly implemented. They also potentially allow a shared evolution of objectives.

In practice, these key moments are organized as a workshop or working meeting, uniting stakeholders from at least two different categories, with the facilitation of a commodian. Although theoretically no one category of stakeholders is essential to these key moments, effectively all key moments mentioned in the collected information indicate the presence of at least one commodian. Key moments are where scientific, technical and lay knowledge confront each other.

These key moments have their place in the dynamic of the dual translation process described above. Table 2.1 below summarizes the types of key moment as identified. Note that they can be included in all stages of this dual translation/interpretation process: from analysing the actual system up to defining an action plan for this real world, via collective exploration of the virtual world (i.e. role-playing game session or interactive simulation session), which remains the most frequent case. Lastly, the objectives specific to commod-ians and academics, such as model comparison, also gave rise to a key moment, which shows the interpenetration of interests of the various stakeholders.

Table 2.1 Types of key moment and case studies with examples

Type of key moment	Case studies
Training	Radi
Collective surveys	Tarawa
Presentation of the virtual world	SAGE Drôme
Co-construction virtual world	Nîmes-Métropole, Ouessant, Pays de Caux, Larzac, Vosges du Nord
Collective exploration of the virtual world	Mae Salaep, Méjan, Nan, Nîmes-Métropole, Njoobaari, Ouessant, Pays de Caux, Radi, SugarRice, Tarawa, Ubon Rice Seeds
Scriptlets and scenario building	Kat Aware, Méjan, Nan, Larzac
Model validation	Kat Aware, Méjan, Nîmes-Métropole, Pays de Caux, Larzac, Vosges du Nord
Collective action plan discussion	Méjan
Presentation/discussion of results of the virtual world exploration	Nan, Ouessant, Larzac
Presentation/discussion of formalizing hypotheses on the real world	Njoobaari, Sage Drôme
Comparison of virtual worlds	Ouessant

Collective exploration of the virtual world is the most frequent type of key moment. These are most often role-playing games, but there are also some cases of interactive computer simulation sessions. This type of key moment is described in Chap. 4 with the associated artefacts. The case study with irrigated systems in Senegal (Njoobaari) has, for example, featured role-playing game sessions to lead Senegalese farmers to criticize the model representing them as well as to elaborate consequences for themselves. In this case, the commodian in charge of the case study called upon the farmers to take part in a game session and then to discuss this session in relation to their activities in the real world. The organization of a key moment starts with the identification of the targeted population. It continues with the selection of a suitable place to gather all the participants. It goes further with the facilitation of interactions among participants. It ends up with the facilitation of interpretation in the light of daily life. Pre-tests of the setting for the key moment are crucial. In the case in Senegal, the commodian had a little control over the choice of participants: the local partner, a leading farmer, was in charge of finding 10 volunteers to participate. It was first a test of the setting and its ability to generate discussion on the real irrigated systems. The place was chosen in order to make the implementation of the key moment easier: proximity but also some neutrality have led to the selection of schools in many sessions. Facilitation of interpretation in light of daily life took place as an open non-structured discussion. This discussion in the Senegalese case was rich and continued beyond the key moment. This caused the commodians to think about a more structured organization of this stage: the debriefing is presented below.

Initial Conceptual Model of Companion Modelling

All these items make up a companion modelling model that we have been able to clarify by implementing the project. It is a representation of the project's common culture at start-up. The follow up to this book shows how the reflexivity involved in this comparative research generated changes in the representation of this common cultural background. This conceptual model, which required clarification at the beginning of the project, is the framework for the design of documents used in completing the case studies. Figure 2.1 below summarizes the presentation of components detailed in this section and includes the relationships between these components.

The companion modelling approach inserts itself into this fairly traditional collective action dynamic, and brings in status and additional types of interaction, related to knowledge of the resource. The two approaches exist side by side and meet during key moments, which rely on mobilizing tools and implementing virtual worlds available to the commodian who provides the approach dynamic. Agents who endorse other statuses contribute to the elaboration of these tools and/ or in their exploration. The typology of tools is not developed in Chap. 4. With this point of view, the companion modelling approach is clearly thought of as having to take charge of the interaction between the two dynamics, as the collective action approach that existed before it has, in most cases, its own dynamic. These interactions also theoretically cause the participants in the two dynamics to modify knowledge of their interactions with the dynamic of renewable natural resources, to modify the power relationships, to modify their ability to schedule the collective use of resources and potentially, to transfer new knowledge to stake-holders only involved in one of these approaches. These types of modification theoretically induced by the crossing, or even a temporary merger, of two approaches form the basis for Chaps. 5–12 of this book.

An Iterative Approach

One of three principles underlying the definition of companion modelling is the principle of commitment over time and adaptability. The commodian commits to following the decision-making process in the ways he changes the object, objective or participants. He upgrades his tools and interventions based on changes in the decision-making process during his support time. When we talk about loops and iterations we are not referring to learning theory loops as postulated by Argyris and Schon (1996). They describe several loops that differ according to the learning register (about the object itself, the values, the learning itself).

The principle of committing to the process over time is a pioneering concept for researchers in modelling. The model is most frequently considered as an object of

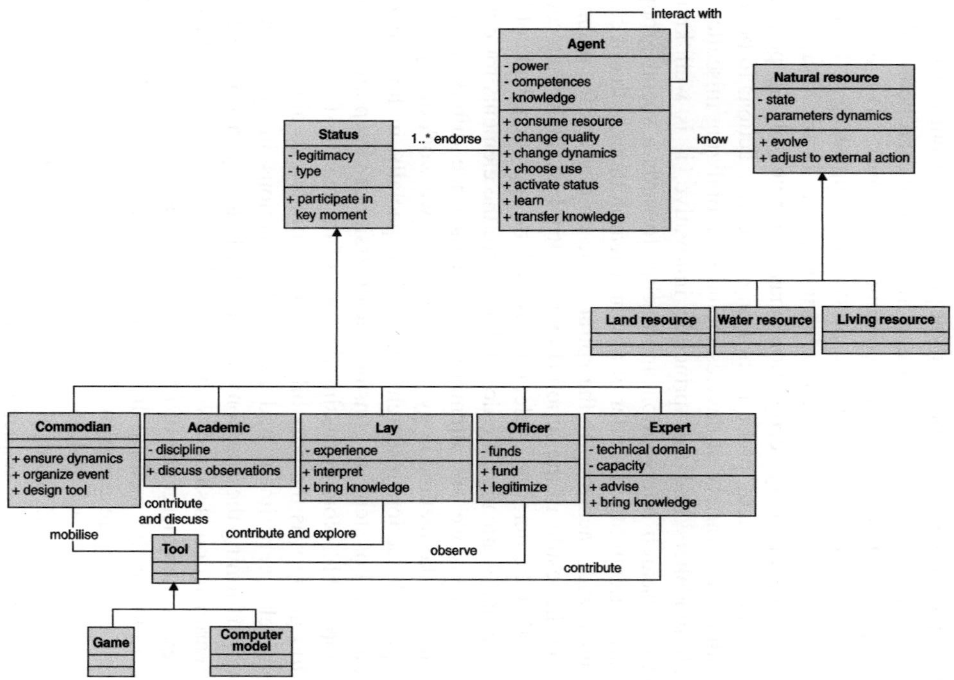

A case study is considered to be a set of agents in relation to each other and with one or more resources. The existence of these relationships structures the whole. The agents are described at least by: status based on their type of knowledge of the system; powers corresponding particularly to the chances they have of controlling or regulating the relationship of other agents with each other or with the resource (or in helping to control the regulations). Every agent uses the resource (i.e. consumption, modifying quality, intervening in the attributes of the dynamic) and has some knowledge of it. Every agent has their own methods of choosing how to interact with the resource and is capable of taking into account their relationships with other agents and reasoning on the system's dynamic. The resource units are described by their state, location and the parameters specifying their dynamic. They change over time based on the usage of agents in relation to them and parameters for their dynamic.

Fig. 2.1 Classes in a generic model of companion modelling synthesis, of integrating acquired knowledge, regardless of the time the process takes. The few researchers who have committed to the participatory modelling processes (Costanza and Ruth 1998; Gonzalez 2000) have frequently underlined that the modelling time should be the decision-making process time. Thus the modelling by Costanza and Ruth lasted seven years. Although ComMod does not stand out for its intervention time, one of the features of the approach is the potential development of a series of models, more or less interlinked. The social dynamic encompassing companion modelling could prompt all participants to develop their questions and thus construct new models to support -consultation and even negotiations.

Thus the ComMod group claims a modelling process set into the decision-making process, which is not the consultation process in itself, meaning that: (i) it is not considered as the social consultation process itself but a 'snippet' based on the modelling within a social process that may have started before the collective modelling experiment and which will continue afterwards; (ii) it is susceptible to changes or ruptures based on the social process encompassing it; these cannot be scheduled; it can also be mobilized at various points in the social consultation process (e.g. issue diversity recognition phase, dispute solving phase, common issue co-construction phase, phase identifying collective actions to be undertaken, etc.), depending on a variety of devices including mobilizing miscellaneous tools.

Although modellers find this principle innovative, it is well known and a constituent of research intervention approaches towards a social group, such as action research. Lewin, the founder of action research thought that it should spiral through stages, with each loop in the spiral made up of steps for planning, action and assessing the results of the action (Lewin 1946). List (2006) in a review of the theme, noted that the iterative cycle, the basis of the approach, had not been much developed methodologically. We do not have here the elements for detailing how the iterative process we are attempting to characterize differs from the action research spiral. In the methodology advanced here, we work on a gradual refining of one of possible action research 'stances', especially in how to design and implement a 'spiral of steps' to consider the diversity of viewpoints in the world in the best way possible. Note in passing that one of the reasons for the emergence of the ComMod group was the possibility of using models to allow stakeholders to plan collectively, experiment and assess their actions in virtual mode before considering performing these actions in real life. In a way 'action research' means that the research action is shared with the stakeholders. Companion modelling was thus involved in the emergence of participatory research processes starting during the 1990s, more in the medical field (Cornwall and Jewkes 1995) or with the development of action research in agronomy.

In the next section, we define loops, before going on to test this definition through various case study analysis methods. We lastly choose a few cases that we feel reflect the diversity of sequences within a ComMod process.

Notions of Loops and Cycles

The first stage in this work on the iterative nature of the approach was to organize a series of collective discussions on the notion of loops and cycles. We relate here briefly a pictorial story, with each graphic innovation reflecting a new vision of the iterative process (see Fig. 2.2). The first proposal came from Barreteau in 1998. It establishes the relationship between the model and the field and introduces the idea that the process can be repeated (Barreteau 1998). A few years later, d'Aquino and colleagues suggested a synthesis in the form of a new figure, featuring the integration of lay knowledge using models that can be either computer simulations or role-playing games (d'Aquino 2002). Barnaud and colleagues then suggested a representation in the form of spirals that were sequences of loops (Barnaud et al. 2006a). This pictorial evolution underlines the discontinuity that exists when moving from one loop to the next: loops are changed when the problem changes or when the stakeholders change. This proposal was adopted by the collective.

D'Aquino and Étienne (unpublished) then went further to consider the breaking down of a loop as presented by Barnaud et al. (2006a) into subloops. These subloops are phases used to deal with the same question, phases which run in-field operations and modelling operations alternately.

Finally, two types of loops were considered and defined:

- 'macro-loops': the macro-loop changes when at least one design phase and one collective visioning phase (scriptlet) have taken place in succession and further investigation is necessary
- 'micro-loops': the micro-loop changes when moving from one phase in the approach to the next, that is, an iteration similar to that of the real world and the model. The possible phases are as presented above.

We decided to test this conceptual model by confronting it with data acquired during the description of case studies.

Diversity of Implementations

The aim here was to consult the empirical data compiled to see whether the macro-loops can be defined unambiguously. Three methods were used for this purpose. The first uses an algorithm for automatic detection based on different phases. This method assumes that moving from one macro-loop to another takes place after a validation phase, a scriptlet phase or a reproduction phase. Figure 2.3 illustrates the result of this investigation. Each vertical bar marks the end of a macro-loop.

The second method involves cross reading the descriptive canvases of the case studies. Apart from the time chart used to produce the previous figure, these canvases contain a literary description of the operation. Two independent readers indicated the number of macro-loops they recognized in a sample of 12 case

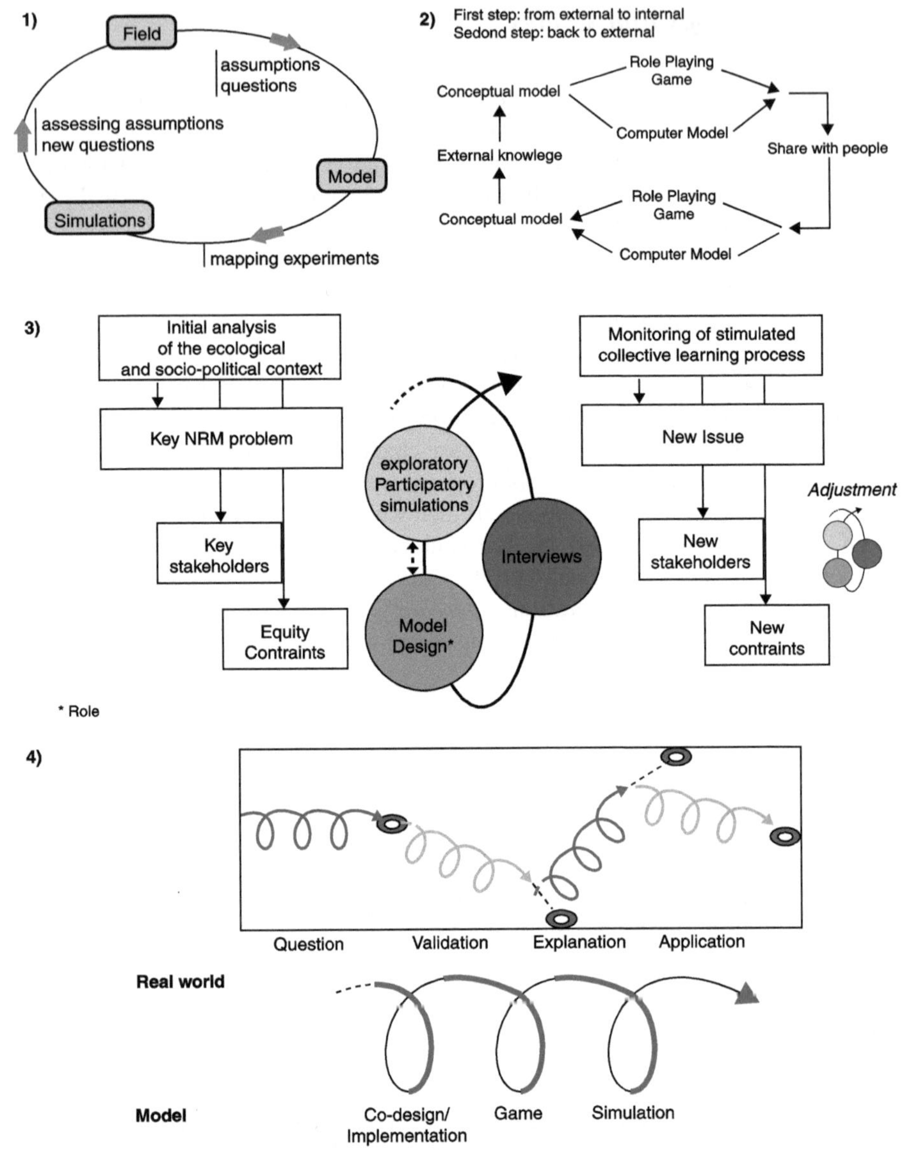

Fig. 2.2 Evolution of figures showing a companion modelling process.From left to right and top to bottom: (1) Barreteau (1998), (2) d'Aquino et al. (2002), (3) Barnaud et al. (2006a), (4) d'Aquino and Étienne (unpublished)

studies. This assessment was compared with the perception of each designer of a ComMod process. Among the 12 case studies, only three show total consistency. These differences were due generally to a different appraisal of the cause for changing a macro-loop. One reader considered that moving from one macro-loop

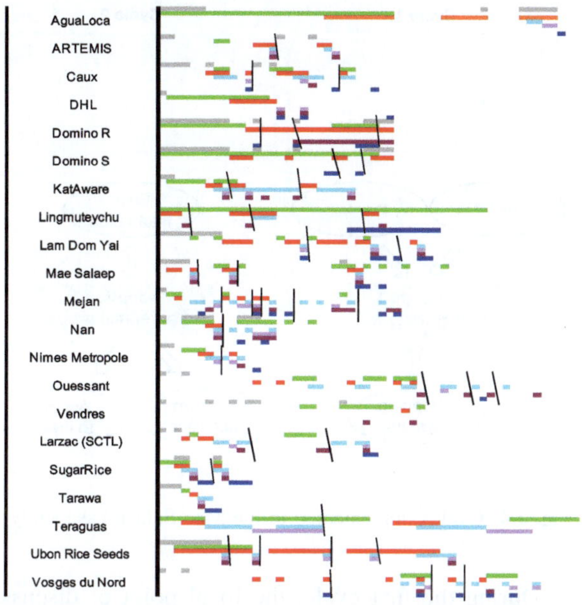

Fig. 2.3 Description of the ordering of various phases for several case studies, with indication of macro-loop ends according to the automatic detection algorithm

to another took place if the perception of the system changed (mirrored in a modification to the model), whereas the other reader considered that the move occurred if the question changed explicitly (e.g. changed objective, change in question dealt with). However, the evaluation makes a clear point that change in issue is barely explicit as it is never mentioned during interviews with participants.

The concept of cycle and loop can be used to construct a model of a ComMod process; this serves as a tool in organizing its presentation by synchronizing the various field phases with the inflections of the decision-making process. This organization of a ComMod process postulates that the loop changes when there is an explicit change in the question dealt with in the opinion of the process designer. This construction is very subjective and cannot, therefore, be presented as the viewpoint of its author(s). The advantage of this framework and rules we have thus established is that the author must be in a position to clarify their model. We shall describe below a case study that is used to appreciate the heuristic range of the model in a companion modelling process.

Evolution of Questions During a Companion Modelling Process

The case we present here covers work in a catchment area in the north of Thailand, the Mae Salaep basin. Figure 2.4 shows the whole sequence for this case study. There are three distinct loops.

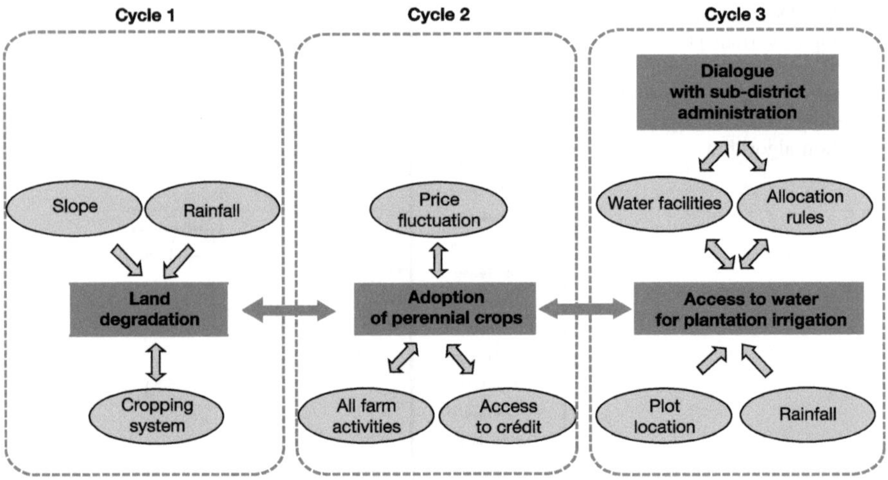

Fig. 2.4 Cycles implemented in the Mae Salaep case study (from Barnaud et al. 2008a)

During the first cycle, the focal point of discussions changed agro-ecological topics concerning soil erosion towards the envisaged solution, that is, the adoption of perennial crops (Trébuil et al. 2002a). The participants then asked for the model to be modified to address the social and economic conditions for this adoption.

This was the purpose of a second cycle covering the interactions between formal and informal credit, off-farm work and investment in plantations (Barnaud et al. 2006a). The participants formulated scenarios for changing the credit rules to offset the problem of unequal access to perennial crops. One proposal was to increase the duration of loans allocated by the government under a decentralized rural credit policy. However, such decisions are taken at government level. The villagers then explained that several villages had already alerted the government to this proposal (independently of the ComMod process) and, if the government agreed to it, new role-playing game sessions would be useful so that they could adapt collectively to these changes. The government was overturned in the meantime and so this proposal could not be pursued.

The villagers expressed two hopes for the future during assessment surveys in this second cycle: bringing the question of irrigation water into play and involving the representatives of the Tambon Administrative Organization (TAO)[3] 'so that they knew what was happening in the village'. The third cycle, therefore, addressed the interinstitution consultation between village and subdistrict. The villagers linked the water question to TAO participation in the ComMod process, as the TAO could finance hydro-agricultural development projects. The third ComMod cycle thus aimed to stimulate a collective learning process on water management in the under-catchment area between villagers and the TAO.

[3] Administration at the subdistrict level, which most often includes 10–12 villages.

Invariants Noted During Implementation

Apart from the overall organization and the use of the concept of cycle (inherent in the approach even though quite heterogeneous across case studies), other invariants testify to a common ComMod culture. An invariant is an element found frequently in documents describing case studies. We discuss these invariants successively in the remainder of this section: sources of legitimacy for the approach; stakeholder involvement; regular recourse to a complex system model combined with simulations or long-term exercises; amount of time allocated to debriefing, especially during role-playing games.

Local Anchoring of the Commodian as a Source of Legitimacy

Analysis of canvases has shown that over 60 % of the case studies investigated were initiated and carried by commodians. The question arises in these situations as to the legitimacy of implementing a ComMod process: what makes participants in collective action dynamics accept companion modelling dynamics, which comes in interaction, up to taking part in them? The collective key moments, such as interpreting observations from the virtual world to the real system, are times when stakeholders could refuse the interaction between the two dynamics. For that, we have noted the importance of local anchoring of the commodian, especially in case studies that have worked well.

Local anchoring often seemed afterwards as one condition for the trust placed by the local stakeholders in the implementation of the approach. This local anchoring either comes from the social capital of the commodian due to extended work experience shared with some of the participants in the accompanied collective action process, or from recourse to an intermediary with this social capital and ready to mobilize it to set up the ComMod process, or from indirect social capital. This local anchoring installs a relation of trust in principle. This is particularly useful when implementing the companion modelling approach after an experience of failure by participants in traditional consultation methods that also had a collective decision-making basis. These consultation failures can nevertheless prompt the stakeholders involved to find new ways of doing things and, therefore, to be curious about the method proposed by a known researcher recognized for his skills, all the more so when he has been seen at work locally. Companion modelling, therefore, more easily becomes a participatory approach, perceived as a new method that can be tried out by the stakeholders when the legitimacy of the researcher has already been established. This trust also theoretically endangers any exploratory situation or role-playing game (Caillois 1967): this endangerment is acceptable when associated with trust, like a safety-net for a tight-rope walker. The acquisition of this trust, which is essential, incites individuals to reveal their viewpoints and questions on the system under the cover of simulation.

In many situations, the ComMod approaches have been implemented successfully by researchers during their stay in a country as a foreign national. The personal insertion of the researcher into local academic or professional networks ensures this local anchoring by relying on the social capital of members of this network. Some even organize and formalize a specific strategy to mobilize this local social capital successfully: identifying stakeholders with the power to block, persuading them to accept the opening induced by the approach, and helping existing institutions to maintain legitimacy over and beyond any changes that may be concluded (d'Aquino 2009). In several cases, as in Thailand or Bhutan, the commodians were local and had acquired this strong local anchoring through past experience. The experienced commodians taking part in the project supported them methodologically. Training researchers within academic networks of international commodians has produced new commodians who have adapted the implementation of the approach in contexts familiar to them. This is illustrated by two experiments in Bhutan. In 2002, a first contact at the request of a hydrologist responsible for solving irrigation problems in the Lingmuteychu catchment area, especially when replanting the rice, prompted the interest of the CIRAD in this question. However, it was the involvement of a senior researcher from the Renewable Natural Resources Research Centre (RNR-RC) in Bajo, who attended a ComMod course when studying for his Master's degree in Thailand, which actually launched the first study in Bhutan. Having identified ComMod as a promising approach, he decided to devote his Master's placement to developing role-playing games covering the sharing of water between two villages in this catchment area. The success of this first role-playing game led to other sessions being organized, which a few years later would culminate in the creation of a management committee for the catchment area. Armed with this first ComMod experience, this researcher, as requested by the Ministry of Agriculture, then decided to apply the same approach in another sector, Radi, which he knew well, having worked in the region in the late 1990s. This Radi site was an area of conflict between two communities of herdsmen belonging to different ethnic groups who had been disputing access to high-altitude grazing areas for more than 30 years. The previous experience at Lingmuteychu, considered to have been successful, made a major contribution to legitimizing the use of the ComMod approach in the eyes of the Ministry of Agriculture.

Involvement of Stakeholders

As described at the beginning of this chapter, companion modelling is based on a dynamic of exchange between various categories of stakeholders. It is regularly going to associate or confront 'lay' knowledge (of local stakeholders), 'technical' knowledge (of the development engineers) and academic knowledge (of researchers). This is active involvement, providing the stakeholders involved with a real chance to intervene either by explaining their vision of the world or by constructing intermediate objectives or formulating projects for the future. These exchanges or confrontations of knowledge are especially intense during the key moments that regulate the progress of the approach.

Numerous criteria can be brought into play in the choice of stakeholders associated with these key moments. Based on the situations and sponsor preferences, three types of situation are currently encountered.

A global vision of the system is preferred: the participants will then be 'connoisseurs' of the region where the local experience legitimizes their invitation to speak on behalf of stakeholders with whom they meet every day (technicians) or who they have studied in depth (scientists). Attention must be paid to not forgeting a theoretically decisive activity for the question raised and overrepresenting one activity versus another.

Priority is given to the involvement of local stakeholders, while maintaining a global vision of the system: the participants will then be representatives of local stakeholders chosen for their legitimacy (e.g. presidents of unions or producer groups, association directors, elected representatives) and for the relevance of their activity in terms of the question raised.

The involvement of local stakeholders is always preferred while seeking to appreciate the diversity of the system: the participants will thus be local stakeholders chosen for the diversity of their practices, according to the question raised.

These three ways of forming the involved collective can alternate during the various phases in the process of a single companion approach. For example, priority is given initially to the involvement of local stakeholders to cover the diversity of practice and social status. On the other hand, the collective will subsequently be expanded to include researchers and people in charge of the various administrative levels. Or in the case of fisheries management in Thailand (Don Hoi Lord), all the agents of the industry were interviewed at the beginning of the process. First game sessions gathered fishermen of one village, who requested that the following sessions be opened up to fishermen from other villages, other stakeholders in the industry, local authorities and finally, to policy-makers. It should be stressed that this evolution was controlled by the participants themselves. Similarly, according to the translation/interpretation stage, which includes the key moment, certain stakeholder categories will be preferred as they are found to be more relevant than others. Thus, when analysing the actual system, local stakeholders and researchers will be the dominant elements in the group, whereas priority will be given to local stakeholder involvement during the collective exploration of the virtual world.

The place of researchers, and thus of scientific knowledge, in the process remains variable and is still subject to discussion. Normally, scientists are selected who possess knowledge on the main processes in play. Some will be present, therefore, as soon as the approach is initiated as the question raised is linked to one or more clearly identified processes of which they are well aware. Others will be included during the work on dynamics or interactions, if the participants feel a need for expertise on a particularly important theme to gain a better understanding of how the system functions.

Lastly, the involvement of institutional stakeholders (e.g. elected representatives, administrations), frequently sought when defining an action plan for the real world, in most cases is inadequate. This is either because they have not been involved in the first phases of the approach or because their time schedules and acceptability of the approach require adapting the group to these constraints. Chapter 11 returns to this involvement of people whose decisions concern various levels of organization.

Systematic Mobilization of a Complex System Model and Dynamic or Long-Term Simulations

Another strong common point to ComMod processes is the use of complex system models for simulation purposes. Whether constructed along the approach, as is frequently the case, or imported, they take their place in the social and technical network mobilized by any commodian. They can be intermediate objects (Vinck 1999) or boundary objects (Star and Griesemer 1989). As intermediate objects they convey the viewpoints of a group or stakeholder at a given moment in the shared system. Presented or used by others (even by the same people later on), the model communicates these viewpoints. Although this model language occasionally looks esoteric, Chap. 4 will show that it explains the dynamics clearly and assembles heterogeneous knowledge. We will also show that implementation techniques exist to adapt the communication modalities and that the model construction process can limit the implicit assumptions in the collective decision process. As boundary objects, models support interactions. They concentrate the viewpoints of a group of stakeholders (be they lay, expert, commodian, etc.) on the same object and focus the interactions on a restricted number of domains. In the daily relationships of these stakeholders when they exist, the complexity and size of the system that this object is assumed to represent, as well as the time constraints, often make it difficult to confront viewpoints.

Recourse to these intermediate objects can also go beyond the constraints of field interventions with practical (need for simulation over long periods) or ethical (problems in taking charge of the consequences of a full-scale experiment) difficulties. Simulation can thus complete the arsenal of experimental approaches (Bousquet et al. 1999). On this point, companion modelling thus revives a long tradition of using models as an advisory tool for the state (Saunders-Newton and Scott 2001), by reinstating it in the perspective of a distributed decision-making process collecting together all the stake-holders involved. The second objective explained in the charter is a form of decision-making support. However, it is modified from classical decision-support approaches, because it embodies a collective decision-making process considered as a flow of interactions between individual and/or collective, heterogeneous stakeholders in terms of their political weight and representations of the world (Weber 1995b).

The models used in the case studies bolstering this work are found in three types of usage, which are not mutually exclusive.

- First, sharing viewpoints in the same medium: in SelfCormas, the medium, initially basic, forces each individual to state his representations and to complete his statement with what seems essential to him to characterize the issues at stake (d'Aquino et al. 2003).
- Second, social mirror: in SHADOC/Njoobaari Ilnoowo, the model sends back to the peasants of the Senegal valley an image of the collective they are forming by interacting within an irrigated system (Daré 2005). The role of this social mirror forces the stakeholders into an awareness of their interactions. The social mirror is, therefore, a catalyst for collective learning (Hatchuel 2000; Pahl-Wostl and Hare 2004).
- Lastly, the model encourages an exploratory stance towards understanding the world (Auray 2006; Richard-Ferroudji 2008): the SylvoPast model is restricted to a cohabitation of forest officers, cattle farmers and hunters, and reacts to a catastrophic event threatening their activities (Étienne 2003). Experimentation in a virtual world allows the testing of possible modifications with a controlled endangerment due to the distancing it introduces. Taboo issues can thus be explored, and the consequences of the experiment can always be discredited later on because of the virtual nature of the support. One example is the introduction of water stealing in Njoobaari Ilnoowo (Daré and Barreteau 2003).

Chapter 4 will discuss these key components in companion modelling at more length, by specifying the stages and the diversity of implementation and the technical modalities for their use.

Importance of a Debriefing Time

When acting out role-playing games, that is, interactive simulation, the debriefing is an integral part of a key moment. Debriefing provides the way back from the virtual world to the real world: it supports the interpretation of the virtual to the real, and must be structured as such. These essential stages allow us to understand the link between behaviours noted in the game and the specific situation of participants when it takes place.

In all the case studies featuring a role-playing game, an immediate collective debriefing was arranged on site, at the end of the game session. Later individual, or more rarely collective, debriefings took place to complete these. These later debriefings occurred days or weeks following the original session and were based, in the vast majority of case studies, exclusively on the game sessions. In a few cases, the analysis also took into account results from surveys before or after the game and/or results of other role-playing game sessions (see Fig. 2.5).

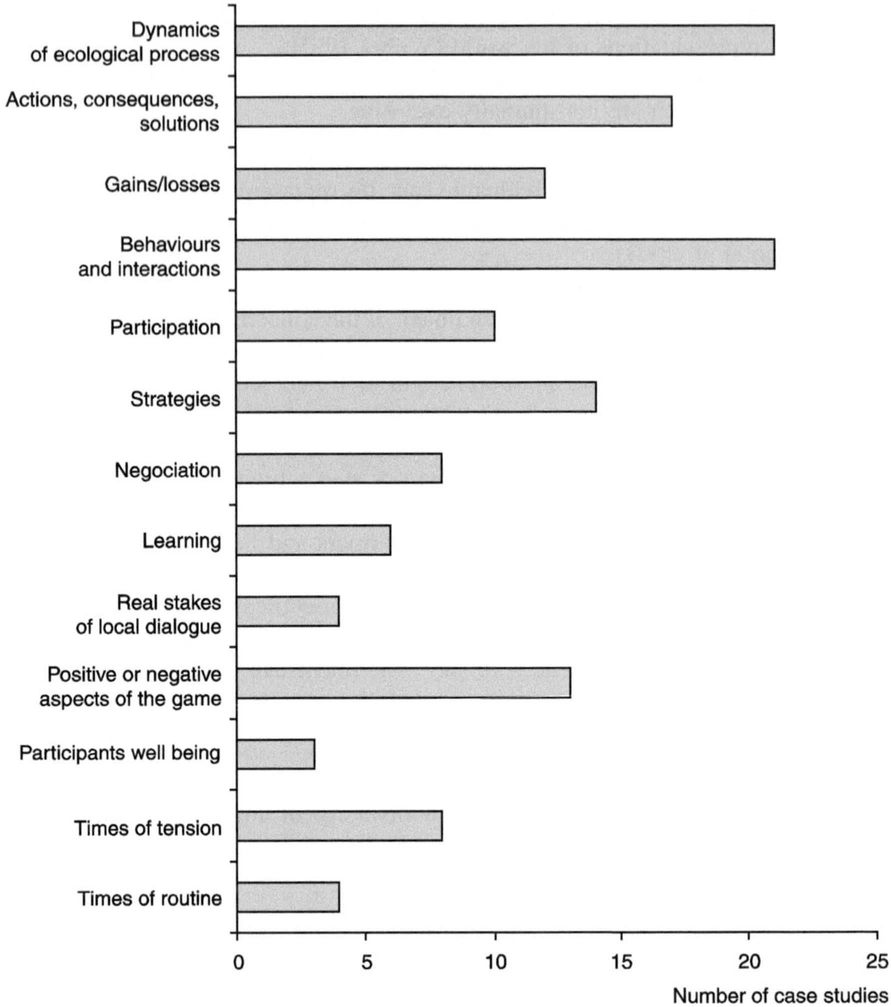

Fig. 2.5 Elements analysed during debriefings

This analysis relies mainly on the participation of players and coordinators. However, any observers present also systematically take part in the analysis of the role-playing game. In very rare cases, the players are not required to contribute and the analysis is provided by the coordinators or observers only. Lastly, also very rarely, non-players are invited to put forward an opinion.

Figure 2.5 features the diversity of items that could be included in debriefing sessions to generate discussion between participants in order to return them to their daily life and facilitate the emergence of a sound interpretation.

Discussions most often concerned items relating to the dynamics of the ecological processes represented and linked most often to changes in land use. The ecological processes analysed varied considerably between case studies: resinous tree planting of the environment or regrowth on fallowland, access to water or forest resources, degradation of grazing land, risk of runoff and erosion, etc. Given that this evolution results from participant decisions during key moments, a major part of the debriefing period is also given over to analysing the decisions made by stakeholders during the simulation, and the consequences in terms of managing the natural resources in question and the solutions provided individually or collectively to offset any harmful effects. Computerization can provide in most cases a quantitative analysis of elements described previously by dynamic monitoring of indicators.

Mathevet and colleagues carefully worked out the organization and setting of the debriefing in the Camargue case study (Mathevet et al. 2007). Working on the decline of reedbeds, they designed the role-playing game ButorStar, with a special focus on social learning and understanding of socio-ecological processes. In game sessions with stakeholders of the Vendres lakes (Hérault county, France) and Scamandre (Gard county, France), they organized a debriefing to discuss and collectively analyse the outcomes of the game, in order to explain the reasons for specific collective or individual decisions during the game session, and to understand what occurred during the game session. Mathevet and colleagues facilitated the debriefing in three stages: (i) individual identification by each player of the rationale, values and behavioural patterns of other players; (ii) collective assessment of results, emotions and understanding of the processes at stake (i.e. perceptions of the social behaviour of various players, suitability and coherence of decisions, group behaviour and dynamics); (iii) discussion of any misunderstanding of the social, economic and ecological processes; assessment of the negotiation of the agreed process and its evolution.

In this case study, the debriefing also included a comparison of the results of the game (e.g. indicators, charts and maps) with those of previous game sessions. This provided an interpretation of the outcomes of another group of players, less directly concerned but with the experience of one session. This comparison pushed participants to discuss the outcomes according to various fields such as integrated management, environmental problem solving, communication, information sharing, ecological processes, group dynamics, public policies and their implementation or negotiation. Finally the debriefing provided the commodian with a basis on which to assess the results from the collective exploration of the role-playing game (Mathevet et al. 2008).

Discussions during debriefings also cover the behaviour between players, how they interact together and changes in their relationships during the process, to understand better the mechanisms of individual or collective decision-making processes. This rather more qualitative analysis relies most frequently on observations during the role-playing game by observers and coordinators. It can be supplemented later by studying video and audio media when *ad hoc* recording devices have been used (e.g. camera, recorder or camcorder). Depending on the case study, the analysis pays particular attention to stakeholder participation levels and clarification of the individual or collective strategies followed. The coordinators and observers also spend time identifying and -understanding the various negotiation phases observed. With whom and why do the stakeholders -negotiate? How do any existing power relationships and the presence of formal and informal leaders guide the negotiations? Does any particular subgroup emerge, for example, conveying the involvement of certain stakeholders in debates, or indeed their exclusion from them? In some cases, the coordinators and observers also try and decide whether learning took place during role-playing game sessions in particular. They rely most frequently on questionnaires that measure this learning, as the same questions are asked just before and just after the role-playing game session.

Debriefing discussions finally consider the feelings of participants as the game session unfolds. The coordinators are keen to know whether the players felt at ease, or if there were moments of boredom or tension. Many questions are raised over identifying positive or negative points of the game, with the specific goal of improving the game and the representation of ecological processes if necessary. The debriefing can then become a way of evaluating the companion modelling approach implemented.

Discussion

Companion modelling case studies feature widely varying implementations, up to particularly identifying components, such as the iterative nature. However, a number of elements are seen as strong components of the case studies assessed in this project, that is, the use of virtual worlds in the models of complex systems, the legitimacy of the approach and its participants, anchoring to local settings and a high level of interaction between researchers and stakeholders.

Although discussing the approach is one objective of this book, at the end of this chapter we return to these key points in companion modelling. We discussed initially the originality in the huge number of participatory research approaches, restricting our investigation to the field of renewable natural resource management and more specifically, those projects undertaking computer modelling. Here we discuss originality in terms of method, not of stance, which will be addressed in the next chapter. We shall then take advantage of surveys carried out afterwards with stakeholders to see the perception held by stakeholders of these key points and what are the consequences in terms of a framework for the approach and sustainability of the dynamic introduced. Another point noted in numerous assessment reports that should be addressed before promoting this type of approach is the cost/effectiveness ratio. Lastly, we end by discussing the possibility of adapting the implementation of companion modelling.

What are the Original Features of the Method?

Companion modelling is a constructivist approach, in that it suggests devices and settings for its implementation. In terms of use of models, these approaches aim for a consensus in constructing representation tools. This provides a means of assisting participants to build up a representation for themselves (Dias and Tsoukias 2003; Tsoukias 2007), which serves to confront and sometimes structure the various viewpoints. The choice of constructivism, however, is based far more on a choice of stance than of method: some methods fall more in line with a constructivist stance than others, but even interactive methods are not necessarily constructivist (Dias and Tsoukias 2003; Tsoukias 2007). This justifies the distinction we make in the book between this discussion on originality in implementing the approach (this chapter) and the discussion on the originality of stance (Chap. 3). It also makes this comparative exercise difficult, as companion modelling is first and foremost defined by a stance and resorting to all the elements described above is not enough to qualify a companion modelling project.

The wide diversity in experiences of ComMod processes relies thus on a set of invariants. These include: the origin of the legitimacy of the approach via the involvement of stakeholders; the use of artefacts serving to represent issues under discussion; the system-atic use of debriefing periods for the collective interpretation of what is happening in the virtual world formed by the artefact.

The first invariant deals with instilling an initial trust between the participants and at least one of the commodians. Although trust is an issue normally considered to be a product of a participatory approach, few other comparable approaches pay much attention to initial trust. Thus a comparison of four participatory modelling experiments only highlighted the question of trust in the results of these approaches by the participants (Hare et al. 2003). The rare works devoted to initial trust within participatory approaches in general conclude that there is a need for participation over time (Höppner et al. 2007), towards merging the decision-making support processes (in the widest sense) with the actual collective decision-making processes. Companion modelling considers that the modelling process interacts strongly with the decision-making process it is supporting: this powerful interaction implies that both processes share timeframes, but that it is always possible to identify within the decision-making process one time 'before', one time 'after' and one time 'beside' the modelling process. The good level of trust installed at the start means that the companion modelling approach can lead participants in the decision-making process towards an exploratory approach (Auray 2006).

The second invariant is stakeholder involvement. This is not original in itself. This view of the approach makes it one method of participatory modelling among several others. In fact we have frequently borrowed the techniques used (e.g. workshop, role-playing game, etc.) from classical participatory methods. Organizing this involvement around collective key moments is also a feature of many collaborative decision-making support experiments, such as the group model building exercises (Vennix 1996; Rouwette et al. 2002). However, they focus basically on business issues where it is easy to identify the population concerned: there is a customer, a company or an organization within an easily identified network of formal relationships. Moving to issues of natural resource management raises problems of fluctuating population or changing natural dynamics, and prompts us to revise the question of participating populations (Barreteau 2007): customers are 'affected people' or stakeholders (Landry et al. 1983). The term 'stakeholder' itself is poorly defined (Claeys-Mekdade 2001). In addition, being affected does not produce customers in the same way: when there is a transaction it is frequently only moral, and the heterogeneity of this whole, combined with its lack of representative organization, does not produce a representative, legitimate spokesperson. Companion modelling take this situation in charge with an adaptive approach used to enable the evolution of this participating population and how it is involved, based on joint evolutions in the supporting and supported processes.

The mobilization of intermediate objects, especially models, is not original. Nor does it particularly relate to the involvement of stakeholders. Decision-making support approaches produce, sometimes co-produce, a number of such artefacts, despite still being fairly poorly perceived in the corresponding scientific community (Kikker et al. 2005). Multi-criteria approaches can, however, be found in the literature, which consider that the result of aggregating preferences aims to provide a starting point for the debate (Hämäläinen et al. 2001). These artefacts often form representations of issues under discussion. The translation process introduced into companion modelling from the world of stakeholder issues to a virtual world supporting reflection is thus encountered in many other methodological approaches. Most of these tools are used to take charge of heterogeneous knowledge. More significantly original in companion modelling is the use of non-computer tools, such as simulation models (this term will be explained in Chap. 4). Companion modelling is even more original, however, in not targeting the convergence of the modelling process. It thus differentiates from approaches, such as the evolutionary system design (Shakun 1996), which aim for 'the' suitable representation.

Lastly, the debriefing period (we have been greatly inspired by work in communities exploring the use of simulation games: Lederman 1992; Ryan 2000; Peters and Vissers 2004) provides interpretative feedback on the collective action process. Although this specific time period is well documented in the use of role-playing games, specifically for questioning the possibilities of exporting what happened in the simulation to the real world, few experiments outside the companion modelling examples implement this interpretation period simultaneously with the translation period specific to the design of the virtual world. This is the heart of the iterative nature of the companion modelling approach: alternating translation/interpretation is specifically included in the implementation.

Stakeholder Perceptions of the Approach

The iterative nature of the approach, substantial to the group is effectively barely perceived by the stakeholders in the field and partners in research or involved in the collective action processes in which the commodians intervene. Thus the words 'loops' or 'iteration' never appeared in the responses of participants to our questionnaires. The word 'cycle' appeared in a few questionnaires but never when referring to the implemented approach. It covered cycles in the world of collective action, that is, political, economic or cultural cycles. Cycle was also used when referring to the tools used: role-playing games where 'rounds' are sometimes called cycles, in a similar fashion to computer simulation time steps. Confusion can also reign between the notions of cycle and workshop in a case study.

The external assessments themselves rarely used these key words. They occurred when it was a question of a quality approach to monitoring and complying with the companion modelling process, therefore, for issues only found within the commodian world.

The notion of cycle/loop/iteration does not relate, therefore, to a perception framework for stakeholders taking part in *ex post* interviews. These interviews lead more to an integrating viewpoint on the effects of the approach rather than on its implementation. More than anything, the stakeholders questioned kept returning to changes in their viewpoint resulting from the entire process, regardless of the number of cycles taking place. They found it difficult to remember cycles clearly, even in case studies like Mae Salaep, which had clearly identified cycles. Traditionally the last cycle wins in the analysis made by the stakeholders questioned.

The stakeholders thus maintain their analysis categories and do not take on the commodians' categories for implementing the approach. Their perception comes across as being more from the collective action process in which they are involved, where interactions become grafted on to the companion modelling process, and its effects on their own positions. This does not involve two sealed dynamics side by side, which are involved independently. Instead, it involves for a limited period a single dynamic created from their interaction, where researchers and stakeholders have two different viewpoints.

Any framework for the approach created implicitly by the methodological choices is scarcely felt by the participants, at least in terms of its iterative nature. Note that this framework, however lightweight, is totally exogenous to the field. The methodological choices, like the iterations between the real world and the virtual world, are not defined jointly with the participants. In current practice, there is normally no prior discussion to initiating the debate on creating companion modelling and deciding how it will be run. This framework is tempered by increasing the interaction formats and adapting them to the local context, thereby limiting the exclusion phenomena due to the formats not matching the way some stakeholders grasp the world, with the methodological choices remaining in the remit of the commodians.

This framework can, however, be counterbalanced in certain cases by key moments combining lay contributors and commodians, the co-construction of intermediate objects and learning phases on the use of these intermediate objects. This is the second key methodological point: the strong interactive nature of the companion modelling between researchers and stakeholders. Participant questionnaires and assessment reports show good perception of this feature of the approach by qualifying it more precisely. A secondary effect of the iterative nature on legitimacy is thus revealed, as the strength of inter-action between researchers and stakeholders is in part attributed by the participants and assessors to the regular appearances by researchers in the local arena. These appearances give the stakeholders the chance to monitor the research process.

As addressed in the participant questionnaires, the interactive nature firstly gives more weight to this relationship, by taking into account the diversity of researcher and commodian numbers on one side and lay, expert and institutional contributors on the other. A few researchers have preferential links with a few stakeholders in several case studies. In their responses, the stakeholders recognize network heads, some even identify themselves as such—'the researchers are always directed to me'. The social network analysis of logbooks confirms these favoured go-betweens, both researchers/commodians and lay/expert/institutional contributors.

These privileged relationships raise questions over the influence of companion modelling on power relationships (see Chap. 5). They also raise the question of case studies where the commodian is only present for a limited period. This is because of the necessary sustainability of the relationship, which is essential in consolidating the legitimization of intermediary stakeholders in their 'risk taking' in relation to the local context, due to their central role in changes induced by the approach, as shown in the external assessment of the Nan case study.

Lastly, the interactive nature of the approach is also recognized for its effects induced in terms of learning, that is, transfer of knowledge on the use of resources or to develop political arguments, mediation and contribution of know-how in the collective decision-making process.

The perception of the interactive nature of the approach is therefore shared: the specific role of the researcher in the network of interactions even manages to be erased in certain case studies, whereas this relationship is institutionalized in others. Nevertheless, most frequently the perception of the relationship gives researchers a supervisory role, confirming the residual asymmetry of the approach in practice.

The Costs of Implementation with Regard to Profits

A great deal of work in assessing participatory approaches highlights the importance of fostering participant development, maintaining a certain equity, trusting that knowledge will be shared and ensuring transparency in the learning method

(Reed 2008). The participatory nature of the companion modelling approach is coupled with a process of co-construction and shared use of a model based on an intense and complex social process. Based on the principle of technical democracy, the co-construction method channels the various types of knowledge involved towards shared interests: solving a problem jointly or thinking about a common future (Levrel et al. 2009). This process can generate problems of organization (when participant availability is limited), fatigue (when the approach does not produce concrete elements quickly) or integration of the process (when new participants join in during the ongoing process).

As the participants must gradually share partial qualitative knowledge on the workings of a socio-ecological system with precise quantitative knowledge on a specific domain of the same system, the process can be heavy and costly in time and information. It generates high transaction costs that only can be compensated by the pleasure of the collective construction and the widening of knowledge of relationships with others and the processes that drive the system dynamics. This constraint forces participants to volunteer immediately for an exercise of variable length, but which always requires intense cognitive investment and an acceptance that doubts potentially will be cast on their knowledge by other bodies of knowledge or that it will not be considered sufficiently precise. It is partially raised by the variety of tools used (e.g. conceptual model, role-playing game, computer simulation) and by the partial effect of surprise they generate when involving local stakeholders more accustomed to traditional meetings for exchanging or reproducing views.

The design, implementation and use of a model are central to the approach but require a mediator, a crosser of boundaries who will gradually lead the stakeholders from a personal expression of their knowledge and practices towards a logical and structured explanation that can easily be converted into computer-speak. Where the transparency of the conversion process is respected, and the levels of uncertainty and lack of knowledge are clearly identified and accepted, the model is validated socially. This process poses the problem of the non-generic nature of the models produced or at least a validity linked to a particular context.

Lastly, many participants underline the difficulty in coordinating this type of approach, as the way in which debates are led and familiarity with tools used are critical to the degree of success of the exercise (Chess and Purcell 1999). Companion modelling requires the facilitator to have coordination abilities, a minimum knowledge of the socio-ecological processes in play, a certain ease with computer tools and an undoubted ability for dialogue and exchanging information. This makes the standardization of the approach and its broader diffusion even longer and more difficult.

A Flexible Approach, A Heterogeneous Implementation

Choosing companion modelling does not, therefore, appear to restrict too much the ways of getting involved in a decision process. The assessment reports show tremendous dependence on the coordinator of the approach. Just like every participatory approach, the diversity specific to companion modelling experiences demonstrates the need to characterize and explain the process: the label 'ComMod' is not enough to let the participants in the approach know what they are in for. Without this explanation, the participants may expect something different from what will actually happen; thwarting this expectation risks discrediting all future implementations of the approach. This does not involve characterizing explicitly the forthcoming process, including the possibilities for adaptation. It can relate more simply to specifying the role taken by surveys in the process, the nature of the interaction devices between participants that may be introduced, and the place given to the model in the process.

Indeed, this flexibility is also a constituent of the companion modelling stance. It offers a contingency in the conditions of each case study, which is not restricted to taking into account the know-how and favourite subjects of each bearer of a case study. Companion modelling is subject therefore to the condition of triple contingencies in order to understand changes proposed by Miettinen and Virkkunen (2005): contingencies in time, stakeholders present and available artefacts (Miettinen and Virkkunen 2005). By adapting to the questions that evolve at each iteration, by involving the available stakeholders and by mobilizing the existing artefacts or those co-constructed during the process, the commodian takes on the role of a DIY enthusiast or craftsman, fashioning the dynamic of collective decision-making by relying on his context and the stakeholders setting it up (Innes and Booher 1999). This is the basis for the entire companion modelling issue: remaining flexible and iterative whilst maintaining specific principles and a common stance. This is the direction taken by the collective work and reflections in the ComMod group, which is reported in the following chapters.

Chapter 3
The Commodian Stance: Interpersonal Skills and Expertise

William's Daré, Cécile Barnaud, Patrick d'Aquino, Michel Étienne, Christine Fourage and Véronique Souchère

As suggested in the introduction, companion modelling is a stance, not merely an approach. By stance we mean someone's moral attitude. By analogy with the analysis of literary stances by Meizoz (2007), we consider that the scientist involved in 'implicated' research becomes part of the social play. So in a companion modelling approach, he goes beyond the boundaries of the scientific field. He enters the public arena and sends out a certain image of himself. His stance presents a double dimension, that is, rhetorical and action based. Stance is conveyed, therefore, by taking a moral, affective, social, philosophical and political standpoint that leads to action. The commodian stance is thus our term for referring to a particular way of considering the position of the researcher in the relationship between science and society.

In Chap. 2 we set out the key components governing our approach (e.g. protagonists, sequences, etc.) in order to state the invariants noted during its

W. Daré (✉)
UR GREEN, CIRAD, Montpellier, France
e-mail: william's.dare@cirad.fr

C. Barnaud
UMR DYNAFOR, INRA, Toulouse, France
e-mail: cecile.barnaud@toulouse.inra.fr

P. d'Aquino
UR GREEN, CIRAD, Dakar, Senegal
e-mail: patrick.daquino@cirad.fr

M. Étienne
Ecodevelopment Unit, INRA, Avignon, France
e-mail: etienne@avignon.inra.fr

C. Fourage
UMR ESO-CARTA, Université Catholique d' Angers, Angers, France
e-mail: cfourage@uco.fr

V. Souchère
UMR SADAPT, INRA, Grignon, France
e-mail: souchere@grignon.inra.fr

M. Étienne (ed.), *Companion Modelling*, DOI: 10.1007/978-94-017-8557-0_3,

implementation. In Chap. 3 we shall explore the stance adopted by the commodian when facilitating a companion modelling approach.

To characterize how ComMod processes operate we need to revisit the fundamental principles of this facilitation or 'animation'. Animate derives from the Latin *animare*, which means 'breathe life into'. Whether or not you choose to restrict yourself to its metaphysical, mythological or artistic sense, the term has long been linked with principles of creation, movement and life. Derived from popular education, group 'animation', or more usually in English, group 'facilitation', has been a professional activity in France since the 1960s, with its vocational classification, governed by a code of ethics, occurring in specific domains (e.g. sociocultural, business, association, training, education) (Poujol 1994). Its functions have diversified, with the group facilitator playing a role in the socialization of certain individuals in a fun or recreative, educative or cultural, orthopaedic or innovative way. However, the understanding of the facilitator's activity is subject to controversy in sociology. Augustin and Gillet (2000) suggested that there are two trends. Some, emanating from Christian or lay circles of influence, consider their actions as liberating, instilling vitality and vigour into the group and the lives of individuals in society. Others take the opposite view, that the sole purpose of facilitation is to establish social control and sustain the relationships of domination and social norms imposed on those individuals with the least social capital. Under this deterministic vision, the facilitators combine to manipulate and condition individuals. Sometimes a collective emancipation movement, sometimes a society preservation instrument, these two visions of facilitation are contradictory. They are based on premises and philosophical perspectives that render them virtually irreconcilable (Gillet 1995). Starting from these contradictory views, the critical or humanist sociologies do not take into account the wealth of stakeholders' interactions in a group and their relationship with the environment, nor grasp the facilitation as *praxis*, a way of acting rooted in the past and looking towards the future, which requires the facilitator to show genuine strategic intelligence in supporting the stakeholders in a given social situation (Augustin and Gillet 2000).

Companion modelling is a participative approach. Participative approaches have been developed with the agreement of international development institutions since the 1960s. This was done in order to take into account local stakeholders as legitimate partners of the research and development projects that affect them (Olivier de Sardan and Paquot 1991; d'Aquino 2002). Yet the many ways of applying the concept of participation (Pretty 1995) have fuelled widespread criticism, including the performance of participative approaches. Barnaud (2008) distinguishes between three major types of limitation: intrinsic, horizontal and vertical. We will consider the first type here: we shall return to the horizontal and vertical limits described by Barnaud (2008) later in the chapter. Firstly, intrinsic limits relating to the polysemy and ambiguity of the very concept of participation, or calling into question the objective of participation in which stakeholders should ultimately be released from exogenous driving forces. These first criticisms question the genuinely participative nature of methods imposed by outside participants, whether or not they are well intentioned. They also emphasize a lack of understanding of the complexity of local social situations, when maximum participation of all stakeholders at all stages in the process is requested (Barnaud 2008). Some authors underline the decisive effect of the designer's behaviour on the progress of his approach (d'Aquino 2002; Chambers 1994a; Scoones and Thompson 1994; Water-Bayer and Bayer 1995). Where does the commodian stand in implementing the approach: do they encourage the local stakeholders to reflect and decide on options for themselves or guide their choice of development? Do they act to shrug off their own perceptions, their own choice of development, frequently discipline based, to help the society to decide its future for itself? Is their action liberating or, on the contrary, manipulative?

To respond to these questions we must return to the stance described in the ComMod Charter (ComMod 2005) to demonstrate how it stands out from other participative approaches. We shall then reveal how adhering to the principles of the charter is not rhetorical but the actual way in which the commodian tackles his field and facilitates the companion modelling approach. We are seeking here to demonstrate that the commodian stance, as described in the charter (interpersonal skills), and its place at the start of the ComMod process, are the forerunner of practices, methods and techniques (know-how), which are activated in order to 'animate', in the sense of give life to and/or invigorate, the approach and the communities involved.

The Commodian Stance: Principles of Original Interpersonal Skills

Principles of the Commodian Stance

The stance of the designer-facilitator[1] of a companion modelling approach was established in the first version of an ethical charter published initially in English in the *Journal of Artificial Societies and Social Simulation* (ComMod group 2003), then revised and amended in the French version (Collectif ComMod 2005). This version was widely commented on in issue 13 of the journal *Natures Sciences Sociétés*. We indicated in this charter that companion modelling was an 'implied' research approach, imposing a special relationship between science and society. In fact, we consider that our research practices must be assessed according to traditional procedures and criteria in science but also with respect to questions raised in the field. We recognize the uncertainty of decision-making situations in managing renewable and environmental resources and the legitimacy of multiple

[1] We shall henceforth use the term 'designer' to denote the leader or sponsor of a ComMod approach, and the term 'facilitator' to designate their activities in a collective key moment (see below).

viewpoints in dealing with this uncertainty, despite these being occasionally contradictory. One issue in our research action is considering, during an iterative exchange-focused process, the comprehension and analysis of these various viewpoints with all stakeholders in the social and ecological system in question. To achieve this, we believe that this process must be capable of explaining the implicit hypotheses assisting in building up the speeches, representations and perceptions of each individual. The commodians are not bystanders in this implicated research. They participate in the process that they are facilitating and, therefore, must also explicitly render their vision of the world. We view scientific knowledge as just one type of knowledge among all the others present. The companion modelling approach relies on intermediate objects constructed with the stakeholders. These intermediate objects will help explain hypotheses and formulate scenarios of change in their system, thereby exploring the various options and thus dealing with the uncertainty.

Originality Among Other Participative Approaches and Modelling in Natural Resource Management as Seen by the Commodians

From the beginning of the companion modelling approach, researchers took a novel stance to distinguish it from two different practices, that is, natural resource management (NRM) modelling and participative approaches. This attempt at differentiation initiated 12 years ago is continuing and has conveyed since its origins the meeting of two requirements concentrating around a new concept—companion modelling.

For researchers from the world of participation, companion modelling was a means of moderating far more the influence of the subjectivity of the researchers leading the participation. Firstly, inconsistent viewpoints can be retained under the chosen formalization (object-orientated modelling). Thus, it limits the involuntary tendency of the researcher to twist the perceptions of others to suit his own when he attempts to reorganize the diversity to make it coherent. Secondly, it introduces stakeholder participation well upstream of the usual participative approaches and it relies on far more in-depth testing of the proposed viewpoints (diagnostics) by bringing these first diagnostics into play (*cf.* simulation).[2]

The new concept of companion modelling was noted by researchers from the world of NRM modelling for its attachment to a 'post-normal' posture (Funtowicz and Ravetz 1993), whereby a model designed by a disparate collection of scientists and non-scientists could be more valid and relevant than one emerging from purely

[2] Is technical improvement in our practices really going to be enough to preserve resources? Will it really be possible to correct current tendencies with the current rules? Will improved protection of resources really have a significant economic impact on our revenues?

academic theory. As questions continued to be raised over expert systems, these researchers considered that modelling a complex system could only be relevant and valid through far greater integration of the lay stakeholders' perceptions of the system. They also felt that NRM could not be reduced to only technical and scientific issues. The stakes, practices and motives of stakeholders should be taken into account upstream of scientists defining the issues or improvements in the system.

The companion modelling approach was conceived from the start to stand apart from an approach where lay stakeholders are considered as agents in a system where the scientist had to understand their behaviour to be able to include them in his analysis (as was the case in the games theory applied to NRM). We focused more on considering lay stakeholders to be partners with the scientist in a joint understanding of the behaviour and analyses of all concerned (including the scientist).

Once this became a joint issue for members of both communities (participative approaches and NRM modelling), the specific features of companion modelling continued to take shape as experiments and methodological concepts progressed, as with progress in other trends of NRM modelling and participation. It seems today that the originality of the companion modelling posture lies in this initial positioning as well as in the explanation of its ethical and methodological implications.

Ethically, what makes this approach stand out from its neighbours, the participation and modelling worlds, is that the ComMod designer recognizes that his analysis and knowledge are only one viewpoint among many in the reality of the field and its related issues. In the ComMod Charter (Collectif ComMod 2005), which describes the ethical position of the group, we have clearly stated our awareness that our intervention is subjective. Accepting our subjectivity, the prism of our view and the particularity of our viewpoints urged us to control it as best we can, hence the need for the formalization expressed in the ComMod Charter. This still distinguishes the ComMod process from the majority of other participative approaches. They have an objective vision of what is 'good', 'fair', 'shared' or 'primordial',[3] like the critical sociology of participation that proves without question the subjectivity of participative approaches but does not recognize the subjectivity of its own analyses and judgement values.

Methodologically, the 'focal point' setting companion approaches apart is still, and this relates to the point above, the aim to express the diversity of viewpoints in the system, rather than starting by constructing a common representation. Helping to distinguish between and formalize various viewpoints in the system make up the first essential key stage in any companion modelling process. This is done without

---

[3] For example, this also still distinguishes companion modelling from other participative approaches developed around certain environmental issues, which consider them as 'meta-issues' that can only be reconsidered by the partners. Another example can be found in other modelling approaches, which consider the perceptions of other stakeholders in the system to be modelled as less rigorous versions but which must be integrated with (their?) theoretical conceptualization.

giving priority to consistency between these viewpoints (or with the designer's viewpoint). The continuing methodological process is designed to take this diversity into account in the best way possible (i.e. comprehension and validity of the diversity by the assembled stakeholders, conceptual formalization of this diversity and so on). Recognition of the existence of these multiple viewpoints helps build a representation shared by all protagonists, without this inevitably culminating in a common, unique and consensual representation.

Companion modelling is today pursuing its specification and, therefore, its identity relative to the other approaches, by widening its questioning on new themes (see the remainder of the work, for example, Chap. 5, taking the social context into account and Chap. 11, how to comprehend the multi-levels, etc.).

Originality of the ComMod Approach as Perceived by Other Researchers

The originality we have just described is also perceived outside the ComMod community and was expressed in comments made when the ComMod Charter was published in the journal *Natures Sciences Sociétés* (Collectif ComMod 2005). These comments encouraged us to state certain principles we believe to be fundamental to our approach and render the stance of the commodian relatively novel compared with the facilitation methods of other types of participative approach or approaches proposed by groups of researchers working on the design of regional projects (Lardon 2005). We have retained six.

(i) The initial definition of the question and partners to be involved in the approach. Regardless of whether the request comes from researchers, development agencies or political decision-makers, the commodian must state the nature and epistemological status of the questions from the field (Laloë 2005). They must then investigate, in the context of an interdisciplinary research group and stakeholders in the field, of what does the involvement of each discipline consist and the various types of knowledge present around the table.

(ii) The insertion of the modelling activity in a collective thought process. This 'participative modelling' results in a representation of the 'shared' reality, even a search for multiple consensus that adds a new collective dimension (Hervé 2005). The approach has a duty to replace the stakeholders (i.e. scientists, managers and operators) in a well-regulated set design. Depending on their role in society and in relation to the question asked, each individual has the legitimacy to suggest interconfrontational representations. This is realized during a hypothetical deductive process, which ultimately opens up a series of hypotheses accepted by all and developed into scenarios.

(iii) Leading the approach relying on collective key moments in priority. During key moments the commodian uses tools and suggests methods to explain any potentially contradictory viewpoints and put the stakeholders in a situation for action. This involves encouraging better understanding of the relative positions of each individual and their effects on the system dynamics rather than a passive application of recommendations by experts (Hervé 2005).

(iv) The essential contribution of accessible formalization methods. In a complex situation or in contexts where science sheds little light on the matter, the effort of formalization can provide invaluable assistance in conceptual clarification and communication between stakeholders with different knowledge and divergent interests (Lobry 2005). The commodian will, therefore, suggest methods to stimulate reflective thinking on the framework of rules for each individual's activities, and be capable of explaining these rules whilst respecting scientific methodology and the stakeholders involved.

(v) 'Good conduct' in terms of modelling. Part of the originality of the approach, as emphasized by Pavé (2005), lies in the type of objects modelled and the status given to the modelling and the modeller. The models used, which are frequently co-constructed, are never in a stable state, therefore, never truly validated, but they do not play a normative role. They serve basically as mediators in the social dialogue to explore a field of options. The modeller's major concern, therefore, becomes transparency in the field and the model's operating limits, and emphasizing the hypotheses governing its development.

(vi) The refusal to manipulate using the model's outputs. Given the complexity of the situations faced, it is important not to ignore the possibility of the suggestive power of models, their plasticity and the events they are used to produce being used for manipulative purposes (Mullon 2005). The commodian, therefore, devotes part of his energy to keeping a firm grip on the quantitative or spatial validity of the data produced or to preventing the aesthetics or visual aspect of the outputs proposed by the model from obscuring the relevance of reflective thinking. For example, it is important to know how to assess if a simulation proving statistically incorrect remains spatially acceptable, in other words capable of correctly representing the general configuration of spatial structures (Guermond 2005).

The ComMod Charter sets out the constituent components of the commodian's interpersonal skills. Taking into account that he operates in a social and ecological system where others have already built up bodies of knowledge, both scientific and non-scientific, he considers them all as legitimate and does not place science on a pedestal. Aware of his involvement in the field, he knows that as soon as he intervenes, his action is interpreted by the stakeholders, turning him into another component in the system. He attempts to be as explicit as possible about the hypotheses forming the basis for his representation and the milestones along the way in his approach, so that it and the results obtained are considered legitimate by all concerned. As the system studied is viewed as complex, uncertain and changing, there is no ideal solution, simply agreement on the principles for developing solutions.

Companion modelling experiments have been developed over 10 years in a variety of social, ecological, political and economic contexts. The commodians have facilitated these approaches by attempting to apply the theoretical principles whilst adapting them to the reality of the field. Thus, how are these interpersonal skills, this stance, conveyed in the multiple and varied expertise developed in these various areas? This question is analysed below from the perspective of initiating the approach; to what extent is how to tackle a field dependent on the interpersonal skills of the commodian. Later we shall turn to the practical translation of the stance in how the commodian runs the entire process and, more specifically, the collective key moments.

The Commodian's Role in Initiating the Approach

The way social demand is taken into account or, more widely, in the way the approach is initiated is essential when running a companion modelling approach. It will influence the way the issue is elicited and adjusted to the context, the partnership to be established, the stakeholders involved and the definition of roles played by the designer during the process.

Different Ways of Initialization of a Companion Modelling Approach

An analysis of various case studies reveals much variation in the initiation of companion modelling approaches, especially in the expression of social demand behind the implementation of a ComMod approach. However, this social demand, which justifies applied research activities, is frequently vague in the field. A researcher's first task is to deconstruct the expectations of social subjects, their formulations of the joint meaning and build them back up into research issues with which the scientist can then deal (Castel 2002). A companion modelling approach is subject to this same reality. Thus the social demand behind the initialization of a ComMod approach varies between two extreme situations:

- firstly, a specific demand, formulated by an identified sponsor (the commodian thus has an explicit mandate to deal with the subject);
- secondly, a vague, non-explicit demand, without a specific group to deal with it.

In the first situation, the local stakeholders have learnt of the approach (through exchanges with stakeholders in another field where it has been set up or through a simple presentation of a case study) and call on a commodian to set it up in their field. The approach is totally unknown locally in the second situation. As no one is familiar with the companion modelling approach, it is up to the commodian to initiate the approach after a context analysis. A midway solution has been encountered where a local institution has a mandate to work on an issue but has no idea which methodology to use.

This diversity in approach initialization impacts on the legitimacy of the intervention and the designer solicited. Where the commodian is part of a new field, he must construct a necessary legitimacy to invite the protagonists of the approach and construct the group he wishes to lead in it. He normally constructs his legitimacy based on the recognized legitimacy of other stakeholders, persons and resources already in the field. This affiliation to a partner must be carefully thought through as it determines the way in which the approach designer will be perceived (see Chap. 6). The initial analysis of the context plays a fundamental role here.

In these various scenarios, the respective influences of the commodian and other stakeholders to convert a potential demand into a workable question for the companion modelling process vary greatly, more so when the demand is not always clearly expressed. Whether mandated or not, the researcher must make an effort to translate and deconstruct the expectations of local stakeholders before co-constructing this demand with them. This way the commodians act as midwives, by sketching a Socratic method.

From Social Demand to Supporting the Group Taking it on

Once this social demand has been identified, translated and reformulated into something comprehensible to the scientific activity, questions of identification, definition and involvement of the group(s) in charge, that is, stakeholders in the companion modelling approach are raised, as in all participative approaches. Before seeing how commodians address these questions, however, we feel it important to revisit briefly the criticisms of participative approaches regarding their insertion in a pre-existing social and political context.

The first criticism, so-called 'vertical', condemns the localism of participative approaches, which can only claim success locally and for a limited time due to the mistrust of decentralized institutions; yet these institutions are in a prime position to make this success endure and incorporate it into the decentralization process (Barnaud 2008). The so-called 'horizontal' limitations are that participative approaches capable of reducing social inequalities are rare, despite their stated goal. Due in particular to a naïve belief in the existence of a homogeneous community with a strong social cohesion, they have frequently referred to representatives unique in their community, thereby treating with disdain the diversity of local hierarchical structures, interests and strategies and giving a boost to local elites, concentrating new sources of power in their hands (Barnaud 2008).

Noting these criticisms, the commodian is fully aware that he is intervening in places that are not devoid of institutions or power relationships. Thus more often than not they analyse the social context in which they are planning to intervene. This analysis can be more or less advanced (see Chap. 5). Let us now see how that translates from the point of view of inserting the approach into existing consultation opportunities (see Chap. 5 for an in-depth description of the context and more specifically, the formulation of social demand). An analysis of case studies reveals three scenarios.

In the first, the commodian intervenes in a field that has no group to deal with the question that the companion modelling approach wishes to address. In this case, they go with the institution likely to be the focus for reflective thinking, and identify the relevant stakeholders to co-construct the group designated for the companion modelling approach. This cannot happen without analysing the context so that the options proposed by the institution can be discussed. For example, in the Nan case study, in the context of open conflict between a national park and highland communities, there was no pre-existing arena for consultation, no incipient dialogue, and the relations between the various parties *in situ* were characterized by mutual distrust and misconception. In an attempt to establish a dialogue between these stakeholders, the commodian suggested an arena for consultation, for which participants were chosen so that the various interests at issue (especially in villages) could be aired. The legitimacy of such a group was, however, weak. The choice of participants was discussed systematically with the legal representatives of the institutions involved (i.e. village leaders, head of the national park) to give extra weight (Barnaud 2008). However, although the ComMod process improved relations between the park agents and the villagers up to a point, the arena for consultation created did not last beyond its end.

In the second scenario, analysing the intervention context identified the existence of multiple groups dealing with related questions. Thus in the case of Domino Senegal, the issue of water and land management was dealt with in multiple consultation areas (Diop Gaye et al. 2007). These areas nevertheless remained disjointed and it was impossible to address the question of land allocation coherently, with the added complication of integrating several organizational levels. The various social and economic studies identified the institutions and relevant stakeholders with which to address these issues under a companion modelling approach. A user committee was set up around the approach with committee members taken from local, municipal, regional and national institutions, therefore, the issue of land allocation could be addressed as a whole.

The third scenario covers situations where the question the commodian is seeking to address is already mandated to an institution responsible for organizing the consultation on this topic. However, several situations can appear from the point of view of social demand. For example, the urban area of Nîmes-Métropole submitted an explicit demand to the commodian. In Domino Réunion, the demand was instigated by the commodian. In La Réunion, although all stakeholders in the regional development, regardless of their scale of intervention, all noted competitive dynamics in using a restricted territory, the problem of ensuring consistency in multiple choices was evaded, overshadowed by sectoral issues. The commodians, therefore, formed an initial group of researchers and stakeholders from the main sectors occupying the area to construct together a prototype representation of intersectoral interactions. Only then did this first group, with its prototype, lodge a request with the unit in charge of revising the Regional

Development Scheme (SAR) to continue with the companion modelling approach incorporated into the consultation process (Daré et al. 2008).

Finally, these examples suggest that the commodian can sponsor the approach (SylvoPast), sponsor and fund those developed under certain research and development projects (e.g. SosteniCAP, AguAloca, Njoobaari, etc.) or simply be a companion to it (Ouessant, Nîmes-Métropole, etc.). Note also that these various positions can change during the approach cycles. Thus, the first loop in Domino Réunion was led by commodians then sponsoring the approach. They then became companions in the approach when the SAR unit took over the running of it.

Faced with criticisms on the naivety of sponsors of participative approaches, the commodians make no claims in believing they have arrived in virgin institutional spaces or egalitarian social systems where interactions between players are always consensual (see Chap. 6). However, the institutions do not always have the mandate, legitimacy or recognized neutrality required to sponsor a companion modelling approach with regard to its principles. This explains why the commodian has been forced to propose constructing new arenas of consultation grouping relevant stakeholders to deal with a given issue. By so doing, the commodian is aware of creating an arena that will find a niche in interactions with other formal or informal arenas already in existence but still likely to influence the debates within the ComMod arena. How and to what extent do pre-existing and newly created arenas influence each other? The question remains open and the evaluation suggests a few ideas for reflective thinking and improving knowledge on this point (see Chap. 7).

Roles of the Commodians and Facilitation Techniques Employed During an Adaptive and Iterative Approach

We have focused on two levels of analysis to characterize the role of the designer during a companion modelling approach. The first covers the entire approach. The second focuses on facilitating collective key moments, particularly the simulation phases.

Who Facilitates and at which Phases in the Process?

Our analysis first focuses on the entire companion modelling process. We shall initially state who are the study designers and then analyse when precisely they intervene.

To understand the nature of associations between the different statuses of the designers during the various phases in the process we must first return to the question of their legitimacy. The commodian intervenes in a social system that, socially speaking, is not native to him. Companion modelling in this context aims firstly to facilitate collective reflective thinking between stakeholders who have a certain legitimacy in the eyes of the commodian as well as other members of the invited group. (We shall return to the legitimization of protagonists invited to take part in the process when analysing the initial awareness-raising phase). However, we make the assumption that in turning social phenomena into science by collaboration between researchers and their lay counterparts, these stakeholders activate the tools and dynamics to test other stakeholders, indications of their legitimacy that borrow from a range evolving between the two approaches described by Weber (Daré et al. 2004). The first covers the 'subsequent legitimization of relations of domination' and the second considers the legitimate order as a guide, a convention which the stakeholder holds in his own intimacy (Corcuff and Lafaye 1996). By so doing, legitimacy is not a state, but is built up in the interaction with other participants. Legitimacy can be transitive (the designer takes his legitimacy from the institution guaranteeing the process), self-improving (change based on interactions between the participants and/or the designer), circumspect (the legitimacy of a role-play session leader is not necessarily enough to facilitate a project restitution, a sponsor may not possess a sufficiently recognized legitimacy to facilitate the design phase) or, on the contrary, open (the legitimacy of a participant in the approach can be used to boost his legitimacy in other arenas of the consultation).

A Facilitation Team Comprising Members of Different Statuses

Here, we summarize briefly the different types of protagonists discussed in Chap. 2 by distinguishing the forms of knowledge mobilized during the approach, that is, lay, researcher, technician, institution, commodian and student. We are simplifying it to describe the status of designers by considering students—apprentice commodians—as commodians as these two types do not belong to the social system studied (Fig. 3.1).

In most cases, the entire process was led by a team made up of members of different status. Only two experiments were run by commodians alone. The designer of a companion process is never alone whatever the circumstances; the approach is driven by a -facilitating team with set roles and intervention times split between its members.

Who Facilitates Each Phase?

Our companion modelling approach can be divided into six phases described in the canvases (see Chap. 2): awareness raising, design, implementation, validation, restitution and simulation/game. These phases do not necessarily follow each other but they make up a complete cycle. Based on the canvases of 27 case studies, we analysed the status of individuals responsible for facilitation in the main phases.

Responding to a demand they have identified or for which they have been solicited by a stakeholder or institution, the commodian facilitates the first stage of making

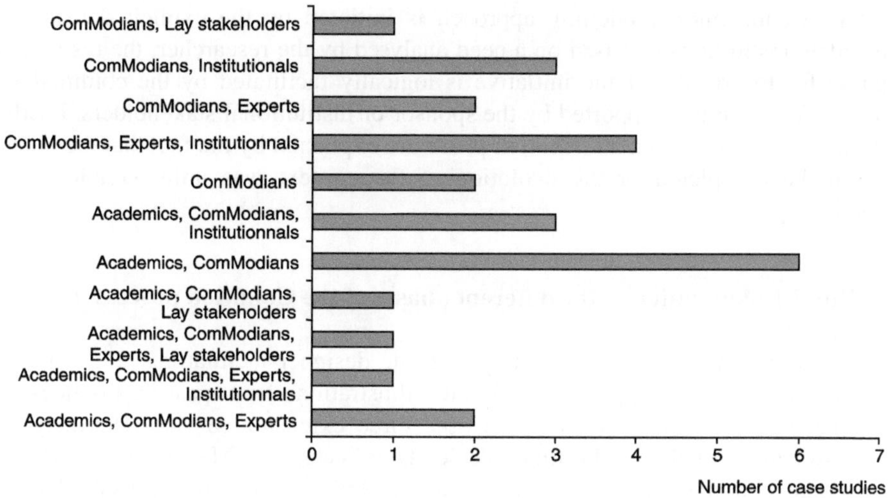

Fig. 3.1 Association of various types of facilitator intervening during the companion modelling process. This figure presents the number of case studies per type of association acting as facilitator

contact with other local partners alone or with a local sponsor. In 10 out of 27 cases he was supported by local stakeholders (i.e. academic, expert or institutional).

Once the social demand has been formalized, the aim of this phase is to raise the awareness of the various protagonists to the issues, constraints, limitations, requirements and advantages of the companion modelling approach. In this phase, the commodian relies on a presentation of past experience, thus building up their legitimacy in the eyes of local stakeholders. The commodian can call on institutions or local stakeholders acknowledged to be relevant to address the topic in question and thereby boost his position. A transfer of legitimacy thus occurs. The rare circumstances that do not require the commodian to organize this awareness-raising phase are when a research project has been defined clearly by non-commodian researchers already installed in the study field (e.g. Luberon, Ubon Rice Seeds and Tarawa). These researchers normally have submitted an intervention request to the commodians. The Luberon and Larzac cases are explained by the existence of strong local groups or by filiation of cases studies, so much so that it is institutions or institutions and experts that presented the proposed experimentation.

The need for familiarity with the tools and methods for formalizing knowledge and model design and implementation explains why in particular only a few academics manage to facilitate the model design and implementation phases by themselves. This was true for the implementation of the Ouessant case study. Note that for Ouessant this was a desire expressed by the project designers, which, given the technically advanced formalization tools used, meant that one participant had to be capable of taking charge of the development/modelling aspects. A special training course was organized for him. These phases are normally facilitated by a commodian alone or in a group of facilitators.

As a companion modelling approach is initiated by the explicit demand of mandate institutions or based on a need analysed by the researcher, the restitution phase for the results of the initiative is logically facilitated by the commodian, more often than not supported by the sponsor or institutional stakeholders. Lastly, those rare cases without a restitution phase are explained by the fact that they have yet to be completed or the evolution of the context was unfavourable to its continuing.

Box 3.1: Animation of the different phases of the ComMod process in the Pays de Caux

During the awareness-raising phase, the designer-facilitator of the companion modelling approach used some illustrations from past experiences. She organized discussions to help the three extension services (Farmers' Association of Seine Maritime, Veules-Dun Watershed Management Committee and the Regional Association for Soil Preservation) consulted to formulate the common question they wanted to deal with in the framework of a ComMod process: how to engage and support a dialogue between stakeholders on the management of erosive runoff at the watershed scale that would include the implementation of both better agricultural practices and landscape structures (e.g. grass strips, storage ponds). The main phases of the approach, the usable tools, the implications in terms of workload, etc., were explained to the stakeholders. The facilitator took charge of running the following steps: design, implementation, exploratory simulation and diffusion. She was assisted by students (in particular for data collection, design and animation of the role-playing game) or by commodians (in particular for the -implementation of the models).

The ComMod processes are organized around collective key moments. The Pays de Caux case study was the object of accurate monitoring of trade during the collective key moments and the data were recorded in a logbook to reconstitute the network of participants present. The analysis of the social network formed around the ComMod process showed the centrality of the commodian, the designer-facilitator of the process (the white square in the centre of the Fig. 3.2).

Four groups can be distinguished in the figure. The first brings together participants involved in the design of the model and role-playing game. The size of the symbols used allows us to identify the hard core of this group. Two other groups show the participants involved in the two sessions of role-playing game organized in July 2007. Finally, the last group is composed of leaders of watershed management committees of the Seine Maritime region who attended a meeting on the restitution of the game results to make them sensitive to the use of this game with stakeholders from their own territory. Stakeholders (light grey circles) played a part not only during exploratory simulation exercises, but also in the design phase. The small more isolated

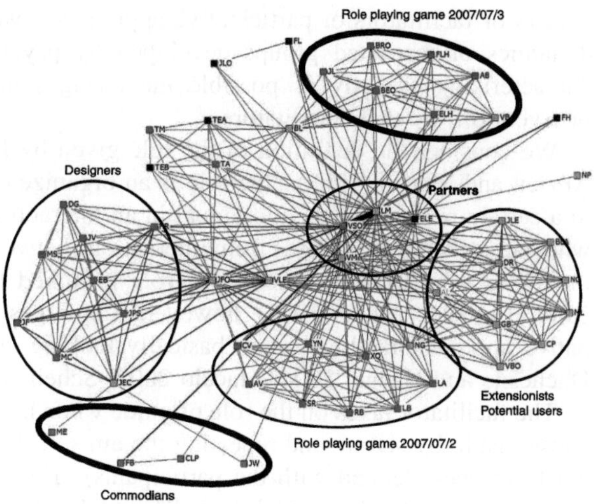

Fig. 3.2 Relationships among participants and the facilitator of a ComMod process. Social network based on the logbook of the Pays de Caux case study. The thickness of the line is proportional to the time shared by 2 participants. The *dot colour* indicates the main type of knowledge beared by each participant: scientific (*red* for Commodians, *black* for other scientists), technical (*gray*) or lay (*blue*)

groups, because they were less involved in terms of duration in the project, mainly included scientists (black squares) belonging to the ComMod network or scientists and students involved in the evaluation of the approach. Occasionally, technical experts (dark grey triangle) were also consulted to gather additional data or stakeholders and students (light grey circles) to test the role-playing game.

With the results of this final phase we return to the issue of the legitimacy of members of the facilitation team. The fact that legitimacy is partly contingent on the approach and partly a result of external interactions explains the varying associations of facilitator status for each of the ComMod phases. The strong presence of institutions in facilitating the awareness-raising and restitution phases is open to question. Is this not a means of supervising the system? Does this not open up the risk of one minority instrumentalizing the entire approach, with the companion modelling thus lapsing into other participative approaches? Or, on the contrary, is the presence of institutions necessary to guarantee the experimentation and sustain the results? Finally, these results show that the commodian, whether alone or supported, occupies a central position in facilitating the phases of the initiative (see Box 3.1). This raises the question of the complexity of the role of facilitator in the process and the abilities required to take on this function.

The Various Roles Played by the Commodian

Now, we analyse the various roles played by the facilitators to inspire and bring to life the groups formed around a ComMod approach. To achieve this, having reminded ourselves of the main roles traditionally performed by socio-cultural leaders or facilitators of participative approaches, we shall use an analysis of the dynamics of restricted groups developed by psychosociologists as a basis for characterizing as fully as possible the changes in participant groups and the interventions by their facilitators.

We consider the definition of the role given by Rocheblave-Spenlé quoted by Anzieu and Martin (1968): 'the role is an organized model of behaviours relating to a certain position of the individual in an interactional whole'. The interactional whole considered here arises from interactions between participants throughout the modelling approach. The notion of role is linked to concepts of standards and cultural modes (Linton 1977), however, it emerges that socio-cultural or participative approach facilitators are basically looking at four roles (Creighton 2005; Duchesne and Haegel 2004; Maccio 2002; Schein 1988).

The facilitator takes on the role of guide when he helps the group remain on the chosen path, by stating and repeating the aims of the meeting (which may or may not have been defined with the participants). They restate the group's operating rules agreed with the participants. They take the decisions on the appropriate methods to achieve the group's aims. They attempt to reduce intrusive interactions that are full of emotion and likely to force the group off course compared with its aims. Like a moderator, they must encourage clear, accepted communication by regulating the contributions and ensuring that each individual feels they are understood and listened to (if necessary by reformulating or summarizing an intervention). They also act as a mediator to prevent and calm tensions and disputes by specifying prohibited behaviours, by allowing feelings and problems to be expressed within a framework and set time (if necessary by reformulating accusatory comments and defusing charged emotions), by suggesting a procedure to resolve emotional tension and by putting forward a range of options when the group finds itself in deadlock. They also guarantee the unity of the group by trying to maintain or even increase its cohesion. Here they must mark the consensus stages to stimulate the group and boost morale, summarize and clarify the direction and propose tools and alternatives when the group gets stuck.

During the companion modelling process the participants are put in situations of action, reflective thinking or exchange, which cause them to interact. This not only encourages the promotion of understanding between the various parties involved but also nurtures or even creates a social link. Henceforth the various participants in a companion modelling approach, regardless of their level of involvement, become temporary members of a group, an 'us', which can be short-lived but very real as long as the exercise lasts (see Chaps. 7 and 10). This 'us' is an opportunity for interpersonal, even interinstitutional exchanges, when they legitimately represent institutions. The stakeholders get to know a topic during these exchanges and question, transform and translate it, projecting their own issues into it to turn it into a shared object, the medium for their discussion within the group. If the participants continue the process despite the social and economic costs inherent in the approach (Levrel et al. 2009), it is because they find (or hope to find) an interest in the collective interaction produced in this action arena.

We therefore consider that regular participants form a genuine group in a companion modelling approach.

Psychosociological work on the dynamics of restricted groups has developed since Lewin (1959) showed that the behaviours and interactions of their members fluctuated between phases dominated by emotions or by the activities causing them to meet (Anzieu and Martin 1968; Bales 1950; Hare 1994).

Analysing the life cycle of a group of individuals participating in a ComMod process shows that the roles of the facilitator change in line with its various stages (Fig. 3.3). For example, when the group is formed, the facilitator's team presents the subject to the people invited, which results in questions being asked by the group. The individuals present watch themselves trying to understand, to gauge the relevance of each other's presence and the suggested organization. In this phase, the facilitator specifies the group's objectives, explains the move between the subject of the invitation and the issue asked of the group, encourages discussion between the members and presents the methodology. Tensions combine during the various stages in the life cycle to perform the task. However, as the approach advances and they meet successively, the stakeholders learn more about each other, a sense of trust builds up between them and they gain a better understanding of others' viewpoints, without necessarily agreeing with them. All this combines to reduce tensions or at least better cope with them.

The facilitator thus sometimes acts as a moderator, sometimes as a mediator and sometimes to ensure the group dynamics. However, he does not guide in the sense that the facilitator of a ComMod approach is familiar with the overall direction (the subject that is going to be dealt with), but he is aware that the group is potentially faced with a choice of multiple paths to achieve a single objective. This is one major distinction compared with traditional facilitator roles and is a direct spin-off from the post-normal and recognition stance of the uncertain nature of changes in the social and ecological system in which he develops his approach.

Intense Collective Key Moments of Exchange that 'Fashion' the Group

Chapter 2 showed that the companion modelling approach undergoes periods of intense collective exchanges and collective key moments, which alternate with more stand-alone periods of compiling/taking data, design and reflective thinking. Seven logbooks were used to record the information required to analyse the collective key moments. Most collective key moments are organized for meetings and discussions with local stakeholders, commodians or between project sponsors. The design and simulation phases are performed systematically with commodians or apprentice commodians who have collected the information exchanged during these meetings. In addition, when participants are put in situations for action, as part of a role-playing game or when creating the conceptual model, the members of a restricted group find themselves placed in particular circumstances. For

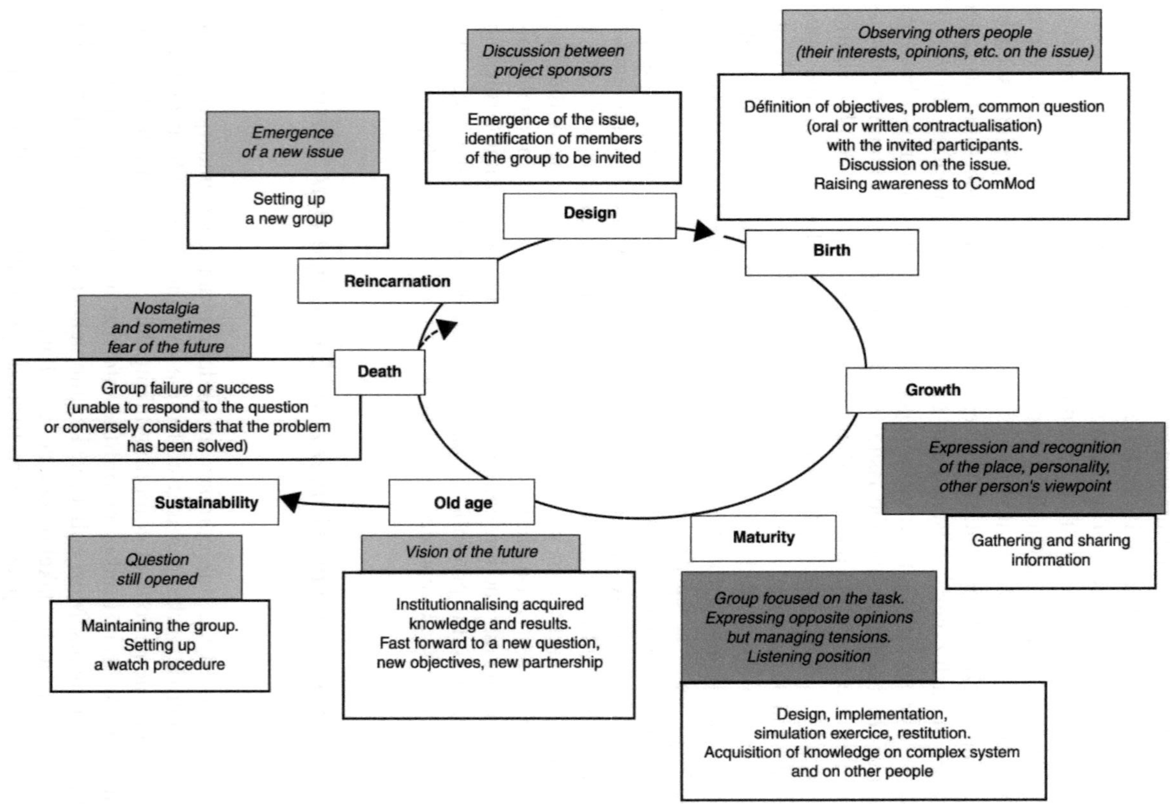

Fig. 3.3 The stages in a ComMod group life cycle. *Key* the *gray boxes* describe the atmosphere between the group members and the modalities for action and exchange. The deeper the colour, the greater and tensions and the more interactions are emotionally-charged the interactions. The group's expectations and objectives are in *white*

example, participants are put in a situation where there is a concentration of time and space as in a role-playing game or a need to explain their representation by following, in a limited time, a methodological conceptualization framework such as ARDI (actors, resources, dynamic, interactions), which exacerbates their interactions. Of course, exchange also happens without a commodian in attendance, but it is very difficult to collect precise information a *posteriori*.

The following analysis focuses on the facilitation of sessions of the role-playing game, which provide the best informed collective key moments in a ComMod process.

In this game of interactions, emotions can be provoked and the facilitator is obliged to deal with them in order to focus the attention of the group on performing the underlying task. The analysis of logbooks and canvases shows that the facilitator is rarely alone when animating the group of participants during role-playing game sessions. The facilitation team comprises a role-play master, technical assistants and observers and can vary in size from 2 to 13 people but, in most of the cases, a buddy system is used for facilitation.

This facilitation structure is explained by the format of the role-playing games, especially the spatial arrangement required. Thus, half the game sessions take place in areas where not all the players are together (e.g. the villages and irrigated perimeter in Njoobaari) or some provide several game areas for simultaneous use in the same venue (e.g. private property, community area and town hall in ButorStar, forest and forester's office in SylvoPast).

The reason for this lies firstly in the multiple functions carried out by the facilitation team. Figure 3.4 presents the functions undertaken by the facilitation group according to the 10 major categories defined in the ComMod role-playing game design documents (Étienne et al. 2008a).

'Invite the participants' or 'record the actions' are the only two functions that stand out from those taken from a bibliographical analysis of the four main roles taken on by socio-cultural or participative approach facilitators.

Participants are identified and invited with partners whose legitimacy is recognized by all the others. This is to try to ensure that they are the most relevant participants in terms of the objectives of the game session and, more widely, the companion modelling approach. This explains why the game sessions are mostly run by a commodian and another status facilitator (e.g. lay, academic, expert or institution).

Given the multiple interactions between the players and the variety of information to be processed (and recorded), managing and analysing a role-playing game calls on a large number of functions that a single facilitator is incapable of performing. It is, however, essential for the success of the exercise in collective reflective thinking. It also depends on scientific needs as the scientific community uses these recordings for analysis, -evaluation and confrontation (see Box 3.2).

These various roles played by the facilitator are converted into a set of principles in terms of his stance. Their goal is the constitution of an 'us' conveying the cohesion of the group. They attempt to create an atmosphere of mutual respect, a user-friendly ambience and a climate of 'psychological security'. The facilitator

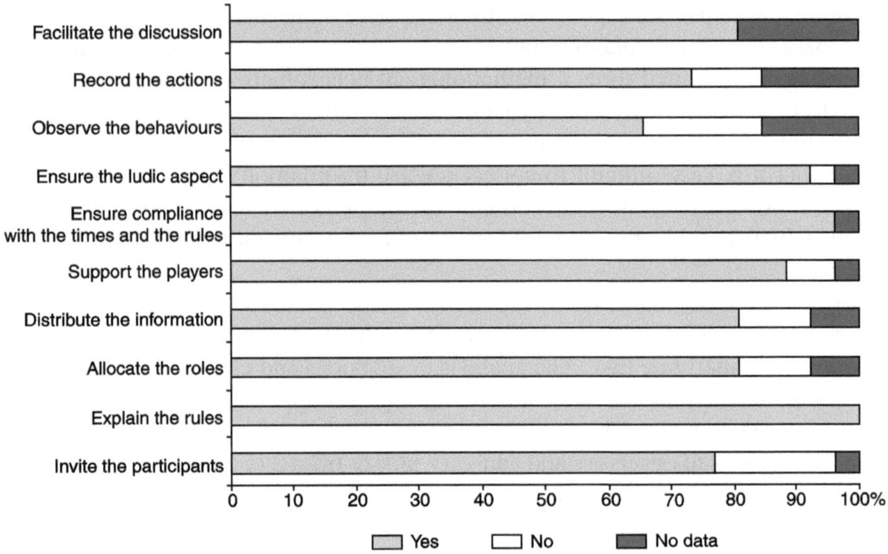

Fig. 3.4 The functions performed by the role-playing game facilitators

does not preside. They propose, suggest, invite and consult the participants who have room for manoeuvre in interacting in a framework defined by the game rules or by the structure of the method used. They do not judge. They thus develop their abilities based on situations of interaction between the players. In so doing, the facilitator works horizontally, not vertically, to develop reflective thinking drawing on collective expertise where, together, the individuals apply themselves to resolving a problem.

In reality, in the companion modelling approach, the design and simulation phases are those where collective key moments feature the most. These particular moments are when points of view are exchanged, opinions aired, actions simulated and so on, elements which all subsequently serve to support the collective reflective thinking. Analysing the evaluations shows clearly that the collective key moments play a fundamental role in the dynamics of collective action and the learning acquired during the companion modelling process (see Chaps. 7 and 10).

Facilitation Techniques and Methods Used in the Intervention Device

Now, we turn to the facilitation of the entire approach.

The range of facilitation techniques used throughout the companion modelling approach as a whole is relatively small. Little information is currently available with which to analyse facilitation techniques. It has only been possible to analyse seven case studies for their facilitation techniques. It emerges that the facilitators use various methods in line with the stage of group dynamics. We present here the most general case, in the knowledge that there are of course context-related variations.

During the ComMod group design stage, the facilitator's methods are to encourage the expression of viewpoints and stimulate the opinions of project sponsors to outline their expectations as far as possible for the companion modelling approach. Brainstorming or brainwriting (normally using post-its) are two of the most used techniques, most frequently combined with slide show presentations. In the brainstorming sessions, the problem to be raised must be simply and clearly stated by the facilitator. They then give the group the floor for a limited period. Creative, original and incongruous ideas must be encouraged. It is forbidden to criticize the ideas expressed. The facilitator lets people speak as and when required, reformulates certain confused or misunderstood ideas and produces interim summaries. The main goal of brainstorming is to list options. For the brainwriting session, each individual writes his ideas regarding the question asked by the facilitator on a separate card. After a few minutes the cards are exchanged and read and each person adds a new idea. The aim is to list the options in a more structured fashion than in brainstorming. Although defining the subject or even the question giving rise to a companion approach is in construction, it is rarely formalized at this stage. The contract between the various sponsors is raised but not systematically formalized in writing. The structure of the agreement is, however, discussed at length in order to identify each party's undertakings as comprehensively as possible.

Box 3.2: Facilitation of the role-playing game in Njoobaari
Description of the game

In this role-playing game, 12 farmers, each a manager of one plot, simulate a simplified rural operation in order to produce two varieties of rice in an irrigated perimeter. The irrigants are organized into two groups of farmers with a person in charge. Each group is in charge of managing the irrigation of a watercourse. Two watercourses are represented in the irrigation scheme. One pumping station supplies water to each watercourse. Players are divided into two villages, with irrigants from both groups. The social interactions for the exchange of work or credit are organized according to social status. The play area is divided into two, one hidden from the other. The first one represents the villages and the second the irrigated scheme. In the village area, players draw 'occasion' cards at random at each round of play, which authorizes them to go to the irrigated scheme area or not. In the second area, players can decide activities on their plot. The limited water resource, the different objective of intensification of rice production, their presence or not in the perimeter area, the different loan payment strategies, etc., create tension in the perimeter and the farmers have to work together to deal with it.

Facilitator's functions

Players are invited by a resource person belonging to the irrigated system. This person sponsors the ComMod approach. He invites farmers in accordance with the objectives of the approach he has explicitly requested: to heighten farmers' awareness of the consequences of a lack of reimbursement of collective loans and the problem of social organization of agricultural production in the irrigated system. Two persons are necessary to facilitate, one in each area. Observers are also present. In the village area, the first facilitator draws cards at random for the players. These cards give them the element of their role: goal of production, social status or repayment behaviour rules. They explain the rules. They take part in recording the activities the players want to carry out in the irrigated scheme before drawing the occasion cards. They supervise to ensure the rules of exchanging work or credit are respected. In the irrigated scheme, the second facilitator presents the level of water for each plot after the activities made by players during the previous round. An abacus familiar to everyone is used to calculate the quantity of water allocated to each plot according to the number of plots irrigated at the same time on the same watercourse. They register the activities really performed on each plot. They note the time so that negotiations about water allocation do not overrun the schedule. A camcorder is use to record the interactions between players in the irrigated area. Moreover, the facilitators have to listen to players and defuse any crisis that could happen, for example, when an agreement on water allocation previously established is not respected. In so doing, they maintain the playful atmosphere. Lastly, during the debriefing, the facilitation team organizes the discussion and regulates the contributions so that every player can express themself. The facilitators precise or reformulate the opinions and viewpoints expressed. They put in perspective the actions and interactions that occurred during the session (relying on recorded data). In so doing, they help the collective reflection to facilitate the passage from learning acquired during the game session to the real situation.

Then, with the formation of the group in the awareness-raising phase, the facilitators tend more to seek to pass on information on the subject and to explain the thinking behind the formation of the group of stakeholders in the companion modelling approach. Presentations at this stage basically rely on explanatory methods, such as a picture, paperboard or slide show to excite the visual and auditory senses of participants.

Once through this stage, the techniques used as the group develops or has even reached maturity seek to encourage exchanges between participants. However, those exchanges must be rich and structured, that is the reason why facilitators resort to brainstorming and round table sessions or developing mindmaps[4] using formalization methods described in Chap. 4. The role-playing game occupies a special position in the facilitation techniques used as it is both the result of a conceptualization/scriplet process (during which several facilitation techniques can be used) and the special moment when participants embark on the situation simulation exercise. It is activated as a long-term/simulation exercise to create surprise, encourage creativity and interaction and even defuse potential sources of conflict.

Other more marginal methods have also been used, such as the palaver or problem tree in Senegal (Fig. 3.5), the land coat-of-arms in Madagascar, facilitation games in Bhutan (Fig. 3.6) or photolanguage[5] in Thailand or at Tarawa, which can be compared with the logical phrases used to bridge the gap between the ARDI interaction diagram (see Chap. 4) and model implementation.

There is seemingly no standard combination of facilitation techniques apart from using role-playing games in the majority of companion modelling experiments. The most important seems to be the adaptive nature of the facilitation and, therefore, the mobilization based on contextual needs of a particular technique. The result is a desire for adaptive facilitation, with more emphasis placed on matching the use made of tools to the companion stance than on reifying tools. However, questions can be asked over the potentially substantial improvement from greater familiarity with facilitation techniques, especially in terms of interaction between participants expressing viewpoints, whether or not they are appropriate for taking power plays into account or more widely to become more familiar with the social effects of the facilitation.

Formulating the social demand or taking over a clearly formulated social demand, facilitating groups where emotions can run high, having the necessary scientific knowledge to fuel the debates or mobilizing it outside, helping to modify the perception of other individuals' viewpoints, analysing and inputting the social context of the intervention—all these skills are seemingly required to facilitate a companion modelling approach based on the fundamental principles. The commodian does not claim to be 'omni-skilled'. They surround themself with individuals with complementary skills when faced with the complexity of facilitating a ComMod approach. They and the members of the facilitation team will build up their legitimacy to intervene in the social and ecological system studied during the process. The commodian must, however, possess two essential skills to initiate the approach and take it to a successful conclusion: familiarity with the principles of

4 Each individual writes his ideas on a card. The cards are pinned to a panel on the wall. The ideas are then put in order. The aim is to produce a structured list of options.

5 The facilitator uses a visual medium to assist expression. He presents illustrations haphazardly (e.g. landscapes, personalities, words, expressions, sketches, etc.), then states the theme of the exercise. Each participant chooses one or more illustrations and must explain what comes to mind for 3–4 min. At the end, the facilitator produces a summary, noting the variety and wealth of ideas, concentrating on the interesting ideas and correcting any errors. The aim is to produce a structured list of options.

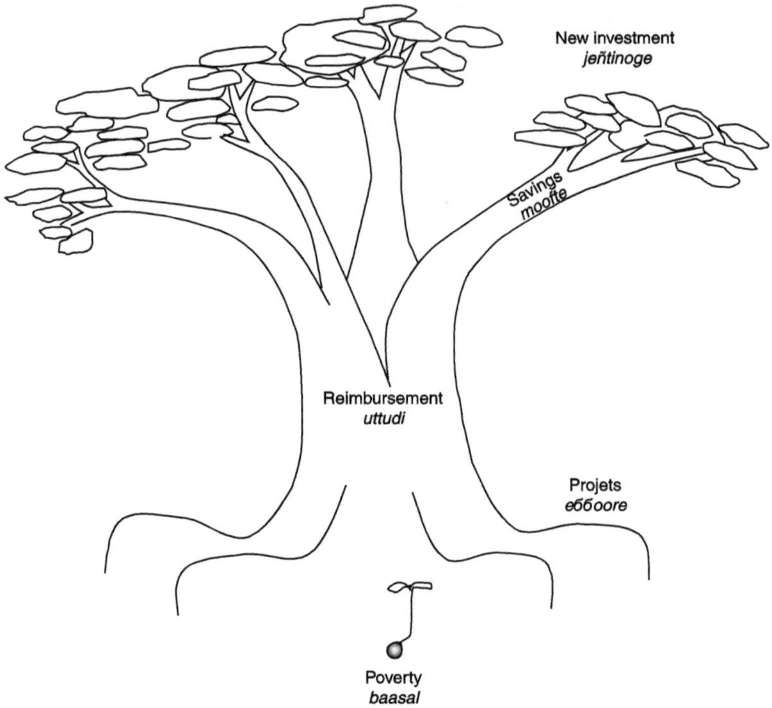

Fig. 3.5 Palaver tree—Village perception of credit/debt (*ñamande)* and how it has changed, Wuro Madiu, Senegal, 2002

the approach and the ability to demonstrate drive in supporting and animating the group that is developing the approach. They must be aware that the results of their facilitation efforts will be interpreted, taken over or even hijacked and improved by the stakeholders.

An Intervention Stance of the Commodian that Cannot be Naive

The participative approach of companion modelling refers to a comprehensive approach of social reality, which involves giving meaning to the practices, actions and projects of communities so that they can become part of the joint perspectives of sustainable development. It presupposes at least bringing together a multiplicity of heterogeneous bodies of knowledge (i.e. scientific, practical, managerial, expert or lay), which are expressed, questioned and answered to define together the conditions for intervention and development of a common knowledge. It involves helping to produce collectively a problem orientation that is not the sum of all the

Fig. 3.6 Facilitation game between the Rhadip (sedentary farmers) and the Merek (nomadic yak herdsmen), Radi, Bhutan, 2005

approaches but the product of a global conception of the question being dealt with, mastered by all participants in the approach (Daré et al. 2007). This ordered levelling is facilitated by producing intermediate objects like glossaries, semantics or model formalism as its assumed abstraction can make its appropriation difficult by other social stakeholders who have played little part in its development or who are far from academic culture. We are here faced with the following paradox: the production of intermediate objects to formalize the approach and make it accessible to all is a mandatory stage in multi-disciplinary and multi-stakeholder approaches, but it could be a source of exclusion of the stakeholders with least cultural and symbolic capital. It would thus risk a facilitation that does not result in collective emancipation, giving participants the chance to blossom, aiming to preserve certain hierarchical social structures. How do the interpersonal skills of the commodian and the know-how we have just described limit this paradox?

In terms of stance, remember that two fundamental principles are the recognition of the multiplicity of legitimate viewpoints and clarification of hypotheses to render the bodies of knowledge accessible or even intelligible to all. Recourse to abstraction can seem difficult to access. Unified modelling language (UML) or ARDI formalism (presented in Chap. 4) have, however, been favoured as they seem the most capable of levelling out the constituent elements of each individual's representations as they involve a representation in terms of relationships and objects (in the most simplistic proposal). This remains far from easy as it requires a facilitator familiar with these tools and capable of reformulating the group's proposals to dissect them in depth. The development of the ARDI method by some commodians (Étienne 2006; Étienne et al. 2008c) has responded precisely to others' concern over potential problems faced by non-computer scientists in the use of UML. The ARDI method sees itself as closer to the spoken language by producing logical phrases before transcribing them into a system of computerized relationships. It has since proved advantageous in multiple intervention situations. Recourse to role-playing games also participates in this logic of producing a viewpoint intelligible to stakeholders. We have thus witnessed a move from the computerized simulation system and role-playing game to a simplification of rules, so that the operating rules of the system can be understood by players not involved in the design phase. Computerization is not always essential; simple cross-ruling in a table and game rules can sometimes create an intermediate object, a discussion medium, of sufficient relevance to the question being dealt with. However, we must not be fooled and it is clear that despite all the efforts to clarify hypotheses, rules and so on, recourse to a third-party object can sometimes cause confusion and, therefore, trigger the paradox described previously. We can, however, be aware of this and still apply a principle of reflexivity to our own approach and thus seek to limit it.

To state our position even more clearly, we now return to the criticisms of participative approaches and facilitation as to the lack of knowledge of the intervention context. It would be easy to imagine facilitating a companion modelling approach with no real knowledge of the system in which the commodian is intervening. The danger here would be to boost local conservatism, whereas the approach aims for movement and support for social dynamics. It is true that in such a situation the expertise mobilized, because it aims to make expressing the various viewpoints present easier, could further the exclusion of absent third-parties, thereby boosting the position of elites or invited stakeholders. The question then is to know whether the commodian adopting this position remains in line with the stance and the ComMod Charter. In practice, when intervening on a site, the commodian more often than not analyses the context so that they can position their intervention and influence the choice of members of the ComMod group set up to carry out the experiment (see Chap. 5). In so doing, the iterative nature of the approach, the development of observation crossed with the expression of multiple viewpoints at various moments with different facilitation techniques encourage the identification of major shadow areas in knowledge and restricts the incompleteness of the system represented. However, we are aware that the discipline of each facilitator has a major influence on the perception of shadow areas, hence the advantage in developing unified, multi-headed and cross-disciplinary facilitation.

As part of the applied research, the researcher also frequently finds himself responding to economic or political questions and measuring the issues and consequences of a critical positioning with respect to the social and economic model founded on the utopia of social progress via growth (or the market). They are responsible for the theories they produce and are of necessity committed to the social life. They cannot cut themselves off from using their scientific work. They are inevitably engaged in producing and structuring the world they are studying and in permanent dialogue with the constituent stakeholders. They construct society at the same time as they attempt to understand it. They cannot therefore be neutral. When explicit reference is made to the support and co-construction of interpersonal skills and know-how, they must of necessity cast an eye over the effects induced by the participation of social stakeholders specifically convened.

Lastly, the companion modelling approach is not an even deal; it intervenes in the social dynamics and can modify them. Even further, it can arouse social expectations, particularly by encouraging the emergence of consultation arenas, by allowing the ignored interests of 'social trainees' to find expression or by altering the symbolic field of the circulation of power. At the service of human communities, researchers can only stay for a limited time and the support periods do not coincide with operations. What happens therefore when the commodians leave the field? How are the expertise and interpersonal skills required to keep the facilitation going transferred? Faced with the complexity of the facilitation due to the combination of all the functions assigned to the designer-facilitator of the approach and even the design and simulation/game workshops, it seems that interpersonal skills must reign supreme. To continue the companion modelling approach under an action programme rather than in a research process, the future designers-facilitators of the approach must first be identified and trained. The problem with this training is the transfer of the stance, not the transfer of the facilitation techniques, which can be reinvented depending on the fields to suit the question being addressed. The charter is only one stage. The methodology guide is another (see the Appendix). Scientific training programmes have been achieved in the main until now. However, in all circumstances, close links with a commodian must be maintained so that the future designer-facilitator can learn by doing, develop through experience and thus acquire those interpersonal skills that take precedence over know-how (see Chap. 12).

Chapter 4
Models for Sharing Representations

Christophe Le Page, Géraldine Abrami, Olivier Barreteau, Nicolas Becu, Pierre Bommel, Aurélie Botta, Anne Dray, Claude Monteil and Véronique Souchère

Companion modelling implementation is based on a network of individuals and artefacts amongst which models occupy a special place. This chapter presents the various models developed in a companion modelling process for purposes of representation sharing. Designed as a way of understanding actual (reference) systems in which social and biophysical dynamics overlap, models represent the evolution of these systems and are used for organizing exploratory simulation

C. Le Page (✉) · A. Botta
UR GREEN, CIRAD, Montpellier, France
e-mail: christophe.le_page@cirad.fr

A. Botta
e-mail: aurelie.botta@cirad.fr

G. Abrami · O. Barreteau
UMR G-EAU, IRSTEA, Montpellier, France
e-mail: geraldine.abrami@irstea.fr

O. Barreteau
e-mail: olivier.barreteau@irstea.fr

N. Becu
AGILE, Université de La Rochelle, La Rochelle, France
e-mail: nicolas.becu@univ-lr.fr

P. Bommel
UR GREEN, CIRAD, Brasilia, Brazil
e-mail: pierre.bommel@cirad.fr

A. Dray
Swiss Federal Institute of Technology, Zurich, Switzerland
e-mail: anne.dray@usys.uthz.ch

C. Monteil
UMR DYNAFOR, INP-ENSA, Toulouse, France
e-mail: monteil@ensat.fr

V. Souchère
UMR SADAPT, INRA, Grignon, France
e-mail: souchere@grignon.inra.fr

M. Étienne (ed.), *Companion Modelling*, DOI: 10.1007/978-94-017-8557-0_4,

exercises involving stakeholders in the reference system. Chapter 2 introduced the different purposes for using models in a companion modelling process:

- make heterogeneous viewpoints visible and open to debate
- question the coherence of these viewpoints and the consequences of their simulation in relation to the real world as lived by the participants
- propose a medium for exploring scenarios collectively through simulations in a virtual world.

There are different ways of translating multiple viewpoints to a reference system in models. This translation is based on conceptualizing the system studied to describe the share of reality perceived as useful by each stakeholder and culminates in specific artefacts being developed and computer techniques and situation simulation exercises (role-playing games) being mobilized. A brief introduction will explain the choice of MAS as a preferred method for representing the domain studied. This chapter will describe different model lines: domain models, conceptual models and simulation models. The knowledge extraction, abstraction, formalization, conceptualization and finally, implementation stages will be presented in succession. The use of simulation models as a medium for exploring future scenarios will then be developed before concluding the chapter with an analysis of the adaptability, complementarity and versatility of these models, characteristics that provide tremendous flexibility in implementing the companion modelling approach.

Introduction

The first section explains the choice of MAS as a main method of representing a studied domain based on actual agents from the reference system. The translation process from domain model to executable model (simulation model) via the conceptual model stage is then presented. Lastly, the elements used to characterize simulation models are detailed. These take the form of a purely computerized simulator, that of a role-playing game simulating the situation of stakeholders in the domain around the actual game media, or appear as a combination of these two extreme modalities.

Ways of Representing Reality

We view modelling as an explicit writing process in which knowledge and heterogeneous hypotheses are distributed in the same artefact, allowing them to function together. The resulting artefact is more or less capable of taking charge of unplanned interactions.

Adopting a constructivist conception of representations (see Chap. 2), we confine ourselves here to representations attempting to account for a given system, which are perceptions (viewpoints) of this system frequently designated as being 'the real world'. To represent it is to create hypotheses about that which seems to characterize it best based on a targeted objective. This involves breaking it down into building blocks so that these hypotheses can interact, then suggesting how to reconstruct it. We briefly review here a few possible structuring approaches from the many options, by giving slightly more detail on the MAS frequently used in the companion modelling approach. These structuring approaches are not exclusive and some work in progress is seeking to use them jointly.

We could, therefore, set out to represent stocks and stock flows. This is the dynamic system approach. It involves highlighting the flow regulation, control and action options and any retroactive loops. The system is described by a set of state variables (the stocks) and equations describing their dynamics (the flows). At stake is identifying what these stocks represent in terms of the system in play. It often involves energy, biomass, water, monetary units and so on. This type of representation is not particularly explicit about the stakeholders linked to these flows or the conditions for their intervention.

Game theory provides a framework for understanding these stakeholders and anticipating their choices by identifying their rationalities and decision rules strategically. It is a static type of representation: all possible interaction scenarios must be specified. The system is represented by a set of strategic stakeholders equipped with an objective function applied to this system. The dynamics of resources combined with these interactions are generally described less accurately.

The description as MAS sets out to conceive virtual worlds made of interacting entities with the explicit aim of reconstructing simplified, yet relevant, situations with respect to the question dealt with. It involves identifying the active entities (agents) who play a decisive role in managing the system, specifying their management entities and their degree of autonomy and stating how they interact with their environment and the other agents. These entities can be objects, items in the landscape, individuals or groups of individuals (e.g. farm, village, institutions, etc.).

MAS are thus understood here as a metaphor of reality (i.e. social, biological and physical) as a set of interacting autonomous entities located in an environment, given an objective and with representations of their environment (Ferber 1995). Note that defined thus as a representation type, MAS make no reference, even implicit, to computer science. In our field of application, the entities make decisions dealing with resources. They therefore interact both with the resource medium and with the other decision-making entities with which they are in contact. Numerous experiments have shown that MAS are well suited to simulation in the domain of resource management (Bousquet et al. 1999; Bousquet and Le Page 2004).

From Reality to Its Representation as a Simulation Model

Companion modelling envisages developing and using models, which represent a form of understanding of a reference system, as a means of sharing simplified representations of this system. The vast majority of models are dynamic, in other words, they include explicitly hypotheses and rules linked to changes in the reference system over time. The modelling process produces concrete, operational tools used for the simulation, the exploratory activity that refines the forms of comprehension of the reference system and draws new knowledge from them by giving effect to long-term visions discussed and analysed collectively. Allowing stakeholders collectively to see the progressive changes in the system under listed conditions and practices stimulates in particular their ability to comprehend the mechanisms of decision-making processes (theirs and also those of other participants). In addition, inciting reflection on the mechanisms responsible for outputs exhibited by the simulations helps to make explicit certain hypotheses that would otherwise remain hidden in a conceptual model.

The translation chain for the reference model into one or a family of executable operational models has already been described (e.g. Fishwick 1998). Drougoul and his colleagues proposed a detailed analysis of the translation chain in the specific framework (which interests us here) of the design process for simulation models based on agents, by considering the roles of thematician, modeller and computer scientist for each of the three stages they describe (Drougoul et al. 2003). The thematician defines the domain model by using the semantics he associates with the reference system. As the thematician's specifications do not provide direct transcription into an executable model, the modeller adapts the domain model into a more formal model known as the design model (or conceptual model); its purpose is to clarify the concepts used, check coherence and delete potential ambiguities. The conceptual model is the result of a co-construction process that closely links the thematicians and modeller. Ultimately the conceptual model could be transcribed by the computer scientist into an operational model or a simulation model. This stage is too frequently ignored, which compromises any chance of understanding the impact of purely computational specific features on the emergence of artefactual structures in a simulation (see the Sect. "From the Conceptual Model to Simulation Models: Implementation" below).

Whereas Drogoul and his colleagues argued in favour of a clear separation between the roles of thematician, modeller and computer scientists (firstly, because each role demands specific skills and secondly, because the need to transfer elements between role sponsors means they must be formulated clearly) (Drougoul et al. 2003), in practice, companion modelling is frequently seen as an accumulation of roles. Thus, the modeller will also create the operational model or will also help to develop the domain model by contributing his thematic knowledge. The advantage of this accumulation is less online loss in communicating between roles and greater continuity in interacting with the thematicians. Companion

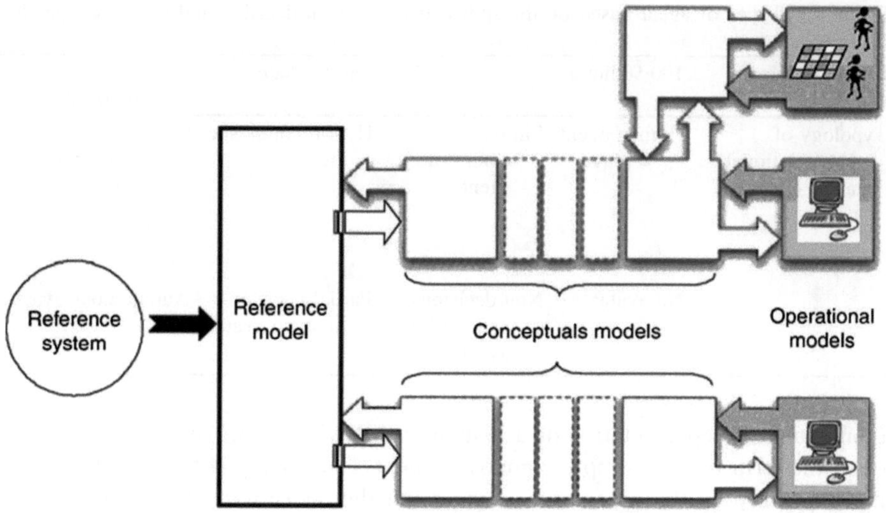

Fig. 4.1 Model lines starting from the same reference model, with each line ending in an operational model (role-playing game or computer simulation model) (inspired by Treuil et al. 2008)

modelling is also an original way of opening up the roles of thematician and modeller to the stakeholders of systems represented.

Presented as successive stages (see Chap. 2), the modelling process, focusing on converting the domain model into the simulation model, seems implicitly linear. Alongside the standard process of refining a conceptual model for a certain time before moving on to the construction phase for an operational model (that can be manipulated to produce exploratory simulations), which is a direct and faithful translation, ComMod sets out to make possible and facilitate the comings and goings between the collective construction framework of a non-frozen conceptual model (a range of conceptual models) and the forming of lines of operational models (see Fig. 4.1).

Simulation Models Based on Miscellaneous Types of Agent

Our conceptual models therefore are MAS. Converting a conceptual model based on the MAS paradigm into a simulation model makes the implicit choice of translating each conceptual entity into a computational agent, which can be classified according to the nature of its decisions. Table 4.1 groups the various types of computational agents used in companion modelling implementation.

When all the decisions of an agent are made by a human being and there is no computerized go-between, the computational agent is a human agent type (commonly called a player). When this same decision is relayed by an avatar

Table 4.1 Types of agent based on the split between human decision and computer-specified decision

Nature of the decision	100 % human		Intermediate	100 % computerized
Typology of computational agents	Human agent = player	Simple composite agent	Hybrid composite agent	Computerized agent = virtual agent

	No avatar	Non-decision-making avatar	Partial decision-making avatar	Autonomous avatar

(computerized representative of a human agent) with no decision-making autonomy, the term used is simple composite agent. Conversely, when the decision is made entirely by the autonomous avatar (all the decision-making processes are created automatically by executing computerized instructions), the term computerized (or virtual) agent is used. The intermediate case of a human agent relayed by a partial decision-making avatar is called a hybrid composite agent.

To characterize an agent-based simulation model it is important to consider all the agents in its make-up. Two major types can be clearly distinguished: (i) simulation models based exclusively on human agents, commonly called role-playing games; (ii) simulation models based exclusively on computerized or virtual agents. A whole range of situations exists between these two extremes where some decisions are human and others are computer-specified (see Fig. 4.2). The term hybrid agent simulation model covers all these intermediary situations.

Alongside the specification of computational agent decisions, the computer is often an effective medium for taking charge of a certain number of other functions unambiguously constitutive of the conceptual model representing the social and ecological system studied. The computerized medium is, therefore, only a component of the simulation model (one of the elements characterizing its structure) and not a computer model *per se* used jointly with another type of simulation model. The five main functions of a computer medium in agent-based simulation models are as follows: (i) inputting decisions by human agents; (ii) calculating agents' performance-related indicators (actions); (iii) simulating the resource dynamics; (iv) displaying the space (i.e. state of resources, positioning of agents, any viewpoints of this space specific to each type of agent); (v) specification using computer language of agents' decisions.

We shall use the commonly used term 'computer-assisted role-playing game' to designate the human agent models calling on at least one of the first four computerized functions listed above. The generic term 'role-playing game' will be reserved thereafter for the simulation models where decisions are all human (including computer-assisted role-playing games) or essentially human (some hybrid agent models). Conversely, by considering the last function relating to the

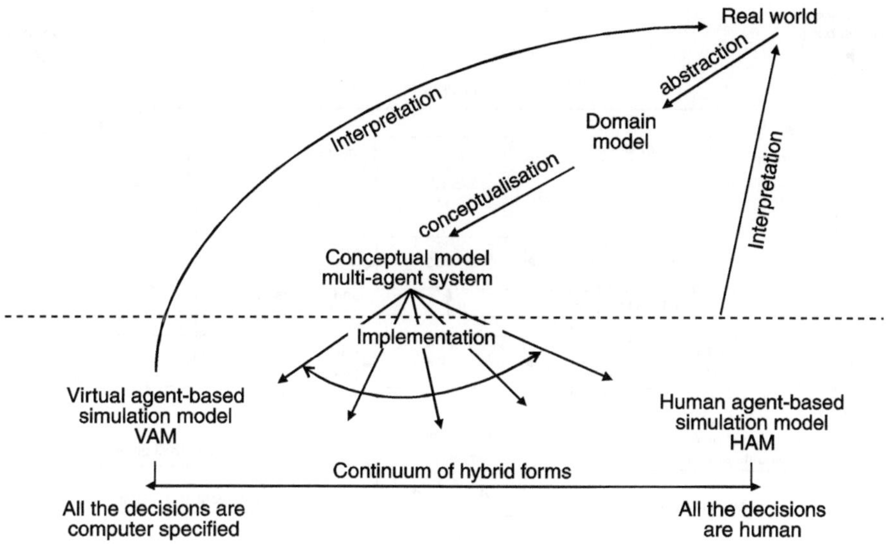

Fig. 4.2 From real world to the implementation of agent-based simulation models

computerized decision specification as decisive, the hybrid agent models where the decisions are essentially computer-specified will be classified with the virtual agent models in the category 'computerized simulation model'.

The analysis of 63 simulation models developed in the 27 case studies analysed shows firstly that computerized simulation models and role-playing games take equal shares and secondly, that the role-playing games are frequently supported in some way by the computer (see Fig. 4.3).

Box 1: The development of models used in the Nam Haen (Nan) case study, northern Thailand

Context

In Nan province, northern Thailand, the government wanted to preserve forest resources in the upper watersheds by establishing a national park. Setting up the park boundaries generated tensions between the officers of Thai governmental agencies (i.e. Forest Department and National Park) and two village communities (ethnic tribes) who feared that agriculture and the gathering of non-timber forest products would be prohibited in the reserved area. Stakeholders' perspectives were integrated into five versions of the same simplified representation of the system under study, derived one from another. These tools were combined to foster the development of a communication platform used to explore collectively and assess prospective scenarios (Figs. 4.4 and 4.5).

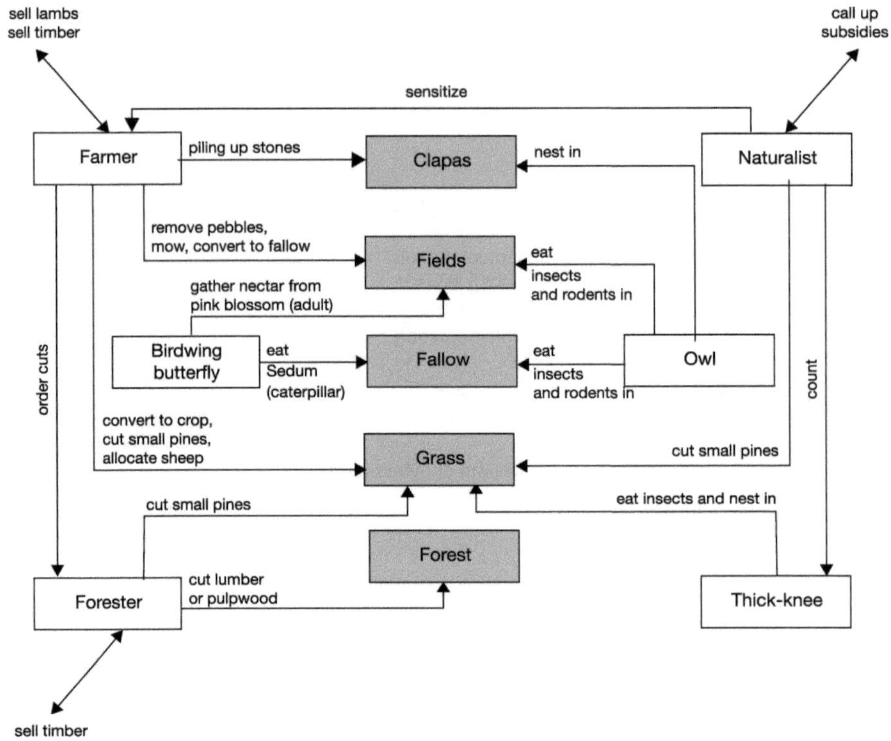

Fig. 4.3 Sample conceptual diagram showing the domain model and the relationships between entities (Étienne, personal contribution) (Causse Mejan case study)

In practice, the adaptive nature of companion modelling produces a wide diversity of simulation models, not only in type but also in their use and association method. To illustrate this point, Box 1 (see page 74) presents the successive versions of the model designed and used in the Nan case study (Thailand): not just one but a family of simulation models (all referring to the same conceptual framework) were mobilized. A useful method of tracing changes in models and revealing their filiation is to position them in the same table based on the typology summarized in Table 4.1. We shall present a few examples in the last section of this chapter to illustrate how the diversity of methods to interlock and associate various models makes implementing exploratory simulations extremely flexible. Before that we will detail the abstraction and conceptualization processes by presenting the methods and tools used for each one.

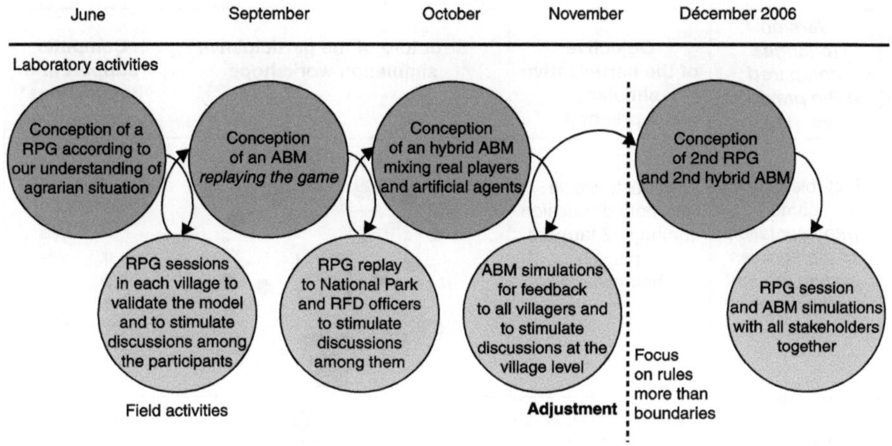

Fig. 4.4 Chronology

Knowledge Extraction and Abstraction: From the Reference System to the Domain Model

Repeating the terminology of the previous section, the first stage is to identify real relevant agents, in other words, make up the domain model. This stage involves pooling the questions and viewpoints from all the 'thematicians' made up of the researchers and stakeholders involved in this stage of companion modelling. This pooling, because it is contradictory, involves mutual learning of the viewpoints of others and changes in questions.

Formulating a Question and a Framework for the Modelling Process

The work starts by formulating a question structuring the viewpoint in which the thematicians will interact to develop the domain model. The framework comes from previous phases in the companion modelling: previous modelling cycles, 'command' from part of the stakeholders in the system or an investigation of a question originated by some of the researchers. One method of initiating the process is to compile the general impressions of the stakeholders who have been invited to participate using a general formulation such as: 'what do you think of such and such an aspect linked to such and such a resource (its management) in such and such a portion of space?'. For example: 'what do you think of the changes in the bluefin tuna population in the Mediterranean?'. It is to be expected that each response refers to a change recently noted (trend or event) and at the same time corresponds to a specific point of view suggesting a sort of appreciation

Version (changes compared to the previous version)	**Objective of the participative simulation workshop**	**Structure of the participative simulation workshope**	**Computer support and its functions***
First role-playing game *(representation of the agrarian system brought by the research team)*	In each village, promote discussion among 12 farmers playing their own role	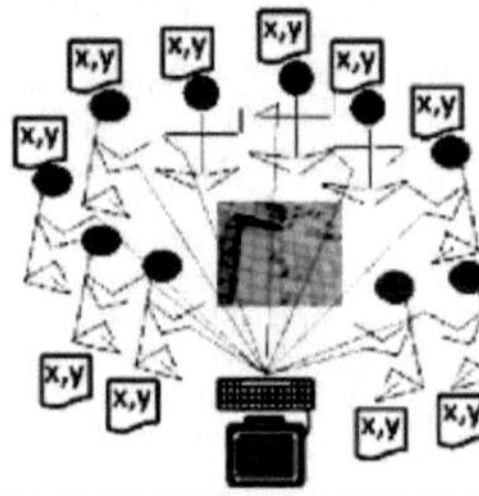	Excel i. yes ii. yes iii. no iv. no v. no
« Replay » of first RPG (computer agents reproduce players' decisions)	Showcase to National Park officers the principles of the RPG and the outputs of the 2 gaming sessions	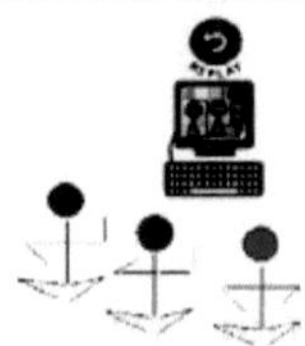	Cormas i. no ii. yes iii. yes iv. yes v. yes
Hybrid computer simulation *(simplified choice of crops)*	In each village, enlarge the group of involved farmers to disseminate information to a larger audience (3 players with 9 computerized agents)		Cormas i. yes ii. yes iii. yes iv. yes v. yes
Second role-playing game (the game board represents a wider space)	All the stakeholders together are able to talk about hot issues and to envision possible options fot the future	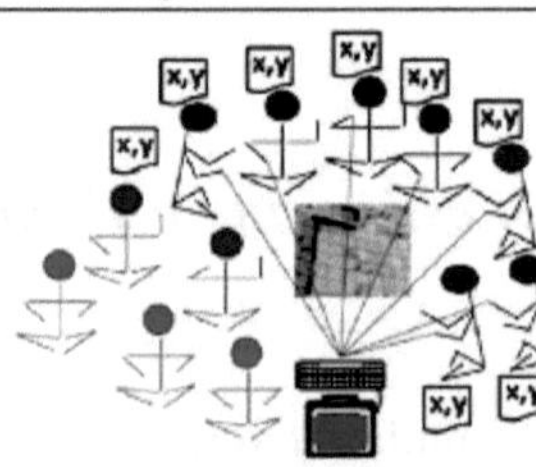	Excel i. yes ii. yes iii. no iv. no v. no
Autonomous computer simulation	To explore and discuss the scenarios identified	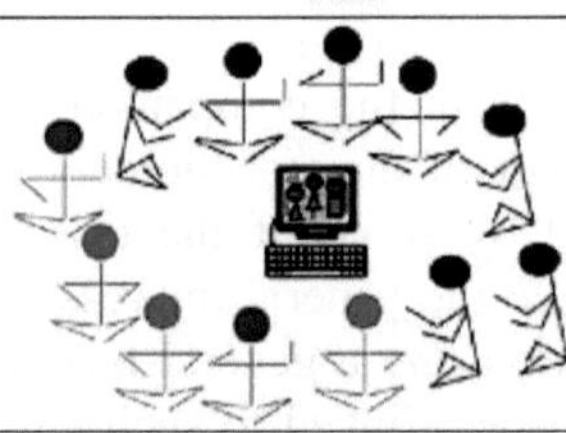	Cormas i. no ii. yes iii. yes iv. yes v. yes

Fig. 4.5 Comparison of the five versions of the model. *Functions of the computer: (i) Inputting decisions by human agents (ii) Calculating agents' performance-related indicators (actions) (iii) Simulating the resource dynamics (iv) Displaying the space (state of resources, positioning of agents, any viewpoints of this space

of this change, for example: 'the species is threatened with extinction'; 'fewer [fish] have been caught over the last 5 years'. For some stakeholders, these changes will be perceived as negative, for example, environmentalists and tuna fishermen, whereas others could take a positive view, for example, sardine fishermen. The confrontation of various responses nevertheless reveals a common character over and beyond the expression of different sensitivities, for example, 'population numbers are dropping'. The model thus considers itself as a means of exploring the combinations of factors that reproduce this characteristic. The framework can be restricted by targeting precisely from the start a certain type of factor, for example, the traditional tools for fishery regulation such as quotas or reserves, which justifies not taking other factors into account, known nevertheless for their influence in reality (e.g. climate change). 'Everything equal elsewhere', what matters is the availability of a means of comparing effects of chosen factors—specifically to be explored.

The first representation of the reference system, called the domain model, is a medium used to gather and assemble the knowledge of the thematicians involved in the modelling process. Based on the assumption that a thematician expresses their knowledge spontaneously, partly tacitly and in disorganized fashion, techniques must be implemented so that they can reveal it, specify and make it as explicit as possible so that it can be 'captured' and formalized during a process also involving a modeller. We call this eliciting knowledge, based on a combination of two operations—extraction and formalization.

Eliciting Knowledge

There are many eliciting techniques. For example, monitoring the process where an individual, considered an expert in a particular domain, is placed in a simulated situation to resolve a problem and is asked to explain the actions he undertakes, or analysing transcriptions, which involves a lexical analysis, analysing a piece of writing or an interview transcription, thereby identifying and organizing all the elements constituting knowledge of a domain.

Applied to companion modelling, eliciting techniques are a chance to identify and formalize the knowledge of stakeholders in the systems represented, their reasoning methods, their decision-making rules or their strategies. However, these techniques have been developed for use by experts in a well-defined, controlled environment close to the one encountered in a laboratory experiment. We have developed our own eliciting techniques to transpose the methods of eliciting knowledge to the specific context of companion modelling (an uncertain, fluctuating environment difficult to delimit). We review three of them.

Note that these eliciting techniques sometimes call on formalisms in varying stages of development, whereupon they refer, at least implicitly, to preconceived conceptual entities acting as frames for the occasionally collaborative formalization of knowledge. We shall then consider that the work of conceptualization sets out to conceive these frames whereas the work of abstracting or eliciting sets out to 'instantiate' these frames (in knowledge engineering, this is the difference between defining an ontology and populating an ontology). It is not of course always possible to detect these two modelling work phases: in a collective workshop (with thematicians), where defining conceptual entities is not stabilized and open to discussion, conceptualization and abstraction will be closely linked, with abstraction used directly to test the conceptualization, which changes on the spot. Most of the time the workshops are turned towards one or other option but it is sometimes difficult to make a distinction, as seen in the ARDI case (see below). However, in that case it is stated clearly that the discussion does not cover the nature of entities and it is, therefore, more a question of elicitation.

Extracting Knowledge from an Interview Transcription

The principle of this technique is to carry out an initial series of interviews with stakeholders from whom we hope to elicit knowledge and to analyse the semantics of interview transcriptions, in order to extract the elementary objects making up their knowledge (individual or collective). Then we reconstitute the logic of all this knowledge (called the cognitive model by some), often as an entity-relationship type diagram. The two phases in this approach, acquiring verbatim accounts during interviews and—extracting knowledge from these transcriptions, are linked and equally important.

Although acquisition is confined to a given subject, for example, the collective management of water in a given perimeter, the social and environmental interactions addressed are such that it is often impossible to predefine the scope of the domain to be treated before the interview. Therefore, the open interview technique, based on an interview guide made up of neutral, open questions, is the most appropriate. In addition, the representation of a stakeholder and the actions he undertakes depend on the context in which he finds himself when he is being interviewed (Suchman 1987). The interview must, therefore, be sited in its action context to be able to capture the empirical knowledge of a stakeholder. For example, to identify how a farmer represents water management systems, the interview should be held on his plot of land as he is irrigating it. The interview can also take place at several places in succession, for example, during a field visit (Abel et al. 1998). Under this same logic of contextualization, it is advisable to start the interview with subjects relating to the activities or events in progress. Thus, if you wish to identify how a farmer represents his interaction with the environment, interview him on his plot and start with questions about his agricultural practices on this plot, before moving on to the links he sees between his practices and the environment. Lastly, do not forget that the quality of the acquisition depends on the more or less direct method in holding the interview (conversational or question and answer) and the relationship of trust between the interviewer and the interviewee. The interviews are then recorded and transcribed word for word for the extraction phase.

Table 4.2 Cross-references between knowledge objects and semantic expressions (Becu et al. 2003)

Knowledge objects	Semantic expression
Concept (object, person, etc.)	Equivalent to names: 'forest', 'river', 'soil'
Process (operation, activity)	'Build a house', 'fish sardines'
Attribute and value	Attribute: 'cost', 'age'; value: '120 kg', 'heavy'
Rule	'If..., whereas...', '... up to ...'
Relationship	Equivalent to passive verbs: '... is part of ...'

The extraction phase involves identifying in the transcription and recording (e.g. highlighting) all the words or semantic expressions relating to the concepts linked to the domain studied. The identification is then repeated for the other types of knowledge, namely, the processes, rules and relationships. Lexicographic analysis software can be used to facilitate this work (Dray et al. 2006). This software encodes semantic expressions and analyses semantic networks, but identifying concepts, rules and relationships cannot be totally automated. The results of the extraction phase are largely dependent on the various types of knowledge, also called knowledge objects (Newell 1982), which you have chosen to identify. There are various extraction structures or grids that cross-reference knowledge objects and semantic expressions (Table 4.2 is one such example).

Extracting Knowledge by Situation Simulation Exercises

Eliciting through a situation simulation exercise involves asking a thematician (who is frequently a stakeholder in the system represented under the companion modelling approach) to explain his projected actions under the various situations presented to him. This technique is used to elicit tacit knowledge from the person by stimulating introspection, which is the ability to raise awareness of vague or dissonant areas between emotions, thoughts and actions, thereby making his representations more coherent (Ferber and Guérin 2003). Various situation simulation techniques can be used (e.g. story, photo slide show, role-playing game, etc.).

Thus the playable stories technique (Becu et al. 2005) only uses one situation simulation exercise per story. The story is divided several times reproducing, for example, the various moments in a crop-growing season. At each point, the facilitator orally describes the situation context (e.g. economic, climatic and social context of a growing period) and asks the stakeholder (sessions can be individual or collective) to explain the activities he would embark on based on this context. The session continues by alternating context description by the facilitator and description of activities undertaken by the stakeholder. In this technique, the actual extraction of knowledge (i.e. the transcription of semantic expressions into knowledge objects) is simultaneous with the stakeholder describing the activities undertaken. The information extracted is presented to the stakeholder on cards on which are written, for example, the name of the entity or process, or as photographs. As the session continues the information is assembled in front of the stakeholder who watches the domain model being constructed. Under this process the stakeholder can intervene directly in constructing the domain model (to enrich or correct interpretations made during his presentation), as in the participative construction of the diagram.

The role-playing game, another situation simulation technique, can also be used to extract knowledge. In this process, acquisition is made by direct observation of player behaviour during the session. Observers (normally several) of the role-playing game note the various actions undertaken by the players and the game situation in which the action has been undertaken. At this stage it is not stakeholder knowledge that is extracted but their actions in a given situation. It is only subsequently, during individual or collective debriefings, that the knowledge applied by the stakeholders to undertake their actions is elicited. To achieve this, each player is asked to explain why he embarked on such and such an action in the game and the information he used to reach his decision. In this way stakeholder introspection is simulated and his tacit knowledge elicited.

Extracting Knowledge by Constructing Diagrams

The knowledge we seek to extract does not only cover knowledge objects viewed separately from each other. It also applies to knowledge on the relationship between these knowledge objects. Drawing diagrams is both a means of eliciting this type of knowledge in the interactional structure of the reference system and a first attempt at the domain model. It seems to us that the domain model corresponds more or less to this first abstraction stage, which could be summarized to advantage as a simple entity-relation type diagram, with the resource(s) and selected stakeholders positioned in it, and where each relationship between stakeholders and between a stakeholder and a resource is signalled by arrows labelled with a verb. This informal format based on 'natural' terms (no semantic restrictions or modelling jargon) provides a first visual glimpse of the model's constituent elements and their interactions, with the advantage of it being easy to comprehend by thematicians who have never been involved in modelling.

This means creating hypotheses for a given question on the elements of the world that are of importance in dealing with it. The three phases traditionally proposed in stakeholder analysis methods can be considered in succession here: (i) stakeholder identification; (ii) differentiation and categorization of identified stakeholders; (iii) specification of relationships between stakeholders. It is essential to consider during each phase the relevance of the planned choices by referring to the question raised. A resource is at the heart of the question considered in the majority of companion modelling implementation experiences. In practical terms, it seems relevant not to abstract this resource during the initial stakeholder identification phase.

Thus, as suggested by the ARDI methodology developed by Étienne (2006), the stakeholders can be categorized according to the more or less direct nature of their actions in the resource and for each type of stakeholder, the nature of the management entity specific to his action can be stated. There is a whole range of tools for use in specifying in a more or less structured fashion the relationships between stakeholders (e.g. as a matrix made up of + and – signs to signify the positive and negative influences between the stakeholders taken two by two).

Figure 4.3 gives an example of this type of diagram created as part of a modelling exercise for a case of disputed use in the Causse Méjan grasslands. In the diagram, the entities considered as the main stakeholders are shown, as well as the resources and also the special case of a passive entity (pebbles). At this stage, reflecting on the nature of entities gives little reward. The aim is to communicate, by presenting as simply as possible, an understanding of the key elements in the operation of the system concerned. Any other format fulfilling this function can be envisaged. Even if the format used is not very formalized, compliance with a few major underlying principles will greatly assist in comparing different versions suggested by several groups working in parallel. It is especially advantageous to implement this first phase in the conceptualization stage collectively, with participants who can be stakeholders in the system, scientists, decision-makers and so on. Comparing the various versions produced is a first step towards sharing representations.

This stage ends by stabilizing the domain model relating to an agreement at a given moment between the participants on the basis of cross-checking areas and cross-references between the various representations in the study domain. It is important to highlight any inconsistencies, contradictions, divergences and singularities remaining in these representations and to agree on how to deal with them, with possible and non-exclusive options including:

- avoiding the issue: the domain model is restricted to what has been agreed
- seeking a negotiation-based compromise
- using scenarios underlining and exploring ‘disputed’ areas (Dray et al. 2006), or taking charge of inconsistent viewpoints when agreement is lacking in an examination to decide between them, such as a common observation in the field
- uniting viewpoints by limiting them to those not generating inconsistencies.

Formalization of Knowledge and Conceptualization from the Domain Model to the Conceptual Model

Once the domain model is set, the conceptualization itself can start. Although formalisms, usually simplified, are frequently used from the first stages of abstracting the domain model, one or more formalisms must be strictly adopted for the conceptualization phase, which describes fully the different aspects necessary to convert it into a simulation model. Conceptualization, therefore, involves specifying the various components of the domain model in the selected formalism(s) and stating the different aspects, especially dynamics, to achieve a full description. Just like the abstraction, the conceptualization can be participative, in which case the processes are often closely linked. However, it can also be semi-automatic. We shall see lastly the large families of entities and processes found in our conceptual models. It is important to ponder the nature of entities and processes when seeking to conceptualize, as it is the way to relate them to available generic concepts.

Formalisms for Conceptualization

A formalism is a formal expression method based on a set of words obeying rules and conventions (called formal grammar or syntax) and underlying semantics. The most frequently used formalisms to conceptualize models in the ComMod processes are as follows.

- Mathematical equations (e.g. for a biological process like the growth of a resource the logistical equation could be chosen). They are used to describe and specify particular processes in the evolution of certain entities.
- Pseudocode, a language close to natural language, resorting to elements in logical syntax (if, whereas, etc.) and a stabilized lexicon to designate entities, attributes and actions. It has the advantage of being easy to use in collective groups, including lay groups, whilst imposing a certain discipline in respecting the specific language of a given group and to a given moment in the life of this group.
- UML, which offers a stabilized grammar (unlike pseudocode) and which is, therefore, understood by all who know it, whether or not they took part in constructing the model. UML can formalize an object-oriented model as well as an agent-based model. It proposes in particular a panoply of diagrams (called 'views' used to present a conceptual model from different angles), such as class diagrams to describe the model structure as a set of connected entities, activity diagrams to present the behaviour of entities and transition-state diagrams to characterize the dynamics of changes in stages (Le Page and Bommel 2005). Although creating UML diagrams can often be arduous, the aim is to achieve clear diagrams comprehensible to all, even those who have not taken part in the conception, which can be the case of the computer scientist who will be in charge of implementing the conceptual model.

The Conceptualization Process

Sequence of a Standard Conceptualization Process

The conceptualization process normally comprises the following activities, which must not be seen as distinct and sequential as they are more often interlocked and iterative.

(i) Model structure specification: 'translation' of domain model entities in the chosen formalism. What are the entities? What are their relationships (which entity 'knows' which other and from which viewpoint)? What are their characteristic properties and the main actions they can perform? This phase is normally a chance to group the entities with similar behaviours by defining the most generic entities. In this case we talk about generalization, in which the match is called specialization. It is at this stage also that elements from previous modelling work can be usefully reused (e.g. generic social simulation models) or include elements from the thematicians' theories (thus the theory of agrarian systems is based on a typology of farmers according to a certain number of criteria that could be used to specify an abstract 'farmer' entity). By analogy with knowledge engineering, this phase may be compared with an ontology construction phase (Bommel and Müller 2007).

(ii) Specifications of dynamics specific to the entities: intrinsic resource dynamics, operations capable of being performed by one entity (behaviours) and influencing its evolution or that of the entities to which it is connected, decision-making mechanisms mobilized to choose the behaviour according to the present state of the entity and the context (its environment). The mobilization of elaborated decision-making mechanisms is normally specific to social entities.

(iii) Lastly, specification of the ordering over time (scheduling) of these processes during a simulation time step.

During this conceptualization work, it is important to maintain a vocabulary that belongs to the domain modelled. The diagrams produced are basically intended for thematicians so that they can understand, grasp and criticize them.

Participative Conceptualization

In most cases, the workshops that could be qualified as participative modelling are simply elicitation objects. The media are used by participants to explain how they perceive their system or simply to help them in reaching agreement. We talk, therefore, about participative conceptualization or conceptualization workshop when the stakeholders and thematicians are actually involved actively in the conceptualization work, be it in defining the formalism or conceptual models or when translating from the domain model. Ubon Rice Seeds case study is an example of co-conceptualization with local experts: under this application, both researchers and regional development stakeholders genuinely worked together to design the UML diagrams. The conceptual model of role-playing games used in the first phase of the study is based on this (Vejpas et al. 2005).

As a general rule, from the moment both the stakeholders in the reference system and the thematicians working to understand this system are involved, it is virtually impossible to separate abstraction from conceptualization. The two mechanisms are closely linked and part of the same process. By relying on a formalism to remove ambiguities, this in fact involves systematizing the description of the model from different viewpoints, by gradually refining the domain model.

Within the framework of participative conceptualization workshops for a model that unites researchers and experts in the domain in an exercise of interdisciplinarity, the modeller proposes a formalism (often UML) to translate the domain model(s) into a conceptual model. His role is to facilitate the translation of concepts conveyed by the experts. Despite the existence of tools (e.g. computerized tables or simple blackboard or post-its) for use by all participants in modifying the model directly, the modeller is frequently the 'pencil holder' who coordinates the discussion in an attempt to incorporate and synergize the concepts. The formalism must, therefore, be comprehended as a language for bringing the disciplines closer together. Morel (1979) expressed this over 30 years ago: 'this is why collaboration between physiologists from all horizons on the one hand and computer scientists and biometricians on the other will remain necessary and desirable for many years to come. But to be fully effective, this collaboration assumes that the people involved in both camps are both taking a part of the path which too frequently separates them. It is in fact essential for each individual to know how to express himself in a language accessible to his partners, if only to comprehend fully the limits and possibilities of each other's respective approaches'.

Considering UML as a tool for interdisciplinary dialogue is also recognizing with Morand (2000) that 'the diagram is at the heart of the cognitive process, not teetering on the brink'.

With the reference system stakeholders, the participative modelling workshops can be considered more widely open: the stakeholders are asked to take part in constructing a representation according to a predefined and normally little discussed 'grammar', such as in the ARDI methodology developed by Étienne et al. (2008c), described above.

Conceptual Model Entities

The reference to MAS representation mode quite naturally gives food for thought, in terms of description, on the entities of the system to be modelled—stakeholders and objects—and how they act and interact. Compared with the domain model development phase, it is a question here of determining the conceptual entities that will be useful in representing the entities considered as part of the domain model. Remember that the model objective is essential, as this overview is useful for anyone encountering the model for the first time to understand why certain aspects of the system studied have been ignored (Grimm and Railsback 2005).

Whether or not an entity will be translated into a class of the model is, therefore, a choice made at this level. The choice will also be made at this level as to whether a group is represented as a single entity or as a composition of agents.

Distinction can be made between different types of entity in the conceptual models. It can be useful to question the nature of the entities modelled as this can help in regrouping and encouraging conscious reuse. Various criteria can be used to distinguish between entities, but the breakdown we present here is the one normally used in the companion modelling processes and reproduced in the Cormas modelling platform.

The social entities, sometimes called agents, represent individuals, groups of individuals or institutions involved in resource management. Their role is normally to manage and/or exploit the system's resources, communicate with the other social entities and act on and/or receive spatial entities and other physical objects. The social entities are the bearers of decision-making abilities and thus of cognitive structures. These cognitive structures can be more or less advanced and depending on the question asked, there are reusable concepts in the multi-agent computer literature, be it in the perception, imitation and learning or intentionality abilities of social entities. In role-playing games, human players directly use their own cognitive abilities to make decisions, only constrained by the rules of the game.

The models we use include a spatial medium almost systematically. This medium is made up of entities (e.g. plots, regions, rivers, etc.), which configure the space and structure the network of spatial interactions. These spatial entities are also sometimes management entities linked to each type of actor, each one with a vision of the space specific to their type of activity.

The other model entities are physical, biological or intangible. The resource entities are a class apart from the others. These entities generally follow specific rules for their distribution and regeneration (e.g. water cycle, population dynamics, evolution and transmission of genetic traits). In computerized simulation models, these functions are often derived from simplified disciplinary models. In role-playing games, the same functions are frequently used as modules to support the game. It may happen that in abstract role-playing games, physical artefacts are used to simulate resources dynamics, such as marbles used to represent water flowing in (Lankford and Watson 2007). The evolution of these resource entities relies on their interactions with social entities (e.g. intakes, modifications, exchanges, relocations, etc.). Reversely, the perception and management modes of social entities are focused on the dynamic of these entities.

The models can include entities representing other physical objects capable of being manipulated and/or exchanged by the social entities or acting on the resource entities. This wide category can group concepts as diverse as infrastructures (dykes) or money as well as communication media. The physical objects can curb the model's interactions in all circumstances.

Lastly, a final category is formed by the communication entities. This covers intangible elements (e.g. information, knowledge, beliefs, emotions, etc.) explicitly shown in the model as they make sense for the social entities and as they manipulate and exchange them. These communication entities are conceptualized most of the time as simple 'messages' containing a contextualized piece of information (e.g. information about water levels on plots in SHADOC, information on seed requirements and availability in Ubon Rice Seeds, etc.). However, it can happen that the information exchanged by the agents needs to be structured and differentiated better, whether socio-technical (technical itineraries) or psychological (emotional states). This type of entity is associated with agents using differentiated information to advance their representation of the world, their emotional states and to influence and possibly mislead the other agents. This is often the case of models addressing resource management problems from the angle of diffusion and impact of beliefs in society (diffusion of innovation or solidarity). Note also that the communication protocols under which these entities circulate (e.g. centralized diffusion, diffusion by mutual agreement, diffusion within social networks, etc.) have a strong influence on the operation of the system. It is often a topic for investigation in different scenarios for models showing information.

From the Conceptual Model to Simulation Models: Implementation

One of the specific features of companion modelling is to encourage the development of several operational models referring to the same conceptual model (this is detailed below). In all circumstances (role-playing game or computerized simulation model), a simulator is developed to 'run' the model. Implementation accounts for this operationalization phase of the conceptual model in a concrete tool.

First, we deal with questions relating to time management and the choice of order in which the model entities are activated. For computerized simulation models in particular, these aspects have a significant yet frequently ignored influence on the behaviour of the simulated system. The practical aspects of manufacturing simulation tools are then addressed. A special place is given to spatial simulator media, essential tool components dealing with resource management issues and targeting shared representations. Finally, the three-way calibration-verification-validation is discussed by considering the specific features of companion modelling.

Managing Time, Ordering Agents and Ordering Actions

Whereas, as we will stress, computerized time management forces a certain number of technical constraints that must be controlled when implementing a computerized simulation model, practical time management when using role-playing games also imposes constraints. Unlike a virtual agent, a human agent very quickly gets bored with repetitive activity. A role-playing game session must be thought out to ensure a ludic side with no down time. The number of time steps in a role-playing game session will inevitably be less than the number of time steps potentially implemented by a computerized -simulation. However, there must be enough to produce something decisive in the game and the period represented by a time step must remain relevant to the processes underlying the resource dynamics. From this point of view, it can prove useful to couple the role-playing game with a computerized simulation model (Barreteau and Abrami 2007).

In computer science, time is either driven by events (the system manages an agenda) or by a clock (the time is segmented into regular time steps of identical length; this is called discrete time simulation). Both approaches have strong and weak points, but it is widely held that the discrete time approach considerably simplifies the development of the simulator and the understanding of a simulation sequence (Treuil et al. 2008). This explains why this time approach is systematically adopted in companion modelling.

We have already pointed out that the simulation horizon (i.e. total number of time steps) relates directly to the definition of the question asked. This dimension is normally determined before considering the granularity (i.e. length of a time step) of the time, which in turn depends on the periodic intervals of all the dynamic processes in play. The granularity is normally chosen as equal to the shortest periodic interval.

The stakeholders in a reference system operate simultaneously. In a computer, the virtual agents who represent them are activated sequentially. Similarly, the sequence of elementary actions executed for each agent at each time step is sequential. It is especially important to specify clearly the chosen ordering rules when the model is implemented as they can have a tremendous impact on the simulation results.

Spatial Media

The analysis of 63 simulation models developed under the 27 case studies reveals that space is omnipresent in these models: 90 % of them have spatial representations. The representation of space is obviously computerized in computerized simulation models, which account for half the simulation models (see Fig. 4.3). Spatial representation is generally not computerized in role-playing games (74 %), despite the vast majority being computer-assisted (see Fig. 4.3). In this case, space is represented by a game board that can take several forms. Three-dimensional blocks are one possibility when the land relief is a major component to be considered (e.g. Mae Salaep, SugarRice or Ubon Rice Seeds), more or less virtual maps representing a realistic space, such as a catchment area, the territory of a park or several municipalities (e.g. Nîmes-Métropole, Kat Aware, Méjan or Pays de Caux), more or less abstract regular grids (e.g. Radi, Lingmuteychu, Lam Dome Yai, Ouessant, etc.) or pictures fixed to the wall to outline an irrigated perimeter (e.g. Njoobaari, Larq'asninchej). Even when a computerized representation of space is provided, a game board is still frequently used during a role-playing game session. This association is complementary in certain cases: all participants visualize the two media, each one contributing specific information. Thus in the AguAloca case study, the game board localizes the hydrographic network and the pumping points (arches/nodes network), whereas at the same time the players can see the same space on the computer monitor from different viewpoints (e.g. soil occupation, municipalities, under-catchment areas). In other circumstances, the various spatial media are allocated to different types of players. Thus, in the MéjanJeu role-playing game, the farmers only had a local view of the space (close-up of their farm) as a printed map, whereas at the same time the naturalists and foresters could consult viewpoints of the entire space directly on a computer monitor.

Implementation

Implementing role-playing games involves identifying, if necessary manufacturing if they do not already exist, and preparing the game media. A game medium is an element that delivers information. A good medium is the result of a frequently difficult compromise between stimulating the ludic aspect and using user-friendly tools whilst avoiding excessive technology and timescales. It is also important to give thought to the form most suited to the type of information that has to be transmitted. Thus for money or material flows, pawns, post-its and notes can be used. 'Chance' cards and dice can be introduced to bring the hazards into play (note that it is sometimes important to be able to control the hazard so as to reproduce the same conditions, especially when you want to be able to compare game sessions). Sheets or *aide-memoires* can be planned to impart information at the start of the game, leaving them available during the game session.

In the case of computerized simulation models, the computer is the principal equipment used to construct the operational model and the computer programming languages are used to 'encode' the model as a list of instructions that can be interpreted by the machine. Several options exist.

- Specific programming: all aspects delegated to the computer must be encoded from scratch. The advantage is total control over the translation chain; the disadvantage is the time spent in reprogramming things that had already been programmed.
- Targeting a specific software program to take charge of a precise aspect (e.g. a geographical information system (GIS) software programme is well suited to representing and analyzing a space).
- Only using a single software program to integrate all the 'encoding'. This software is qualified as a 'generic platform'. A certain number of generic platforms have been developed specifically to facilitate the implementation of the computerized simulation model, but a commercial software program like a spreadsheet can do the job: this is true of the SylvoPast role-playing game developed by Étienne (2003).

The advantage of using a generic platform for implementing a computerized simulation model is unquestioned today. The main reasons are as follows: (i) the generic nature of numerous components (e.g. the space representation module); (ii) the availability of facilities providing efficient implementation of experiments with the model (e.g. launching simulation batteries by specifying the variation ranges for a set of parameters, displaying indicators, etc.); (iii) constitutes an applications library, a source of inspiration for modellers who are not computer scientists.

Cormas[1] (Bousquet et al. 1998), a generic platform developed by CIRAD since the mid-1990s to facilitate computerized agent-based simulation model implementation applied to resource management, is used frequently in the implementation of companion modelling. One of the major strengths of Cormas is its reliance on the Smalltalk programming language; modellers who are not computer scientists find this very easy to use and it makes it easy to reuse extensions already developed in other models. NetLogo,[2] Swarm[3] and Repast[4] are among the other widespread generic platforms developed to implement computerized agent-based simulation models.

When the reference system combines several levels and dimensions, a model can help in making representations specific to each level or dimension coherent. In Chap. 11, the two main options are presented: either the goal is to integrate the diverse representations into a single one or to coordinate them. In terms of generic platforms, the first option requires an integrative tool, whereas the second one requires a toolkit providing tools suitable for each level. An integrative generic platform should also provide functionalities to support the activities specific to all the sequential stages of the modelling process. Mimosa[5] is seeking such an ambitious objective, combining a simulation kernel based on the discrete event system specification (DEVS) with tools to specify ontology and an extensible set of formalisms.

Calibration, Verification and Validation

Calibration

Calibration involves adjusting a model so that its outputs correspond to the expected values (most frequently, a set of empirical data). If not, the model is adjusted by modifying the values of certain parameters. This process is similar to adjusting a measuring instrument using a calibration standard. The parameters

[1] http://cormas.cirad.fr/indexeng.htm

[2] http://ccl.northwestern.edu/netlogo/

[3] http://www.swarm.org

[4] http://repast.sourceforge.net/

[5] http://sourceforge.net/projects/mimosa

selected for this adjustment have a slightly peculiar status. In practice, a parameter is chosen that has both a significant influence on the model outputs (two distinct values produce different results) and for which the value is not known with any certainty. Certain elements of the model must be modifying if adjusting the parameters is not enough to obtain satisfactory results. This results in a more or less in-depth re-examination of the operational model or the design model or even the domain model. Calibration is no longer the issue at this stage it is more a new learning curve through modelling and evaluation.

The design of a role-playing game used for situation simulation exercises sometimes uses calibration to give the model the best chances of achieving its specific objective, namely stimulating exchanges between participants. As the original question for implementing the approach normally causes a problem for certain stakeholders represented in the model, a suitable calibration will be sought to reach a marked situation provoking reactions from participants and stimulating discussion after a few simulation time steps.

Verification

Verification sets out to judge the correct performance of the simulator, in other words, the faultless implementation of a model in a machine. Put another way, it involves knowing that the model is constructed correctly: 'building the model right' (Balci 1988). However, how can one be sure that the simulation outputs flow only from mechanisms that are thought to have been developed in the model? There are many ways of revealing artefacts related to programming or calculation errors, to approximate management of simulation time and interactions between agents or to any other erratic behaviour of the simulator unrelated to the conceptual model it is supposed to express. Miscellaneous small defects or approximations can be amplified into more significant errors, which risk having a strong influence on the overall behaviour of the system.

For relatively complex models, such as those simulating the operation of socio-ecosystems, ensuring the total absence of bugs is virtually impossible to achieve. Thus, according to Gilbert (2008), the number of bugs in a computerized simulation model follows a negative exponential function: after a rapid drop, it never reaches zero, even after a long debugging process. This observation compromises the fundamental scientific principle of reproducibility of results. Several recent studies seeking to reproduce published simulation results have confirmed this problem (Edmonds and Hales 2003; Rouchier 2003). Now aware of this, researchers using computerized simulation models are joining together to produce recommendations intended to make it easier to discover recalcitrant bias and bugs (Gilbert 2008).

Models developed under companion modelling are often very simple and are sometimes called toy-models. They are not necessarily designed to be reused outside the context for which they have been specially designed (throw-away models). These characteristics tend to render them less sensitive to the point raised earlier; not that they are less prone to bugs or artefacts, on the contrary, but more because the consequences of these errors are not decisive in relation to the approach. ComMod does not seek to suggest finely calibrated, expert solutions but to trigger dialogue between the participants to the point where the model can be considered as a pretext, an intermediate mediation object. These models can, therefore, be modified during use (not just the parameter values but potentially also the structural elements) by inputting suggestions from participants during the participative simulation workshops. These modifications are applied ‘on the fly’, without taking the time for a conceptual rethink or to check for the introduction of any computational bias. It makes the models difficult to communicate to anyone other than the people who modified them, which raises the problem of their transferability.

Validation and Validity

According to Balci (1988), validation consists in comparing the behaviour of the model with the ‘actual’ system it is supposed to represent. When this comparison is satisfactory, the model is validated. To put another way, we are seeking to know if we built the ‘right model’ (Balci 1988). A ‘correct’ model is, therefore, often perceived as a model which ‘tallies’ well with the data. Of course, depicting simulation results with variables that can be measured in the field is an essential dimension in model validity. However, it is not enough to consider this dimension alone. Without even mentioning accidental correlations, there are a vast number of theoretical or practical problems in comparing outputs from a model with empirical data (Amblard et al. 2006). Good correlation with data can stem from an external factor not taken into account by the model. The model can also exhibit results consistent with data, whereas the modelled mechanisms are found to be totally erroneous. In addition, if the data used to calculate this correlation have also served to calibrate the model, referring to this single criterion to declare its validity is nothing short of deceitful.

As companion modelling comes under the tradition of constructivism, knowledge built up by the modelling experiment do not in this context form either normative principles or predictive theories. They take the form of ‘generic proposals’, intended to ‘enlighten the reader, arouse his thought processes and his questioning and stimulate his imagination and his creative action’ (Avenier and Schmitt 2007). Thus, the know-how is not ‘validated’ in the traditional sense but ‘legitimized’ by the consistency of the construction method and by the subsequent use made of the co-designed know-how.

It is nowadays widely acknowledged that the concept of validation makes sense within the scope of a given modelling approach: it is useless to discuss it in general (Pala et al. 2003). Companion modelling, considering that the issue tackled by the model is not definitively framed from the beginning but rather that the modelling process contributes to enlighten it, relates to ‘soft operations research’. The model is used as a support to debate the issue along a continuous learning process.

According to Checkland (1995), in such a context, the validity of a model relies mainly on its ability to generate learning. As soon as the model seems 'credible' to users and moreover, the users assess that they are learning through its design and use, then the model is validated. This kind of 'social validation' is quite different from the standard vision that strictly relates the quality of a model to measurements of its distance to the reference system it is representing.

Simulation Models to Explore Collectively Possible Futures

Chapter 3 discussed the human dimension in coordinating participative simulation sessions (and more especially situation simulation exercises using role-playing games) and detailed the roles to be shared by the facilitation team members. We present here a supplementary viewpoint that describes in practical terms the implementation of collective scenario exploration workshops based on role-playing games and/or computerized simulation of a virtual world.

Setting up the Simulation Workshop

Prior to introducing the model to the participants, a participative simulation workshop starts with a general presentation of the context governing it, to provide elements of response to the following questions. Who took the initiative to implement the approach and with what goal? Who selected the participants and based on what criteria (e.g. stakeholders whose actual activities are represented in the model, legitimate stakeholders for representing a group, taking relationships between stakeholders in everyday life into account, etc.)? What is the specific goal of the workshop and which programme will be used to move it forward?

In the case of role-playing games, configuring the space in which the workshop will take place must be thought out based on the spatial characteristics of the reference system, so as to position the players in identified locations, that is, key places with a clearly defined status (e.g. market, public meeting place or private place) and thus reproduce certain major properties like neighbourhoods and distances.

Introducing the Simulation Model

Presenting the model is a difficult phase. It must be short to allow participants to become actively involved as quickly as possible, but it must at the same time provide a whole range of information on comprehending the model structure (i.e. representation of the space and types of agent making up the model), the decisions to be taken by the players in each turn or actions by the computerized agents at each time step, the indicators available to account for the consequences of these decisions and actions, the resource dynamics and lastly, the scheduling of the simulation (periodic intervals for processes, period representing a time step and total number of time steps simulated).

The amount of information communicated to the participants during this model introduction phase must be minimal but nevertheless enough to trigger the learning mechanisms efficiently according to the principles of active pedagogy (see Chap. 10). The information is often asymmetric in role-playing games: the participants received elements specific to their role, which are not divulged to others.

To ensure that the model is understood correctly, a role-playing game session can act out a 'dummy run'; for computerized simulation, checks can be made to see if the participants anticipate correctly the gradual changes in state (in step by step mode) of the simulated entities. The complete disentangling of a first scenario can then be envisaged.

Identification and Formulation of Scenarios

Participative scenario planning stimulates the creativity of participants through simple tools (e.g. short stories, diagrams, etc.), which envisage the trajectories towards possible futures of a socio-ecological system (Peterson et al. 2003; Evans et al. 2006).

In the context of companion modelling, 'scenario' is understood more as an operating mode for the simulation model or more precisely a set of factors that are going to modify its operation: a certain stakeholder behaves differently, certain ecological dynamics are disturbed, a certain variable of social or economic clamping is changed. Changes in how interactions are organized are also frequently envisaged (e.g. new exchange systems, new negotiation protocols).

The scenarios are often defined collectively when the simulation results of a first scenario are discussed, which instigates suggestions for alternative scenarios. It can happen that the first, so-called 'baseline' scenario (established with reference to the current situation of the system studied) has been put together by the workshop designers as a starting point for the exploration process.

The role-playing game technique is effective in generating scenario proposals but far less so in exploring them. During a game session, producing a time step (game round) in fact requires a great deal of time. Whilst taking care to maintain the ludic nature, a method must be found to link the game rounds sufficiently well to get close to the relevant simulation horizon in terms of the question asked and the evolution speed of the processes represented. Here is where computerized simulation really comes into its own.

Exploration and Observations

An advanced exploration of the model's properties is always useful in an experimental approach; it ensures its robustness and measures its sensitivity to the various parameters likely to be mobilized to define scenarios. When run a great many times the model produces thousands of observations. It is important, therefore, to conceive experimental designs so as to produce the targeted information at minimum cost. The use of simulation platforms like Cormas provides access to a whole set of functionalities, which helps considerably in producing this experimental design.

One of the basic principles of companion modelling stipulates that the exploratory simulation of contrasted scenarios, by structuring the exchanges between involved stakeholders, lets them validate the interactions between the different representations and system dynamics integrated in the model. The simulation takes part in a co-learning process (or mutual learning between participants and workshop coordinators) in the system studied, strengthening the interaction with, and between, local stakeholders (see Chap. 10).

Indicators and Viewpoints to Monitor Changes in the Simulated System

Indicators used to monitor changes in the simulated system and compare the proposed scenarios are calculated from the model variables. In agent-based models, these variables correspond to entity attributes, be they spatial, active (agents) or passive. A given attribute may be considered directly as a relevant indicator, but functions calculated from several attributes to develop the most synthetic indicators can also provide the basis. During the various stages in the approach, a panel of indicators mentioned as relevant by certain participants is gradually compiled and supplemented. These indicators, which correspond to what each stakeholder is accustomed or wishes to consider in his activities, orchestrate their perception of the virtual world.

As an agent-based model is often made up of a significant number of entities, it is tedious to use graphs alone to monitor changes in all the indicators. A practical means of observing the indicators in a whole range of entities is to define the viewpoints specific to each type of entity as a function of visualization, which allocates an image of a specific shape and colour to each value or interval of values of the indicator. Applied to all entities defining the spatial medium of the computerized simulation model, called here 'viewpoint' and corresponding to a 'theme' in the GIS, it offers a dynamic spatial representation of the simulation on which can be superimposed the dynamic representations of situated entities.

Thus under the ComMod implementation in the Causse Méjan (Étienne et al. 2003), a series of viewpoints was constructed to facilitate understanding of the process of grassland encroachment with pine trees by making a clear distinction between the physiognomic (the pines are seen in the landscape) and functional (the pine seedlings have become established in the plot) aspects. Another was constructed to localize the heritage issues of fauna, flora and the landscape to produce a synthetic representation. Another accounted for the work carried out and its localization, as seen by either the silviculturist or the naturalist. Some viewpoints have sought to convey a particular aspect as, for example, changes in the risk level of encroachment from adult pine trees located on the ridges.

There are several ways of providing the observable elements in a simulation to the participants. The entities represented according to different viewpoints like the graphs can be printed or displayed on a computer monitor (so that the information can be specific to the participants it is intended for) or projected directly in the room (information shared by all the participants).

The diverse viewpoints are presented to participants who observe (computerized simulations) or experience first-hand (role-playing games) changes in the simulated system. Participants find it easier to grasp the viewpoints when, under identical conditions of simulation for the same scenario, they comprehend them at the same time as those more familiar with them. It is, therefore, easier to share representations.

Behaviours and Interactions Between Players

Situation simulation exercises for people taking part in a role-playing game session influence the way in which they make their decisions and interact with the other participants. Recording arguments put forward during discussion phases between players is a way of gleaning information on the rationalities mobilized for these decisions. Duplication of roles verbalizes the reasoning, which makes it more easily accessible, but hedges the exploration of more standard behaviours as shown in situations of experimental economics (Bornstein and Yaniv 1998). Lastly, observing player attitudes and behaviours is another source of information. This activity has to mobilize assistants capable of performing this function (see Chap. 3) and totally devoted to the task (one assistant stationed in each strategic place). The resulting observations can often prove very rewarding. They are analysed collectively during the debriefing phase of a role-playing game session. They can also be used to activate the changes in the model used by questioning the domain model, the conceptual model or the simulation model. It is the first driving force in the iterative process.

Analysis

After exploring the various simulation scenarios, the participants make inferences from what they have observed and what they felt was revealing. It is fundamental to allow sufficient time for collective discussion of the conclusions that each individual has drawn from the experiment, so that these conclusions can be shared and potentially invalidated given the distance between the virtual world used to reach them and the actual world that could be subjected to them. This collective discussion of conclusions is even more important when they involve individual participants personally. Participants must be given a chance to re-establish their identity to avoid losing face before their group (Richard-Ferroudji 2008). This is achieved through the collective debriefing presented in Chap. 2.

In role-playing game workshops, this analysis is hugely beneficial in making available observations on the progress of the session. Thus, if the decisions of players are recorded on a computerized medium, decisions judged problematical can be explained through speeded-up sequence replays or the various changes shown by replaying other sessions organized elsewhere. If an observer has followed the negotiation phases, the presentation can be analysed and the arguments discussed. If the session has been filmed, the attitudes can be analysed or an attempt can be made to interpret players' movements in the game space (who has been taking the initiative to trigger interactions with the others, and so on).

The analysis phase is when the return of the virtual world to the real world is addressed. The participants are invited to say if they have spotted the links between what the simulation experiment has exhibited and what goes on in reality, or whether, on the contrary, certain aspects exhibited during the simulation experience are never seen in reality. This is the second and main driving force in the iterative process: the participants can raise doubts over the representations mobilized, their implementation or develop the questions asked during this analysis. Finally, over and beyond identifying similarities, the analysis must include a form of workshop assessment by the participants, especially asking them whether they perceive any type of advantage in participating compared with the question asked (see Chap. 7).

Singularity, Complementarities and Versatility in Modelling Tools

This chapter has presented the various stages and methods in constructing virtual worlds for the purpose of shared representation and collective exploration of possible futures. The stages have been presented in a certain order, which makes you think that the sequence always advances in this order, with the end of one stage heralding the start of the next. This is not true in practice: there are frequent interferences between the various stages. This feature is reinforced again when the stakeholders are involved, to varying degrees, in a certain number of stages and when they are also given the chance to steer the process. The companion modelling approach has a duty to be adaptive and its tools must be flexible. Through the diversity of experiments analysed, we discuss the duality between the degree of singularity and the degree of genericity of the models, which can be related to their degree of abstraction/realism. Then we review the diversity of modes when combining the two main types of simulation tools used in companion modelling, namely computerized simulation models and role-playing games.

Singularity/Genericity of Modelling Tools

During each ComMod process, a singular model, specifically developed to represent the reference system under study, is co-designed. Conversely, a generic model, undoubtedly useful as potentially suitable in different contexts, is not co-designed (or co-designed just once). Being a methodology that highlights the co-design process, what are the implications for companion modelling regarding the status of generic models? Sometimes we use the term 'disposable models' to state that a model is just a snapshot of representations and questions at a given stage. Using it makes the questions evolve and designing it makes the representations change, therefore, it is only relevant in catalyzing the process at a given time. To overcome this narrowness, making, from a peculiar case, a model more generic would allow widening its representativeness. Linking such a model to a given reference system is a way to disclose its type. Serving as a reference, a generic model can also keep track of the diverse adaptations (disposable models) that were specialized to make it more relevant to specific contexts.

The singularity/genericity of a model is often directly related to the degree of realism of the representation of the reference system, which is linked to the degree of complexity of the model. Three levels can be distinguished.

When there is an explicit linkage to a specific reference system, the actors, the resources and the spatial configuration are straightforwardly specified in the model from the corresponding characteristics of the reference system. This is often the option preferred at first by the local partners. To recognize in the model some particular features of the system under study make the participants confident about the ability of the tool to represent real-life issues. It may happen that this quest for realism is mainly justified by the need for participating stakeholders to become confident in the model under development. Thus, in the Domino Réunion case study, a detailed demographic module that was first included in the computerized simulation model was not discussed later on in the prospective scenarios. Generally, modellers seek more realism. This bias restricts the ability of the model to distance its users from their reality, which can prevent the consideration of some scenarios that would let sensible real-life aspects surface again.

When there is an implicit linkage to a specific reference system, the model is an archetype based on a realistic simplification of the actors, the resources and the spatial configuration of the real system. For the participants to legitimate the simplifications, some key features, independent from any peculiar details, have to appear in the model. For instance the SylvoPast (Nîmes-Métropole) case study (Étienne 2003) provided a representation of a typical Mediterranean forest based on mean proportions of the vegetation layers observed in that region.

Finally, when there is no linkage to any particular reference system, the model provides an abstract representation that simply aims to deal with an issue. The users may repel a tool too abstract as being insignificant for them, hindering their ability to consider simulation outputs as plausible anticipated situations. However, it may also happen that the participants themselves reckon that a simple and abstract tool best suits their needs to share representations. Thus, during a project entitled 'Levelling the Playing Field' carried out in the Philippines, the abstract version of the CherIng role-playing game, usually used as a pedagogical tool in training sessions about the companion modelling approach, was finally more appreciated by the participants than a contextualized version.

Combinations of Various Simulation Models in Implementing the Approach

In the majority of ComMod case studies, several simulation models are combined. Following a first typology of benefits related to the combined use of role-playing games and computerized simulation to address negotiation issues (Barreteau 2003), we review here the advantages of each type of combination in supporting modelling activities, such as design, communication, exploration or validation.

Filiation links between models exist in all the ComMod case studies combining several simulation models. Thus even in the most prolific case study (seven simulation models developed in 6 years), Mae Salaep, common elements are found between the various models developed to address the problem of erosion, access to credit and sharing of water.

When the role-playing game precedes the computerized simulation, it supports the communication of the conceptual model, whereas the computerized simulation enforces and extends the prospective dimension by allowing the exploration of more scenarios than the role-playing game. In such a configuration, two scenarios stand out.

In the first one, the computerized simulation model is a direct transcription of the role-playing game and is the most often used in the continuity of (or only a few days after) the role-playing game session, with the same participants. The Self-Cormas application in Senegal opened the way to this type of combination (d'Aquino et al. 2003). The understanding of the direct link uniting the two tools is made easier by the proximity of the interface between the users and the model, particularly the spatial representation: what appears on the computer monitor is a faithful reproduction of the game board. Drawings on cards used in the game can also be digitized and displayed. This is the ideal combination for participants to understand clearly the structure and principles of the conceptual model when playing it, and to propose and subsequently monitor scenarios on the computer fully aware of the status of the computerized simulation model. This does not appear as a complex tool issuing recommendations but a more efficient role-playing game equivalent for exploring the scenarios.

The second use scenario for a computerized simulation model following on from a role-playing game, less frequent than the first, refers to a virtual agent model that is not reproducing a role-playing game, but often a more developed representation of the reference domain. Role-playing game sessions are then supporting the process of designing the computerized simulation model. An application in Thailand (Lam DomeYai) demonstrated that this association method can prove more effective in developing relatively complex computerized simulation models with local stakeholders who can then take it over: at the end of this project, the villagers who had taken part in the virtual agent model development process (which was based on three role-playing game sessions) went to present 'their' simulation tool to a seminar at the university. In this scenario, unlike the first, the designer of the virtual agent model capitalized on the analysis of several role-playing game sessions so that he could base defining the virtual agents on typical rather than special behaviours.

When the computerized simulation model precedes the role-playing game, the role-playing game is frequently a simplification of the virtual agent model, which participants not very familiar with this type of computerized simulation model can find useful in understanding its structure ('open the black box'). For instance, Njoobaari (a role-playing game) and SHADOC (a computerized simulation model) are two related simulation tools representing the operation of irrigated systems in the Senegal river valley (Barreteau et al. 2001). This type of combination also relates to situations in which the role-playing game includes certain modules created during the development of the virtual agent model, mainly those linked to natural processes, like the pine dissemination module in the Méjan case study (Étienne et al. 2003).

The recent trend towards hybrid simulation models, which by their very structure incorporate the specific properties of role-playing games and computerized simulation models, demonstrate that both types of tool are very useful in companion modelling implementation. The hybrid simulation models offer interesting possibilities in managing the time constraint of role-playing games most effectively. The avatars thus take over from players certain repetitive actions (MéjanJeu) or with shorter time periods than those taken in the role-playing game framework to make key decisions (Pieplue). Adding virtual agents to a reasonable number of human agents is also achievable with a hybrid agent model (AtollGame), giving the simulation model enough agents to cope with the question asked.

Chapter 5
Contexts and Dependencies in the ComMod Processes

Raphaël Mathevet, Martine Antona, Cécile Barnaud, Christine Fourage, Guy Trébuil and Sigrid Aubert

The ComMod approach involves the interaction of local actors, heterogeneous social groups with different rationales and interests, and researchers and institutional stakeholders (i.e. donors, sponsors, administrators and experts), whose socio-political rationales and biophysical intervention framework are just as varied. The project, or intervention within this 'group of actors' (Henocque and Denis 2001) gives rise to dialogue, confrontation, combined construction of the posed problem and its definition, and a description and understanding of processes involved, be they social, economic or environmental. The process may produce new knowledge or technical resources. The pooling of each other's knowledge and the exchanges that take place in the workshops focus on how to access and manage resources, monitor practices, and limit or encourage some of their effects. Here, the expansion of both decentralization and community-based development, the diversity of actors and power asymmetries involved, the quest for legitimacy of the actions of certain stakeholders, the desire to promote the participation of these

R. Mathevet (✉)
CEFE-CNRS, Montpellier, France
e-mail: raphael.mathevet@cefe.cnrs.fr

M. Antona · G. Trébuil
UR GREEN, CIRAD, Montpellier, France
e-mail: martine.antona@cirad.fr

C. Barnaud
UMR DYNAFOR, INRA, Toulouse, France
e-mail: cecile.barnaud@toulouse.inra.fr

G. Trébuil
e-mail: guy.trebuil@cirad.fr

C. Fourage
UMR ESO-CARTA, Université Catholique d' Angers, Angers, France
e-mail: cfourage@uco.fr

S. Aubert
UR GREEN, CIRAD, Antananarivo, Madagascar
e-mail: sigrid.aubert@cirad.fr

M. Étienne (ed.), *Companion Modelling*, DOI: 10.1007/978-94-017-8557-0_5,

actors in the analysis and design of the management systems, and the possibity of - empowering local stakeholders, together raise a critical question: what role should be given to context?

The ComMod approach creates or is involved in an action arena where participants (individuals, organizations and representatives) interact within a specific context, which is affected by exogenous variables at the time it is implemented. These interactions produce outcomes, decisions or actions, which in turn affect the participants and the action situations as well as some or all of the exogenous variables (see Fig. 5.1). The action situation may be defined by a set of variables: the positions of participants, the potential outcomes, the relations between actions and results, the ability to control participants, the type of information generated, and the costs and benefits of the interaction products (Ostrom 2005). The action situation is nothing more than the social environment in which participants interact. The action arena thus defined by the participants and a given action situation is dependent on exogenous factors that may affect its structure. These exogenous variables can be characterized by three dimensions: the institutional dimension (the rules in use), the biophysical dimension (the biophysical attributes involved), and the social dimension (the attributes of the social community in which the action arena takes place). Within this analysis framework, we thus consider that these exogenous variables form the socio-environmental context and the action arena is the intervention context. This gives rise to three sets of questions. When and how should the ComMod approach consider the socio-environmental context? When and how should the ComMod approach consider the intervention context (action arena)? In return, how does the consideration given to these two types of contexts affect the process and its outcomes?

The ComMod approach can be seen as a process of social interactions, which gradually takes into account the socio-environmental context and, through its flexibility, alters or establishes the intervention context in the process (see Box 5.1). However, the question of whether to consider the initial socio-environmental context and the intervention context before starting the process is the subject of much debate within the ComMod group, as opinions and practices vary. Some members of the group consider that it is important to consider both these contexts from the outset during studies prior to implementation. This initial stage is likely to be highly strategic, as the choice of participants and decisions regarding the intervention procedures are dependent on it and give impetus to, or affect, the process outcome. Furthermore, simply focusing attention on the context issue again raises the question regarding the feasibility conditions of the ComMod approach: is it still appropriate or desirable? Others, however, consider that this initial consideration is pointless because of its relative subjectivity and incompleteness and that the approach is a process which, according to the participants involved, is used to set up the socio-environmental context parameters needed to deal with the problem identified and to define, modify and adapt the action arena accordingly.

So consideration will first be given to the social and environmental context and intervention context in the theories relating to the management of socio-ecological systems and in participatory approaches to development and conservation.

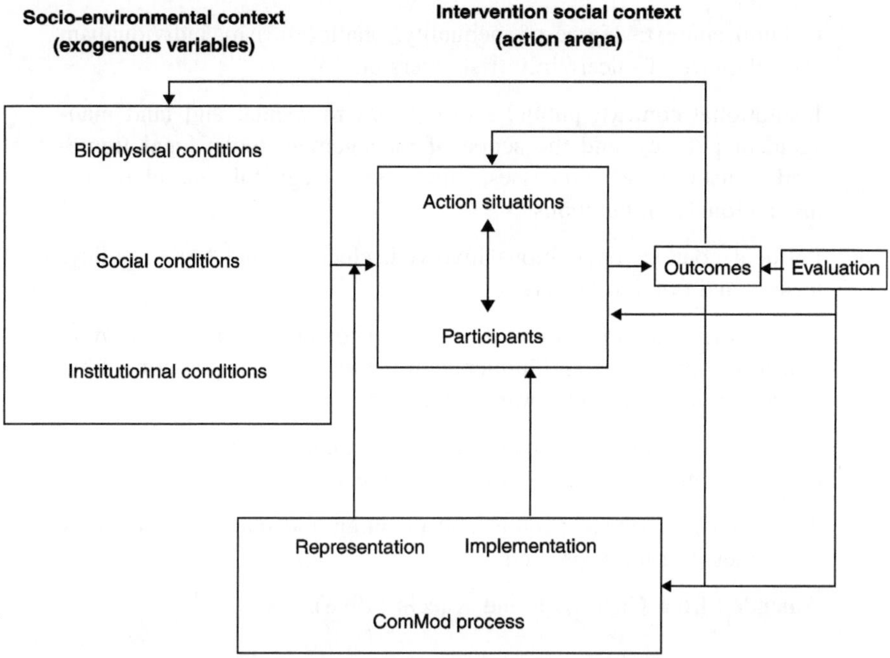

Fig. 5.1 Context analysis framework (adapted from Ostrom 2005)

Then, the tools and methods used to characterize the context in participatory approaches and ComMod approaches will be reviewed. After explaining the nature of the materials analysed and the methodology used, we describe the ComMod approach contexts studied and then discuss: (i) the effects of including or not including the context in the definition of objectives; (ii) the choice of participants; (iii) the influence on project dynamics; (iv) the consequences for decision and action. We finish by discussing the perspectives these results open.

Box 5.1: The Context as Generally Construed

Territorial framework: definition and description of the site scope, its environment, administrative and functional divisions and natural resources.

Historical context: regional history, site history and history of conflicts in the territory concerned.

Socio-economic context: population and land tenure dynamics, dynamics of economic activities and unemployment, and dynamics of the main equipment and infrastructure.

Cultural context: degree of inequality, phallocentrism, individualism and tolerance of uncertainty (risk aversion).

Institutional context: public, sectoral, environmental and land management policies, and the scope of management tools; legal framework, institutional processes, and local, regional, national and international combinations.

Political context: oppositions involved, elections and terms, policy-makers and opinion leaders.

Stakeholders: identification of key stakeholders, interests and relationships, importance and influence of actors and their position within the social network, and power asymmetry.

Social, economic, political, environmental, institutional issues: directly or indirectly linked with the project.

Territorial perspectives: main land planning and management policy guidelines, territorial projects.

Amended from Cicin-Sain and Knecht (1998).

Trends in the Consideration Given to the Socio-Environmental Context and Intervention Social Context

Trends in the Understanding of Interactions Within Socio-Ecological Systems

Several theoretical changes relating to the management of ecosystems and renewable natural resources argue in favour of taking greater account of the socio-environmental context and intervention social context. The shift from the paradigm of stability and equilibrium as the main regulator of ecosystems (Odum 1983) to that of adaptability calls for consideration to be given to the spatial and temporal heterogeneity in the functioning of socio-ecological systems, and the need to include the governance of space and resources. Disruptions are common and widespread: ecosystems are open, non-linear and interconnected in the landscape (Barbault 1997; Kareiva and Wennergren 1995; Holling 2001), and socio-ecological systems considered as natural are in fact strongly influenced by human activities on various scales (Allen and Star 1982; Forman and Godron 1986; Callicott et al. 1999; Folke and Holling 1996). The 'flux of Nature' paradigm stresses that we need to know and understand the past and present effects of interactions between human activities and ecosystems (Leopold 1949; Bertrand 1978; Mathieu and Jollivet 1989; Pickett et al. 1992; Primack 1993). Adaptability in ecosystems is based on maintaining genetic diversity, biological diversity and landscape heterogeneity (Holling 1978). It also requires the existence of institutions and networks that learn and memorize knowledge and experimentation, create flexibility in solving problems and help to balance the power of interest groups (Santos 1997; Berkes and Folke 1998; Berkes et al. 2002). For this purpose, it is important to give consideration to the social context and institutions (Folke and Carpenter 2002; Gumuchian et al. 2003). Ostrom (1990) and Berkes and Folke (1998), in particular, have studied their role in the management of socio-ecological systems.

The analysis of socio-ecological systems systematically faces difficulties in identifying an appropriate, relevant level of analysis to address a given problem (Ostrom 2005). Understanding the sustainability or resilience of an ecological or a social system involves understanding its background and origins in order to be able to incorporate them and transform them by 'unplanning' and 're-functioning' parts of them (Burnouf 2008). This consideration of heritage is facilitated by a combination of tools providing a representation of spatial dynamics on different scales, implementing systemic schemes drawn up with local stakeholders and discussing spatial, social and environmental interdependencies (GIS participatory mapping and MAS). However, the breakdown of decisions and actions on different hierarchical decision-making and spatial levels remains difficult to represent and take into account (Berkes et al. 2002). While this may be due to an unclear definition of the scope of the problem and study, the socio-economic contexts of socio-ecological systems are generally very complex, and their scale, limits and content undefined.

So rather than referring to a vague, unspecified context, the conditions of stakeholder actions should be analysed in terms of their perceptions and beliefs and seen in their social network and interactions with others and institutions (Ostrom 2005; Raynaud 2006; Bailey 1969). Beyond the general biophysical and geographical context, the social context is characterized by (Moulaert and Mehmood 2008; Nguinguiri 2008; Froger 2006) (see Box 5.1):

- the diversity of stakeholders: diversity of actors and stakeholder groups, and diversity of social values, concerns, justifications and interests
- the plurality of norms: official norms (state), local norms (so-called traditional or customary) and international norms (conventions), etc.
- the aggregation of authority and decision-making centres wholly or partly within or outside the system in question
- 'borderline' actors at the interface of representations or stakeholder groups.

In this social context, social change and adaptation to a changing socio-economic and/or ecological environment thus depends on the stakeholders' ability to handle various systems of rules and establish a new basis for institutional innovation without necessarily removing the old ones (Smouts 1998; Ostrom 1990). The understanding of governance thus involves understanding the norms actually implemented (Ostrom 1990; Olivier de Sardan 2001). To identify and determine the role of such norms, analysis of the overall socio-political context also increasingly includes a micro-sociological analysis of local actors and organizations in terms of networks and conflicts through an analysis of practices, strategies and interests (Lavigne-Delville et al. 2000). Similarly, issues involving the modes of collective action, regulation and control implemented within organizations according to available resources are studied (Kalaora 2003; Nguinguiri 2008).

In this approach, the cultural dimension seems unavoidable and is increasingly included in research or action research projects (Long 1990; Jiggins and Röling 2000). As the social learning process takes place in a particular technical and social context allowance should be made for this. Pahl-Wostl et al. (2008) suggested an analytical framework for the cultural context in order to adjust the involvement of local stakeholders. This framework can be divided into four main aspects: perception, rationality, morality and prescription. This analytical framework performs four functions in the project: it identifies a reality that is not spontaneously apparent; gives meaning to this reality; provides value judgements; gives recommendations on how to manage the cultural context of the interactions of local communities with their environments and resources according to the context considered.

The context can thus influence the action context in three ways (Raynaud 2006):

- the context determines (e.g. attributes of actors directly influencing the action situations)
- the context conditions (e.g. previous actions, margins of manoeuvre reduced because of variations in data outside the action situations)
- the context acts as a general reference (e.g. influence unspecified as dispersed, etc.).

Trends in Participatory Approaches

Field projects have produced many methods aimed at understanding the socio–environmental context and the intervention social context. These methods, which were first focused on technical expertise, have evolved towards consultation of local stakeholders and then towards true social engineering.

In development terms, one of the first approaches was a representation based on the technical development of the green revolution of the 1960s. The input of the context established at the time was essentially technical. Experts and researchers identified the technical variables and elements that could increase or improve agricultural productivity (Lavigne-Delville et al. 2000). Limiting factors were then overcome by technical solutions and a transfer of technology with a varying degree of success. On the basis of the many failures observed in this transfer, the vision of taking the context into account through a diagnosis that was too focused on technology was questioned by many authors in the development research field (Lavigne-Delville et al. 2000). This intervention model was replaced in the 1970s and 1980s by a socio-environmental context approach based on a systemic approach to farming system research (Collinson 2000), which identified the social, economic and cultural context of the farmers concerned by development projects. The context study then attempted to understand the functioning of agricultural systems in order to adapt research technologies to local conditions (Collinson 2000). More disciplines were involved and allowed the analysis of socio-economic and agro-ecological dynamics on different scales (e.g. from the plot, to the farm, land and territory). Despite this progress, little consideration was given to the intervention social context. The role of stakeholder strategies remained marginal. Despite the participation of local populations in the change process, choices were still largely dictated by the technological and economic dimensions and were made by researchers, experts and central policy-makers. Finally, since the 1990s, the widespread use of participatory discourse has led to the intervention social context being taken into account more comprehensively (Chambers 1983; Chambers et al. 1989). A gradual reversal of the approach has been seen. After the widespread use of rapid rural analysis methods that allowed the views of local people to be taken into account in order to scale innovation transfer more to the local context, approaches have tended towards participatory rural appraisal (McCracken et al. 1988; Chambers 1994b). Communities are empowered to define their needs so that the diagnosis is the result of a trade-off between stakeholders and experts acting as information providers and/or facilitators of approaches (Olivier de Sardan 1995). Finally, approaches extended participation to project planning and actions with tools such as participatory learning and action, where the learning of both local stakeholders and researchers is emphasized and objectives are defined by the group rather than beforehand (Pretty 1995; Scoones and Thompson 1994), thus contributing to the empowerment of the actors (Pretty 2003; De Koning 1995).

The approaches to biodiversity conservation and management of renewable natural resources[1] showed a similar pattern over the same period (Pimbert and Pretty 1997). Having promoted the exclusion of local populations from areas to create protected areas (Rodary et al. 2003), these approaches considered the conservation area as the ‘society area’ (Mathevet and Poulin 2006), and this called for the social context to be taken into account (Olivier de Sardan 1995). The traditional institutions were reconsidered and the local communities were seen as the most likely actors to respect the resources because of their dependency and proximity (Berkes et al. 1991). Community approaches developed, particularly in the management of forest resources (Ostrom 1990) and participation in the implementation of decentralization policies for the management of natural resources with highly heterogeneous results.

Since the 1980s and now under the order of international donors, participatory approaches have thus become widespread (Pimbert and Pretty 1997). These

[1] Guided by the visions and techniques developed through research and development, they are based on the same observation of failures in the nature conservation measures taken and the centralized management of natural resources by governments.

approaches are based on several assumptions. From the normative standpoint, encouraging individual and social learning is considered to be good for society and citizens in general (Ludwig 2001; Latour 1999). The second substantive assumption considers that encouraging the mainstreaming of multiple viewpoints provides a better understanding of issues and improves the identification and selection of suitable solutions (Van den Hove 2001). Finally, from the instrumental standpoint, encouraging cooperation is considered to facilitate greatly the implementation of solutions and reduce conflicts (Brandon and Wells 1992). Participatory approaches are thus likely to overcome the opposition between general and specific interests and recognize the role of the procedural negotiated dimension of general interest (Trom 1999). The participation of local populations thus brings 'experts' and so-called 'lay' visions closer together (Callon et al. 2001). This brings about a change in the very conception of local actors from that of a uniform, indiscriminate and sometimes ignorant and irrational public (the 'villager' or the 'community') to that of a system of composite, plural actors with know-how, knowledge and skills related to their experience and capabilities, and who have the ability to act as 'lay experts' (Boy 2003; Lascoumes 2005). These changes are profound because on the one hand, they call into question conventional top-down approaches in the formulation and implementation of projects and, on the other hand, they recognize the ability of local actors to manage their own resources and determine their own development.

The Identified Limits

Many criticisms have been levelled at participatory approaches, with regard to the inadequate consideration they give to the social contexts of approaches (Reed 2008; Faysse 2006; Salafsky et al. 2002; Margolouis and Salafsky 1998; Brandon and Wells 1992). The criticisms taken into consideration with regard to the issue of including the intervention social context in ComMod approaches are the choice of participants and the results of these approaches.

The Choice of Participants and How They Express Themselves

If issues and mechanisms of social differentiation are not analysed, social inequalities are accentuated and there is a risk of one stakeholder manipulating the others (see Chap. 6). Such approaches can thus accentuate the positions of strong players in the socio-ecological system, giving significant weight to dominant players at the expense of weak or absent stakeholders. Such power asymmetry accounts for project benefits being hoarded by some local elites. The representativeness of participants is particularly questionable in situations of conflict and gives rise to confusion in the benefit of the approach in terms of the empowering of local actors. Finally, problems concerning the validity of the information gathered undermine the relevance of the solutions identified with these participants.

The Limits of the Results of these Approaches

Any insufficient underpinning of these approaches in an institutional context has three consequences: (i) these approaches do not go much beyond the scale of community intervention, and their effects dwindle once the intervention is completed; (ii) the frustration of stakeholders who, having identified either the problems or the solutions to these problems, cannot solve them because of interdependencies with other regional or decision-making levels; (iii) there is a tendency not to address the actual causes of the situation but only its effects, without questioning the foundations forming the basis of inequalities and social relationships. The lack of a detailed inventory thus often leads to a risk of being unable to ensure either the inclusion of decision levels or fairer processes (Holmes and Scoones 2000).

The issue in defining the social context is thus to provide a better definition of the implementation of the social intervention process or give consideration to the cause and effect relationships within the socio-ecological system examined. However, analysing the context involves more than just describing the territory and the stakeholders, as indicated in box 5.1 of Fig. 5.1. It also makes it possible to define the action situation at the beginning of the project (box 5.2 in Fig. 5.1). Did the ComMod approach arise out of a conflict, an identified problem, a territory project or a local territorial dialogue initiative? If it arose out of a conflict or a problem, who identified it and how was it qualified? In all cases, the attention given to the initiators and to the start of the process helps to provide an explicit definition of the process issues and objectives within the action situation.

The ComMod Context Process

The previous sections showed how conceptions of context have changed and, in particular, how the social context is now seen as heterogeneous, hierarchical and non-uniform, and how networks of interacting individuals, institutions and power asymmetry are highlighted as explanatory factors. This change helps to clarify causal relationships. Having outlined the theoretical and empirical evidence in favour of including the context in a participatory approach, the role of the socio-environmental context and intervention social context in ComMod approaches should now be considered.

The Tools and Methods used for Mainstreaming the Social Context in ComMod Approaches

In addition to reviewing the literature on the problem identified by the representative, for example, and the situation of the project on the ground, a few techniques and methods are commonly used in ComMod approaches to define the socio-environmental context and intervention social context (Box 5.2).

Box 5.2: The Diversity of Methods Used in ComMod Approaches to Characterize the Socio-Environmental Context and Initial Intervention Context

Farming systems diagnosis with a production system entry (e.g. Mae Salaep)

Pro-active-reconciliation tool (PACT) method (e.g. Nan)

Analysis of actors (e.g. AguAloca)

Analysis of rules (land tenure, right of use and institutional analysis development (IAD) and 4R (rights, responsibilities, returns, relationships) framework, e.g. Domino Réunion)

This first involves meeting with local actors, who are initially the easiest to meet in meetings with local leaders, officials and other stakeholders, to reduce or remove any suspicion by explaining the reasons for the project (Mikkelsen 1995). Building trust and good relationships calls for a careful choice of locations and times of such meetings so that they are as convenient as possible for stakeholders (Jackson and Ingles 1998). Thereafter, semi-directive exploratory interviews make it possible to determine the situation, major issues and main stakeholders (Margolouis and Salafsky 1998). Interviews with resources people who have particular knowledge of the subject and intervention field are often very enlightening even though their views may not necessarily be representative (Jackson and Ingles 1998). Semi-directive interviews can be usefully supplemented by participant observation, field visits with resources people to determine local culture and customs and discuss the actors involved and any major changes observed (Guijt 1999). Finally, participatory mapping with stakeholders helps to clarify the issues and interests, in particular those related to land, customs or environmental problems. The many techniques (Mikkelsen 1995) are also involved in the analysis of actors (Grimble and Wellard 1997; MacArthur 1997) aimed at, on the one hand, identifying the key players in relation to a given problem, situation or particular project and, on the other, clarifying the respective interests of these actors and the nature of their interactions (Allen et al. 2002). This initial analysis thus identifies how the interests of actors are likely to affect the process outcome, either by contributing to its success or bringing it to a rapid halt (Margolouis and Salafsky 1998). This analysis is thus part of a survey into the feasibility of implementing the approach.

An analysis of actors usually identifies the individuals, social groups and organizations whose practices or lifestyles are likely to be affected by the project. It shows who has influence or who is able to affect the project and its dynamics. This obviously includes those who are likely to go along easily with the process as well as those who are likely to oppose it. It is sometimes useful to make a distinction between the stakeholders who are likely to benefit directly from the process, those who will be involved in the process but will not necessarily benefit from it, and those who may have an interest in the outcomes of the process but who are not directly involved or concerned (Margolouis and Salafsky 1998). What are the expectations? Why participate in the process? What are the benefits they could gain from it? What are the possible implications? What are their abilities to participate in it? What are the conflicts of interest? What are the attitudes towards other stakeholders? To these questions can be added, on the one hand, an assessment of the positive effects of the process, negative effects, neutral effects or uncertain effects on each of the identified interests and, on the other hand, a definition of the involvement priorities of actors according to their level of influence and their power, in particular as part of a strategic approach to participation.

Other analyses focus more on analysing the institutions (Ostrom et al. 1994). This type of analysis involves looking at the action situations and actors focusing on the rules in use and how they are conceived and implemented (Babin et al. 1999). Together with the institutional analysis development (IAD) developed by Ostrom et al. (1994), the 4Rs (rights, responsibilities, returns, relationships) method (Vira et al. 1998 in Barnaud 2008) augments the analysis of the usual actors by providing information on the rights of access and use, the types and levels of responsibility in the management of resources, the returns likely to be generated by the resources, and finally the relationships with other stakeholders. When the project objective or problem has not been identified by the approach representative or in earlier work, it may be best, as in the approach adopted by Barnaud (2008), to make use of the pro-active-reconciliation tool (PACT) method developed by Jésus (2001). The advantage of this method is that its aim is to define precisely the problem stakeholders want to see addressed. Another advantage of this method is that it allows actors to be analysed according to their own perceptions and interactions. This may make it easier later on to analyse the effects of participatory approaches in terms of learning and negotiation on the basis of observed changes from the standpoint of the perceptions of issues and other actors and interactions, as well as how actors see the future (Jésus 2001).

This actor analysis greatly facilitates the analysis of the initial situation and facilitates the incorporation of power games. It thus makes it possible to evaluate the feasibility of the approach and also reduces the risk of unintended effects in terms of social interaction and balance of power.

Materials and Methods

In the ADD-ComMod project, the characterization of the socio-environmental context of the experiment analysed is based on the outline of each of the 27 case studies. However, the diversity of this material remains vast. Some authors included everything that was known in the field concerned so that anyone reading the outline would know what this field involved. However, most focused on what was needed to understand the implementation of the companion modelling process. This greatly limits the harmonization of data for a detailed analysis of contexts. These outlines will thus only be briefly characterized on the basis of the compilation work carried out by Étienne (2008).

Regarding the effects of context on the approach and results, 18 ComMod experiment assessment reports were available covering a wide range of materials. These outside assessments often involved a reconstruction of the process on the basis of semi-directive interviews. We conducted a survey among the designers or assessors of 13 case studies to collect more detailed information regarding the initial context and its influence in the process.

At a risk of underestimating certain aspects and overestimating the importance of others, it is stressed that, despite the lack of sufficiently accurate data gathered on the context subject, the aim was to select relevant material and not to make a value judgement on the experiment considered.

A Diversity of Socio-Environmental Contexts

The diversity of social, economic, environmental situations in which these case studies took place was highlighted. These case studies present a wide range of socio-environmental contexts. Nearly half of them were implemented in advanced or developed countries and were a continuation of earlier research or development projects. Half of the case studies focused on natural heritage conservation issues (e.g. habitats, emblematic species and genetic diversity) or the control of natural hazards (e.g. erosion or forest fires). Eight studies focused on water management and four on problems of agricultural productivity or hunting. Three made no mention of any relationship with environmental problems. Four case studies were a direct result of an old research programme firmly located in a given area, seven were related to development projects and twelve were associated with the implementation of public policies. For the latter, two-thirds involved decisions taken at national level (e.g. decentralization, creation of national parks or distribution of cultivars) and one-third at local level (e.g. water management scheme or soil protection). In the last seven case studies, reference was essentially made to changes in use (e.g. agricultural abandonment or land privatization) that have led to the future of the territories concerned being put into question. Almost half the case studies concerned conflicts over access to resources or sharing a territory between economic activities, recreation and nature conservation. The other half mentioned the importance of the involvement of social groups or the method used to organize local societies in the issue concerned. Finally, five case studies focused primarily on 'traditional' practices of local actors (farming or use of water).

With regard to the scales and levels of resolutions used, half the case studies related to clearly defined territories and half to areas with undefined boundaries. According to the question raised, the organizational level adopted ranged from the most meaningful biophysical entity in the system functionality (e.g. agglomeration, river delta, catchment area or geoterritory) to administrative entities within which public policies apply or for which statistical data were available. This level can be reduced to a municipality or cover a whole department. The resolution level chosen to address the issue was highly variable and different levels were often used in a given case study, with the operating unit often used.

Mainstreaming of the Context Effects in ComMod Experiments

This section aims to clarify the relationship between knowledge of the socio–environmental context and intervention context and the implementation of the ComMod experiment.

Diversity in the Formulation of Social Demand in the Initialization of the Process

The analysis of the 27 case studies highlighted the great diversity in the initialization of companion modelling. Of these case studies, 16 involved almost exclusively researchers, even if the work of the latter was only possible if they found allies and local go-between actors beyond academic circles. In 11 cases, institutional stakeholders were directly involved, and NGOs and associations played an active role in three studies.

Two scenarios were thus observed:

- a relatively precise request is made by a representative, often a community or a local or central authority; this representative usually has legal legitimacy to address the problem
- a rather vague social request is made, without a group initially identified to deal with the issue, generally defined by a team of researchers, either within a strictly research context or associated with the implementation of public policy.

In the first case, local actors can directly ask a researcher to carry out work in connection with an issue concerning them. This was the case, for example, of the Cévennes National Park in 2000. Concerned by land abandonment and the threat this posed to most of the natural heritage issues identified in the Causse Méjan area, representatives of the scientific department asked a researcher to carry out a prospective analysis. This researcher then suggested the setting up of a companion modelling approach with all the local stakeholders in order to predict how this area would evolve according to various management strategies, while simultaneously including natural heritage and productivity issues and facilitating dialogue between them (Étienne et al. 2003). If the companion modelling approach is carried out successfully in response to an issue in a specific location, in certain situations there may be a spin-off through the emergence of a request for the approach to be implemented by local stakeholders. This is particularly well illustrated by the civil society experiment with land in the Larzac region following on from the success of the Causse Méjan study (see the Larzac case study). In other situations, the outcome of a case study may give rise to several organizations seeking this type of approach to be tailored to similar concerns. As a result of the SylvoPast project (Étienne 2003), proposed in 2000 in response to a request from the Ministry of Agriculture based on forestry development and the prevention of forest fires, the agriculture and forest service of the Gard department thus asked the researcher who carried out this work to tailor this approach to the problems of forest-fire prevention at the interface between urban and natural areas (see the Nîme-Métropole case study).

In the second scenario, commodian researchers alone or in association with other researchers suggested the idea of setting up this type of approach to local stakeholders in response to a local issue identified by the researchers but also likely to mobilize local stakeholders. Researchers most often focus on a specific case connected with their research subjects. They may also be interested in implementing the approach in a country and/or in a situation that they had not encountered so far. In some studies, these two types of interest can be combined. The Mae Salaep case study, which at the beginning addressed the analysis of interactions between the risk of land degradation and agricultural diversification with the aim of reducing this risk in the future, thus enabled the researchers involved, not only to capitalize on the data from several years of fieldwork on the agrarian situation and the risk of soil erosion, but also gave them the opportunity of testing the ComMod approach in Thailand (Barnaud et al. 2006a; Trébuil et al. 2005).

Whatever the scenario described above, the respective influences of the commodian and other actors in translating a possible request into issues that can be addressed by companion modelling vary widely, particularly as the request is not always clearly expressed. In many situations, the researcher must make an effort to translate and break down the expectations of local stakeholders, and then reformulate this request jointly with stakeholders.

Diversity in Mainstreaming the Initial Context

For three-quarters of the case studies, the initial context was mainstreamed within the approach framework (i.e. in setting objectives and selecting participants and the participatory scheme). It appears that in situations of interculturalism in poor, developing and emerging countries, the ComMod intervention was generally preceded by an initial analysis of the intervention social context often based on previous research in the same field. It frequently appeared to be detailed on biophysical functioning and local stakeholders, but seemed less detailed with respect to institutions and decision-making processes beyond the local level. For the remaining quarter of the case studies, no explicit consideration to the initial context was given in implementing the process. The context aspects were essentially identified during the process and gradually taken into account in activities according to the needs emerging during the various stages.

Out of the 13 case studies for which more context mainstreaming aspects were available, only two experiments did not follow on from past research or development work. One experiment resulted in a strict framework approach (a very detailed initial context study provided the main guidelines for the process), and the other a gradual approach that involved a local authority mandate.

Out of four case studies involving a community or local authority mandate, only one gradually took context into consideration, while in the other two case studies the context was considered as a guideline and in the last case study context was considered as a strict framework. When case studies were initiated by researchers, the proportion of the study in which the context was considered as a guideline changed little (55 %), but contexts providing a strict framework were higher (33 %) than for gradual cases (11 %).

When the commodian does not know the local situation or is not familiar with the issues before carrying out an initial analysis, the context provides a guideline or a strict framework. It should be noted that in three-quarters of the case studies the local situation or issues were fairly well known to the ComMod designers and coordinators and when this was not the case, a more detailed context analysis was carried out.

Thus, when the context was considered by designers or assessors to provide a strict framework in the process, a thorough initial analysis of the socio-economic context was carried out, and the analysis of stakeholders and biophysical context was well known but less detailed than the other dimensions. In processes where the context provided a guideline, the socio-environmental context was detailed but the analysis of the intervention context was very general. Finally, when the context was gradually taken into account, the process could begin without any prior knowledge of the biophysical context but in both cases there was a brief analysis of the socio-economic context. Rough monitoring/assessment of power relationships was generally implemented during the process in studies where the context provided a guideline or a strict framework.

In short, the context was mainstreamed and provided a framework for the approach in most case studies whether covered by a mandate or not. Moreover, in view of the wide range of context mainstream possibilities, the decision to define the initial context in detail (and use this knowledge to provide a framework for the approach) depends more, it would seem, on the researcher's stance (see Chap. 6) and the existence of previous studies than the researcher's familiarity with the issue or field of intervention. Note that it seems such familiarity can lead to implicit recognition of power games and power asymmetry (see Chap. 6).

Let us now try to analyse the consequences of context mainstreaming on initializing the ComMod process. However, for reasons of heterogeneity of both the initial studies and assessment reports, it was not always possible to distinguish the effects of taking the context into account on the results and approach. We thus propose to adopt the main consequences mentioned in the assessment reports where an explanatory link can be established with the fact of the context being taken into account or not.

Consequences on the Definition of Project Objectives

The analysis of the initial context had a significant influence on the definition of the ComMod approaches. For example, for the designers of the Domino Réunion experiment, the analysis of the initial context made it possible to carry out a true feasibility study for implementing the process. For the other cases, this type of analysis made it possible, in projects covered by a community mandate as well those instigated at the initiative of the researcher alone, to refine and reformulate project objectives, and clarify the nature of any open or latent conflicts. The Nîmes metropolitan area experiment was the only one not subject to a preliminary analysis before specifying the approach objective, although previous work carried out by a member of the project administration did in fact contribute greatly to the setting out of the issues and characterizing the main stakeholders concerned.

Despite this general observation, it appeared in several assessments that project objectives were not always clearly understood by those involved. Research projects not followed up by concrete action through a development project were often the cause of this confusion. The research and development issue is often identified upstream of a ComMod approach by or with an institutional representative or following on from a previous study. In this case, the institutional problem seemed to dominate all other issues (e.g. relations with territory, use, etc.). The approach may not always allow this initial objective to be challenged or make stakeholders true players able to define their relevant issues, and knowledge to be mobilized. In the case of the AguAloca project, for example, while neighbourhood associations were not involved in identifying the problem, they nevertheless identified new issues relating to waste management, land speculation or the financial difficulties they encountered to improve the situation.

With hindsight, it is clear from ComMod experiments that the analysis of the initial context makes it possible to identify the range of issues and problems, and realistic goals can be set for the project, thereby ensuring that it is not too ambitious, thus limiting the risk of the process failing.

Impact on the Choice of Participants

The identification and selection of participants is important in terms of the approach's transparency, representativeness and legitimacy. There are typically four types of participant selection (Steyaert and Lisoir 2005):

- open on a voluntary basis: anyone who wants to participate to ensure the broadest possible involvement
- representation: selection of participants representing a standpoint, system of values or knowledge of the various stakeholder groups
- demographic: sampling to provide a representative cross-section of the population
- limitation: the number of acceptable participants given the methods and tools used.

In the outline and assessment reports, experiments were often quite vague about the criteria for selecting participants at the various stages of the approach, even when the initial context had been analysed in detail and taken into account. The participants were selected based on their knowledge of the system or for their influence on it by the researchers and institutional players running the experiment. Participants were generally invited as representatives of other actors with common interests with regard to the problem. These representatives, however, were often not recognized representatives designated by the actors in the same socio-professional category, for example. If the process is part of, or initiates, a group decision-making process, discussions thus need to be broadened to include all the stakeholders concerned (Barnaud 2008).

In the case of experiments where the context provided a strict framework, as in Thailand, in particular, participants were invited on the basis of a nominative selection after a survey. Such a selection seeked to strike a balance between socio-economic categories and issue categories, but also depended on the self-expression ability of stakeholders and marginal players, in view of the local balance of power analysed in advance. The final choice of participants was then discussed between researchers and authorities. In other situations, the key players involved were mobilized through the knowledge of the representative or resources people. The actors could thus be mobilized in stages by researchers and the institutional network involved in the project (e.g. Camargue, Pays de Caux or Nîmes-Métropole case studies). When participants are selected by the researchers alone or with institutional stakeholders, the question of their representativeness needs to be raised. While the working circle can be broadened at the proposal of the experts or stakeholders invited to attend a first meeting, some choices seem questionable and have been challenged retrospectively by a few stakeholders in several projects. In the case of the Nîmes-Métropole experiment, for example, several stakeholders questioned the absence of players such as the National Forestry Commission because the socio-ecological system considered was of primary concern to forest management. The expertise of this institution was called for by some stakeholders but for reasons of administrative competency fields, they were not, in the end, asked to take part in the experiment. Others wondered about the overrepresentation of hunters and the fact that land planners, residents and associations did not participate in the joint conception phase.

In Brazil, as in other countries where there is a huge divide between social groups, and where wealthy individuals live alongside people living in conditions of extreme precariousness, relations between state services and populations are tainted with paternalism on the one hand and vote-catching on the other. Excluded and disaffected populations are obviously not affected. The representativeness of neighbourhood associations may also be questioned as they inherit groups set up under dictatorships for the distribution of food baskets and exert some social control. For AguAloca, the actors involved were thus primarily the institutional actors of water management structures and an environmental NGO; at Ter'aguas, representatives of municipal services, professional public health associations, NGOs and neighbourhood associations were involved. In addition, many actors adopted militant stances and were engaged in education and supported actions in favour of the disadvantaged populations living in the catchment areas. Only the most motivated and already mobilized individuals were thus involved in the project. The socially excluded were consequently not involved despite their being the primary actors concerned.

While the context analysis for the project in Tarawa focused on local stakeholders, the institutional context upstream was apparently not really considered. This was reflected in the lack of involvement of key players: government representatives, members of the Smart and Aware Pervasive Healthcare Environments (SAPHE) project steering committee and donors, who did not consider work products willing to move the project forward according to the initial timetable because of a lack of confidence of these -meta-players in the approach. The approach designers and the assessor considered retrospectively that it would have been better to involve them to make sure they agreed with the approach and its results. Unlike this difficult situation, the Domino Réunion project identified the stakeholders on the basis of a large-scale institutional analysis. However, the team encountered difficulties in mobilizing key players for the main issues identified. On the other hand, in the case of Domino Senegal, the mobilization of key players was effective but old conflicts re-emerged in favour of debates on the legitimacy of the stakeholder representatives with discussions on the ethnic-economic dimensions of land issues.

These examples show that the question concerning the choice of participants is difficult to clarify retrospectively and remains a difficult stage in the process. Initial analysis makes it possible to identify key players, interests and potential conflicts. Although it contributes to defining the participatory scheme, it is still sometimes difficult to show a causal relationship retrospectively between the choice of participants and the analysis. The choice of participants is a trade-off between representativeness and the availability of actors. In this context, co-option is relatively frequent and occurs in networks of process researchers or representatives. In other cases, the choice to mobilize representatives of associations instead of inhabitants is often the result of a trade-off between representativeness, representation and interface with direct actors. The mainstreaming of the initial social context is important in clarifying the choices of participants but does not guarantee that such choices are the most relevant retrospectively and does not guarantee their mobilization. Despite this, it does not appear, according to the available assessments, that the ComMod processes analysed are challenged by participants. While instrumentalizing risks are mentioned, they are not, on the basis of these case studies, likely to be very tangible.

Consequences on Project Dynamics

The initial identification of conflicts helped to adapt the approach workflow by gradually and not necessarily simultaneously selecting participants during role-playing sessions as in the case of the work carried out in the Ter'aguas project in Brazil. The same applied in the case of Domino Réunion. While the initial study of the context had identified the players that it would be useful to involve in the model design stages, it was not possible for technical reasons to include all of them at the same time. The process thus gradually included stakeholders in various problem definition, representation and discussion workshops. This caused frustration among some participants. Moreover, the existence of divisions between some personalities representing divergent interests that were not clearly identified initially, led to an imbalance in dealing with issues in favour of the best-organized stakeholders. Similarly, but dealing with issues of water quality and water infrastructure, the interests of domestic users and small farmers were not addressed in the case of the ComMod process in South Africa (Kat Aware). The social and education level greatly influenced the involvement of local actors in this experiment. Large farmers were more familiar with the modelling and better able to discuss it. The study of the initial context had not considered these dimensions, which involve power and knowledge asymmetry. The team thus concluded that the model and role-playing were ultimately more useful in discussing the needs of large farmers rather than dealing with all the issues around the distribution of water resources. In the case of Domino Senegal, the initial analysis identified a committee of users on the basis of both representativeness and availability criteria. However, the players who made themselves available were not necessarily the most representative. Nevertheless, the approach made it possible to rebalance the group and include players who were initially marginalized.

In the Asian case studies, the mainstreaming of the initial context also allowed the approach to be tailored. In the case of Lingmuteychu, the initial analysis selected two villages, the conflict between which was representative of issues throughout the catchment area. The exemplarity of the first phase in the process, regarded locally as a success with these first two villages, made it possible, during a second phase, to widen the project to all the other villages in the catchment area. In the case of Nan, the initial analysis of the context provided a great deal of knowledge about the people and allowed participants to be selected with the village chief. It also established a working method to mainstream the balance of power between the national park and villagers, in other words it supported the stakeholders in a weak position compared with very influential actors and thus gradually allowed them to participate. The process was thus initiated by meetings within the village without the presence of the representatives of the national park so that the villagers could share their views on the park issue and define a strategy. Another meeting was attended by park officials, without the villagers, so that they could discuss the issue between themselves. Finally, a workshop brought together all the protagonists. Since everyone had been able to think ahead about the issue, nobody could withdraw from discussions on possible agreements on the pretext of having to discuss matters with members of their community or institution. In the case of Mae Salaep, the intervention social context was reassessed at the end of each stage and, in the same way as in Nan, made it possible to tailor the method and choice of participants in various workshops. So during the third ComMod cycle, the stakeholders representing institutions with an organization level higher than that of the village were able to be included in the process. This was done at the request of villagers who needed to be reassured (gradually during the first two cycles) prior to attending such a meeting (Barnaud 2008). In this case study, the initial analysis also facilitated the identification of a Christian leader whose charisma and the respect the community held for him made it possible to bring together the villagers around a common project. The designers and assessors of this experiment considered that the identification and inclusion of this actor in the process, whose status and social network were such as to make him a catalyst in the change process, were instrumental in the participation of certain social groups and its success.

Impact on the Decision and Action

Most ComMod projects are led by researchers and are not intended to make the changes, but rather to improve the quality of the process leading to the decision to make changes in accordance with the principles of post-normal science discussed in the introduction to this book. Let us now turn to the context mainstreaming process on initializing experiments and to what is likely to facilitate the move forward to taking decisions or action.

First, it is clear from the assessments that the action arena invested or set up within the ComMod process is generally not the right forum for discussion aiming to move towards decision-taking. It is located alongside this forum, but a stakeholder learning or empowerment process is systematically required to define a new action context to include or create a decision forum. When processes are included in the right forum for decision-taking, it sometimes helps to change the socio-environmental context and rules, etc.

Thus, the intervention field of some projects has been widened to other spatial perimeters or administrative levels. The Nîmes-Métropole experiment has been extended throughout the Gard department. In addition, several mayors drew up a risk prevention plan. They had already begun to think about this but the ComMod approach gave them better arguments with which to discuss the implementation of the plan. In the process at Lingmuteychu, the creation of the catchment area management institution arose from the mainstreaming of interests of the stakeholders and institutions involved. The agreement reached with the villages in conflict was considered to be exemplary and helped to extend it to other villages. In other situations, the experiment was halted because of the unwillingness of some influential players in the social system to change their practices or enter into further conflict (as in the case of Radi or Nan). In the case of Larq'asninchej, the change in the social organization level failed because the integration of new players called for new representations to be included where previously there had been local handling of the problem.

If there is no initial analysis of the context or the analysis is too sketchy, this may result in a lack of mobilization of key players or representatives of institutions whose support is needed to promote social change or the adoption of new group rules or techniques. At the same time, the identification of these key people cannot guarantee their involvement for various reasons as has been shown in different studies (e.g. Mae Salaep, Nan or Domino Réunion). In these studies, several processes showed that to change or innovate, the change stakeholder (i.e. the representative of the approach/the participants in a dialogue) needs a project that will provide a clear breakdown beyond a participatory diagnosis and possible change scenarios to clarify the action context. What action? What action status (legal, etc.)? For who is it intended? In what relational context? To affect whom? The lack of any final outcome and development projects following on from the companion project may frustrate and demobilize people by giving the impression of an unfinished process. In several experiments, it would have been appropriate to anticipate on the basis of initial analysis of context and process monitoring to keep participants mobilized as they often expressed their readiness to engage in new similar experiments. An exit strategy could have been clarified from the initialization with representatives and/or institutions.

Overall, projects still suffer from insufficient analysis of governance structures and insufficient assessment of the abilities of decentralized, local and national institutions to come to an arrangement. There are various consequences:

- insufficient account taken of social organization levels in achieving concrete actions
- insufficient identification and/or involvement of go-between stakeholders between groups and levels of organization to mobilize actors
- change that participants identified retrospectively or intentionally or unintentionally not included in the approach as not involved at the outset.

In view of the information available to us, it does not seem possible to link in further detail an action or change to the mainstreaming of context in the process of a particular experiment. It can, however, be said that the context determines whether or not these actions (as in the Méjan example where the role-playing sessions were then translated into the implementation of a concerted local development plan that was in the pipeline).

Conclusion

The analysis of the initial context makes it possible to define the intervention context and clarify the conditions and procedures for implementing the ComMod approach, during which a group diagnosis takes place on the problem identified by the representative and/or local stakeholders. As we have seen, the context analysis raises many issues that are not specific to the ComMod approach and which have not been fully explored in this chapter. How can an analysis of the initial context identify the factors relevant to the process? How far should the analysis of the intervention social context go? However detailed the initial analysis may be, social networks and the positions of actors will change, and the real issues are only likely to emerge during the stakeholder interaction process itself. A rough initial analysis that provides the approach coordinator with an initial baseline may seem sufficient. However, with regard to the subjects addressed in ComMod experiments, it seems essential that the researcher identifies the social dynamics taking place if he wants to understand the context of his action and its inherent issues. The ComMod approach, while asserting a general aim of creating a mechanism for group thinking around a given problem and for learning to empower stakeholders to participate in group management of resources and their territory, cannot neglect such an identification issue. The desire to contribute to social change cannot result in dominating forces and power games not being taken into account. Neglecting or giving insufficient consideration to social differentiations and local power issues creates a situation for the process to be manipulated by, or turned towards, the interests of stakeholders in social and political competition (see Chap. 6). Group decision processes are generally very context dependent and contingent on the time, people and resources available. Any action or research must thus take this account. It should not be forgotten that the issue involved is the quality of the process, in other words, an approach attentive to the relevance of its mechanism and its results in terms of local issues; the involvement of key actors in the decision-making process, appropriate dialogue forums, a concrete opportunity for each participant to influence the course of the project outcome, no significant divergence between what the project offers (in the case of an institutional sponsor outside the local community), and the aspirations of the community.

We have shown that, despite the difficulty of the exercise, mainstreaming stakeholder power games arising out of the issue considered from the start of the analysis of initial context can be useful in providing the coordinator with a better understanding of the social context and its likely effects in order to scale the process approach by providing the best possible identification of participants in the various workshops. Although the context initial analysis does not guarantee success, it does identify the political and institutional mechanisms that over and above empowering actors to influence decisions, will make it possible to change decision rules and processes and mobilize the funding needed to implement the solutions identified during the experiment or subsequent approaches.

While it is considered that legitimacy of the intervention can be established over time during the process, initial legitimacy may nevertheless be crucial to the project outcome, and raises the crucial question of the intervention social context: who is leading the project and who made the request and raised the question? Who asked stakeholders to participate, and which stakeholders? Are they representative of a particular social group? How is the researcher perceived and legitimized by the stakeholders? These issues can only be put into perspective if the intervention social context is sufficiently well defined to establish the explanatory causal relationships; otherwise they are pointless. Likewise, the representativeness of participants in relation to the issue raised has yet to be considered.

If the decision-making processes in a complex situation are considered to be characterized by great uncertainty regarding the facts and conflicts in the very definition of problems and social issues, how is it possible to improve decision-making with a view to improving the quality of the decision without taking the action context into account? To improve the quality of the ComMod process, the initial analysis of the intervention social context should explicitly take into account the action context involved to:

- allow the commodian team or researcher to introduce themselves, explain the project and raise the awareness of, and reassure, the stakeholders
- clarify and validate the approach's initial objectives
- draw up an inventory, in particular, to assess the process by comparing before/during/after situations
- assess the project's feasibility in terms of risks of the actors misappropriating the approach
- ensure that the selection of participants is not disconnected from the action situation involved
- validate the relevance of the approach in decision-making (forum for interactions and results).

In questioning social relationships, the ComMod approach promotes awareness and the organization of responsibilities. It facilitates the negotiation of resource management principles by gathering and sharing information and discussing possibilities. Even though the portion of the approach and context is difficult to assess in the results and actions taken, the ComMod experiments analysed contribute in various degrees to the pluralistic management of resources and territories.

Chapter 6
Power Asymmetries in Companion Modelling Processes

Cécile Barnaud, Patrick d'Aquino, William's Daré, Christine Fourage, Raphaël Mathevet and Guy Trébuil

As discussed in the previous chapter, ComMod processes, like most participatory processes, are implemented in social contexts characterized by power asymmetries and conflicts of interest between stakeholders involved at different organizational levels. These asymmetries result from complex and context-specific power games, which are often difficult to identify and are dependent on a combination of social, political, economic and cultural factors. One of the most common criticisms levelled at participatory approaches is the lack of recognition of this complexity of the socio-political context in which they are implemented (Lavigne-Delville et al. 2000; Cooke and Kothari 2001; Eversol 2003; d'Aquino 2007) (see Chap. 5). There is now an urgent need for researchers working on participation, including the ComMod group, to formalize their position in a better way in the face of this criticism. However, such an undertaking is difficult because behind the need to take better account of the socio-political context, there is not only the necessity to

C. Barnaud (✉)
UMR DYNAFOR, INRA, Toulouse, France
e-mail: cecile.barnaud@toulouse.inra.fr

P. d'Aquino
UR GREEN, CIRAD, Dakar, Senegal
e-mail: patrick.daquino@cirad.fr

W. Daré · G. Trébuil
UR GREEN, CIRAD, Montpellier, France
e-mail: william's.dare@cirad.fr

G. Trébuil
e-mail: guy.trebuil@cirad.fr

C. Fourage
UMR ESO-CARTA, Université Catholique d' Angers, Angers, France
e-mail: cfourage@uco.fr

R. Mathevet
CEFE-CNRS, Montpellier, France
e-mail: raphael.mathevet@cefe.cnrs.fr

M. Étienne (ed.), *Companion Modelling*, DOI: 10.1007/978-94-017-8557-0_6,

understand it better, but also to think about how to deal with it. This involves taking up the issue of the stance adopted by the designers of participatory processes with regard to the power asymmetries inherent in the socio-political contexts in which the processes are implemented. This issue poses a dilemma for these designers. If they attempt to assert a neutral position with regard to power asymmetries, they will be criticized for being naïvely manipulated by the most influential stakeholders, and de facto participate in merely reproducing or even reinforcing the initial power asymmetries. However, if, to mitigate this risk, they attempt instead to assert a non-neutral position, by taking the initiative to strengthen the voice of the least influential stakeholders, questions will then be raised as to their legitimacy to do so. ComMod researchers are no exception to this dilemma.

The ComMod Charter defines an initial stance with regard to the heterogeneity of the intervention context, namely the need to acknowledge the diversity of existing standpoints with a view to establishing a shared understanding of the situation. This mitigates to some extent the risk of the views of the least influential stakeholders being unheard (Collectif ComMod 2005). Over and above this common general stance, it now appears useful to the group to clarify the diversity of possible positions with regard to taking power asymmetries into account, both in terms of methodological choices and ethical positions.

The aims of this chapter are to provide a method for eliciting the positions adopted with regard to power asymmetries by designers[1] of participatory approaches and then apply this method to ComMod researchers with a view to identifying and analysing the positions in which they recognize themselves. We analyse here the designers' perceptions of their practices, which depend on the meaning they give to their action. The analysis of the practices actually implemented and their effects is a second step, which has not yet been taken[2] As a first step, this chapter formalizes differences and similarities in the positions in which ComMod researchers recognize themselves with regard to taking power asymmetries into account.

Participation, Power Games and Legitimacy: Theoretical Focus

Power Asymmetries Within Society

Power is considered here as a relational concept, that is, we talk about power relationships. It implicitly refers to two closely related concepts: the power to do something and the power exerted over someone. A widely accepted definition of

---

[1] The term designer of a ComMod process, or more broadly a participatory process, is used here to designate someone who designs, leads and coordinates the whole ComMod process.

[2] It is a task that the ComMod group began in partnership with other designers of participatory approaches.

power is given by Weber (1968, 1995b): 'The probability that one actor in a social relationship will be in a position to carry out his will despite resistance, regardless of the basis on which this probability rests'. According to some authors, the definitions of power based on Weber's definition make power relationships a zero-sum game: in a relationship, the more power one person has, the less power the other has (Rowlands 1995). After a range of simplifications, power has gradually become something someone has or has not, which accounts for the frequent dichotomies between strong and weak, dominant and dominated. Scoones and Thompson (1999) deplored this simplistic view of power which conveys a 'populist' view of participation with, on the one hand, rural communities without power, and on the other, technicians or administrations abusing this power. According to Vermeulen (2005), these simplifying dichotomies are problematic because they deny the existence of a form of power among the weakest stakeholders. Vermeulen argues that even in a situation of oppression, the weakest stakeholder always has a certain form of power, as the existence of the oppressor as such is dependent on the existence of the oppressed (Veneklasen and Miller 2002).

To avoid these simplifying dichotomies, it is possible to analyse these power relationships with more nuance. First, it is important to recognize that very influential stakeholders in a given social field (Bourdieu 1994) may be less so in another field. Second, it is necessary to characterize the resources mobilized to exert power. According to Giddens (1984), there are three main types of resources: (i) know-how and knowledge; (ii) the standards and cultural values conferring on some stakeholders a legal, traditional or charismatic authority; (iii) physical resources, such as money, natural resources or human resources. Third, power relations can characterized according to the mechanisms governing them. Boulding (1989) makes a distinction between three different mechanisms: the stick, the carrot and the hug. The stick and the carrot are familiar metaphors, the stick corresponding to a form of power exerted by force or threat, and the carrot reflecting the ability of someone to get someone else to do what he wants by means of incentives, which are mostly economic. Both are forms of coercive power. These are forms of power exerted over someone. Boulding's most important and most innovative concept is the hug, which reflects an integrative or cooperative form of power. This is the power of a group to do something through people coming together to achieve the same goals, in accordance with the same principles and with a sense of belonging to a group. Countervailing power can be considered a form of cooperative power of the least influential stakeholders through a creation of alliances.

Power Games in a Participatory Process

Besides the analysis of power relationships in the initial context, it is also important to follow the dynamics of power relationships in the participatory process. Power asymmetries can have several outcomes in a participatory process (Leeuwis 2000; Faysse 2006). Some stakeholders may use their power to include or exclude other stakeholders in the negotiation, to influence the subject and issues of negotiations, to impose their ideas in discussions while ignoring or dominating the views of others, or to control the implementation of decisions without honouring the agreements reached during negotiations. Some may also use their power not to join the negotiation process and eventually block the process if their presence is essential for formulating a viable agreement.

In contrast, the risk of the views of the least influential stakeholders not being heard is also very present in a participatory process. When a stakeholder lacks self-confidence, freedom of expression, access to information or an understanding of the issues at stake, his ability to defend his interests is limited. Some authors note that, in certain negotiation configurations, the least influential stakeholders may have no interest in participating in the negotiation process, unless this ability to defend their interests is reinforced (Wollenberg et al. 2001). Although this idea is not shared by all designers of participatory approaches, these authors believe that a participatory process cannot work without the empowerment of certain stakeholders considered to be in a position of weakness.

Rowlands (1995) makes a distinction between three forms of empowerment:

- the reinforcing of personal abilities, such as increasing self-confidence, better under-standing the issues at stake, etc. (this corresponds to the power to do something)
- the reinforcing of relational abilities, which is the development of know-how concerning strategic relationships so as to learn how to influence the outcome of a discussion or negotiation process (this is the power exerted over someone)
- the reinforcing of collective abilities: a group of people learning how to cooperate to achieve more satisfactory results than individuals acting individually (corresponding to the power to do something together); this calls for the group to represent itself as a group defending common interests.

This typology focuses on the forms of learning in empowerment mechanisms, echoing the forms of learning identified in Chap. 10. However, the empowerment concept means more than just learning and increasing abilities within a negotiation arena. This learning process (changes in perceptions and modes of interaction) may have impacts on a society going beyond the negotiation arena (e.g., changes in social status, creation of alliances, etc.). The designer of a participatory process should be aware of such social changes (whether desired or not) and carefully monitor them.

Dealing with Power Asymmetries: Dialogue-Orientated Versus Critical Stances

Researchers working on participatory approaches and multi-stakeholder processes are divided on the power issue. Faysse (2006) makes a distinction between proponents of dialogue-orientated approaches and proponents of critical approaches, a distinction that corresponds to two branches of systemic approaches, that is, soft systems and critical systems.

For the former, the main obstacle to the emergence of a successful cooperation between stakeholders with different interests is a lack of communication and mutual understanding. Once these communication barriers are lifted, it becomes possible for such stakeholders to have a shared representation of the situation and reach a consensus, forming a sound basis for collective action (Pretty 1995; Röling and Wagemakers 1998). Proponents of this approach consider collective learning to be the main lever for meaningful social change. They refer to systemic approaches of the interpretative type, also called ‘soft systems’ (Checkland 1981). Based on constructivist epistemology focusing on the various possible interpretations of the system by stakeholders, these approaches have been developed in response to approaches of the so-called ‘hard systems’ type. Hard system approaches are used by the ‘hard sciences’ with strong biophysical and technology dominants and consider the researcher as an objective expert who has to remain outside the system he studies (Flood and Romm 1995).

The 1990s saw the emergence of a third branch of systemic approaches, namely that of ‘critical systems’, in response to the limitations of soft system approaches to managing conflicting and coercive situations in contexts of significant power asymmetries (Jackson 2000; Ulrich 2003). Proponents of critical approaches consider it necessary to manage actively and strategically the power asymmetries in a participatory approach so as to prevent any existing power asymmetries dominating discussions and ultimately resulting in a widening of the initial asymmetries. The facilitator of a participatory approach cannot, according to these authors, be considered as neutral in that he has to intervene intentionally in the discussion arena to give a voice to the weakest stakeholders and let them be heard in discussions. Leeuwis (2000) thus suggested that participation should be considered as a negotiating process and not just as a collective learning process. This conceptual shift considers that the stakeholders in a discussion arena talk and interact according to their own interests, which are, among other things, dictated by the existing power relationships. In addition, while the supporters of the dialogue-based approach consider consensus as a goal and a necessary condition for collective action, the proponents of the critical approach, on the contrary, recommend that a group should not necessarily be pushed too fast towards consensus. Indeed, if such a consensus is reached too quickly, it may often only reflect the views of the most influential participants. In French literature, the concepts of negotiation and ‘*concertation*’ (concerted process) are frequently referred to together. These two concepts have the same theoretical basis, namely negotiation that Dupont (1994) defined as ‘an activity involving several interacting stakeholders who, when faced both with divergences and interdependencies, voluntarily choose to seek a mutually acceptable solution’. Beuret (2006) considered that the difference between negotiation and ‘*concertation*’ lies in the fact that negotiation seeks to obtain an agreement on a decision, which is not necessarily the case of a concerted process. The aim of a concerted process is ‘the collective establishment of visions, goals and common projects, in order to act or decide together’.

Although historically speaking, critical approaches were established in response to the limits of dialogue-orientated approaches, these two approaches are not necessarily antagonistic. Indeed, a number of key writers on dialogue-orientated approaches are open to criticism and instigate changes in their own thinking (Innes 2004). Some critical and dialogue-orientated writers also share a common theoretical foundation, that is, the 'theory of communicative action' of the philosopher Jürgen Habermas. Habermas makes a distinction between three forms of action based on three forms of rationality (Habermas 1987; Leeuwis 2000; Lussault 2003). In the instrumental action, the individual acts to achieve a predetermined goal. In strategic action, he also aims for a specific goal, but considers the other individuals as potential opponents and he takes them into account in developing his own strategy. In the communicative action, individuals seek mutual understanding and consensus. Although each individual and each situation combines these three kinds of action, the communicative action should, according to Habermas (1987), be the goal for the members of a society. In other words, in an ideal situation, the result of a negotiation process should be based on stakeholders' arguments and not on power relationships. These are the only circumstances under which a genuine, stable agreement can be reached. Habermas (1987) defined a number of conditions for establishing such an 'ideal speech situation' in which each stakeholder has the same ability to participate in the discussion (same rights, same information and same opportunities). All forms of asymmetries and coercion in dialogue are seen as 'distortions of communication'. The theory of communicative action is often associated with dialogue-orientated approaches (Faysse 2006) and challenged by some proponents of critical approaches who denounce its utopian nature. However, key authors on critical approaches, such as Ulrich (2003) in the field of systems thinking or Leeuwis (2000) in the field of communication sciences, have also used Habermas' theories to demonstrate that dialogue-orientated and critical approaches may be seen as more complementary than antagonistic. We will come back to this point later.

Legitimacy of Participatory Processes

The question of the position adopted with regard to power asymmetries brings us back to the issue of the legitimacy of participatory processes and those who implement them. We saw at the beginning of this chapter that a designer taking the initiative to empower the least influential stakeholders may be considered as not legitimate to do so. Legitimacy is here taken to mean acceptance and recognition but this concept is inherently subjective and polysemous. As it is commonly defined in terms of justice and fairness, it may be construed in its legal or ethical dimensions. In this section, we introduce the way that this concept of legitimacy, which is central in sociology, is defined by the major authors in this field, before considering the ComMod group's standpoint on this issue.

The concept of legitimacy was coined by Weber to provide an understanding of how an agreement between social agents provides a social value to any practice (Weber 1995b). Legitimacy thus marks the daily existence with a number of collectively accepted benchmarks that guide each society. However, Weber provides two very different definitions of legitimacy without always making a distinction between them: a general definition related to individuals' choices of activities, and a justification a posteriori of a domination relationship. Legitimacy then becomes a concept for considering power and social relations.

Our considerations on the legitimacy of a ComMod process come more within the scope of Weber's first definition. The ComMod glossary refers to Boltanski and Thévenot's (1991) definition, which is along the same lines. According to them, legitimacy is the 'interaction-based and open-ended result of a compromise between principles and values progressively leading to a convention to which stakeholders make reference to justify their choices'. When questioning the legitimacy of a participatory process, we question both its scientific legitimacy (i.e., perception by the scientific community of the validity of the knowledge produced) and its social legitimacy (i.e., acceptance of the process by stakeholders). While this chapter addresses the question of legitimacy through the issue of power asymmetries, this question is much wider. A specific working group within the ComMod network and a few papers (Daré et al. 2004; Aubert 2006; Barnaud 2008) have contributed to the progressive refinement of this question and have suggested to distinguish between:

- legitimacy of intervention—Who formalized its demand? Is the ComMod approach able to respond to it? (see Chap. 5)
- legitimacy of participants—Why some stakeholders and not others? Who invited them? Are they considered as representative of a group? How can this representativeness be guaranteed, if need be? (see Chap. 5)
- legitimacy of models—Has the proposed representation of the system been validated by all the stakeholders considered as essential to ensure its legitimacy? (see Chap. 4)
- legitimacy of the team implementing the ComMod process—Given the profile, the background, the institutional connections, and even the personality of the designers, are they perceived as legitimate by stakeholders? (see Chap. 3).

In addition, is legitimacy seen as given and unchanging, or on the contrary, is it something that may be built or lost over time? When the legitimacy of an intervention is queried, the issue of how the approach was initialized is important (see Chap. 5). For some, the ideal approach would be for the local stakeholders to call on the ComMod researcher to support a group of stakeholders set up by them to

[3] However, the case can be cited of an experiment led by Michel Étienne in which, following a ComMod experiment in the Causse Méjan, the SCTL called on researchers at the Avignon INRA to use ComMod to facilitate collective learning on pine forest management.

deal with an issue that they have identified. In practice this type of situation is rare.[3] Legitimacy is more often progressively built during the process, which is consistent with the iterative and adaptive nature of ComMod processes (Daré et al. 2004). Each time they return to the field, the designer changes the model, the process and the group of stakeholders involved, so as to take better account of the views, concerns and expectations of the stakeholders (see Chap. 2). This way, the designer progressively increases the social legitimacy of the process.

However, the issue of power asymmetries is not fully resolved by these methodological suggestions, and neither is the question of the influence of the designer in power games. To address these issues, there is a need to go beyond a situation where each designer assesses the legitimacy of the intervention according to its own definition of what is legitimate or not in the context in which they operate. Questioning the legitimacy of each designer thus first implies making explicit its perception of the context and position with regard to this context.

Method: A Test for Making Explicit Positions with Regard to Power Asymmetries

To build a shared representation (i.e., shared by ComMod researchers) of the diversity of the positions in which they recognize themselves with regard to power asymmetries, we implemented a ComMod-like process, that is, a collective, iterative and adaptive process.

Method used to Build the Test

It should first be remembered that there was prior knowledge about the positions of a few ComMod researchers (d'Aquino 2007; Daré et al. 2007; Barnaud 2008). We also had the responses provided by 15 ComMod researchers to a questionnaire[4] seeking to state the designer's position in relation to the socio-political context within which the process was implemented. An initial examination of these questionnaires[5] made it possible to propose a first analytical method consisting of four indicators, or axes of analysis, graduated from 1 to 5, and assumed to be relevant for characterizing various aspects of the possible positions of researchers with regard to power asymmetries. From these 15 questionnaires, we were also able to identify four different researcher profiles (corresponding to four combinations of positions along the axis). This initial analysis was then submitted to at

[4] The questionnaire was sent to all the researchers of the ComMod group and ADD-ComMod project, as part of an ongoing comparative analysis on participation issues (ComMod and non-ComMod) coordinated by Patrick d'Aquino.

[5] We used the Atlas.ti software designed for the qualitative and quantitative analysis of text.

least one researcher for each profile to allow them to criticize, validate and/or enrich the proposed axis, indicators and profiles. This confrontation led us to revise and refine the axis of analysis and the proposed profiles. The revised version was then submitted to all ComMod researchers who had led and coordinated at least one ComMod process and had thus been faced with the question of their position regarding the socio-political context. They were asked to indicate their position on the axis and select the profile they felt matched them closest or, if necessary, propose the definition of a new profile. Our analysis was, therefore, based on how ComMod researchers see their own practices, according to the meaning they give to their action.

One limitation was the sampling method: only those who were available or motivated enough to take the test did so (20 out of the 27 asked). However, it can be assumed that the result provides a sufficiently accurate illustration of the diversity of positions existing within the group.

Test: Four Axes of Analysis

Axis 1: neutrality or non-neutrality with regard to power asymmetries

The first axis gives various positions ranging from neutrality to non-neutrality with regard to power asymmetries. This involves the researcher making explicit their position with regard to these asymmetries, and not a neutrality or non-neutrality in absolute terms. Indeed, no ComMod researcher considers themselves as neutral in absolute terms, one of the specificities of the process being to emphasize precisely the involvement of the researcher as a stakeholder within the system, who has, in the same way as the other stakeholders, a particular viewpoint on the system. However, if ComMod researchers agree on their non-neutrality in the sense that they admit to having an opinion on the system, not everyone agrees on intervening in this system on behalf of this opinion. In particular, there is no consensus on the more specific question of positions with regard to power asymmetries, which goes further by examining whether there is any bias in favour of greater equity.

This concept of equity is itself very subjective. Something is regarded as equitable if it seems fair. However, what seems fair or equitable in the eyes of some will not necessarily seem so in the eyes of others. When the concept of equity is used to qualify a participatory process, a distinction can be made between procedural equity and social equity. While procedural equity refers to what happens in the negotiation arena, social equity refers to the impacts on the social system in which this area lies. Although the concepts of procedural equity and social equity are obviously linked, the designer of a participatory process has more control over the former (if any). When the issue of bias in favour of more equity is raised hereafter, it is in reference to that concept of procedural equity. Although no definition of procedural equity has yet been collectively agreed upon within the ComMod network, an equitable concerted process can be defined as a process in which all stakeholders have a chance to voice and assert their interests, with equal opportunity being the ideal (Barnaud 2008). What interests us here is how the

Table 6.1 Meaning of the graduation of axis 1 on the expression of neutrality or non-neutrality with regard to power asymmetries

Position	Definition
1	Neutrality of the laissez-faire type. Deliberately let power games express themselves in the process as they refuse to intervene in favour of more equity beyond the sharing of knowledge
2	Unconditional dialogue-based neutrality. No wish for any bias towards any of the stakeholders or points of view involved. Give a voice to all groups involved in the same way, as dialogue is considered sufficient to promote greater equity
3	Conditional dialogue-based neutrality. Same as 2, the difference being they leave themselves with the option of stopping the process if it becomes detrimental to certain stakeholders considered to be in a weak position
4	Post-normal non-neutrality. Provide a communication arena in which they seek to strengthen the voice of the stakeholders considered to be the least influential ones. Leave any stakeholder free to reject this arena, but consider that the stakeholder's support for the proposed arena is deemed necessary to its legitimacy
5	Strategic non-neutrality. Same as 4, except that the stakeholders' support for the proposed arena is not necessary (e.g., pressure mechanisms on recalcitrant stakeholders may be mobilized if necessary to strengthen the voice of the least influential stakeholders)

designer of a participatory process positions themself in regard to this concept of procedural equity, and, accordingly, how they deal with power asymmetries. Their position is expressed in particular through their methodological choices: the choice of the topics discussed, the tools used, the knowledge mobilized, the stakeholders involved, the communication modes or the timing of the intervention can all be considered as ways of expressing a position with regard to power asymmetries.

Within the ComMod group, at first glance, it can be seen that some researchers adopt a dialogue-orientated stance and consider that, to be legitimate, the designer of a ComMod process should show no bias, while others favour a critical stance arguing that such neutrality is neither possible nor desirable. To overcome this dichotomy, which, while certainly practical, is simplistic and insufficient to provide a more nuanced analysis of the range of possible positions, we have identified a range of five possible positions (Table 6.1). To the extent that we want our analysis framework to be applicable to a whole range of participatory approaches wider than ComMod, the positions described in this range are not necessarily all observed within this group (this point is also valid for the other three axis).

Position 1 occupies a special place in this range as it is the only position that has no explicit aim of promoting greater equity. It may be considered to be the position that expresses most strongly its neutrality with regard to power asymmetries, but we should not overlook the fact that deliberately allowing asymmetries to persist is a form of bias in itself. Positions 2 and 3 correspond to dialogue-orientated positions. Researchers corresponding to these positions express a bias towards equity, with dialogue and sharing of views as the key levers for promoting procedural equity. However, they assert neutrality in the coordination of the process, considering that they have no legitimacy to intervene in the communication arena by increasing the voice of certain groups of stakeholders. Position 3 (conditional neutrality) differs from the position 2 (unconditional neutrality) in allowing the possibility of intervention in this communication arena by stopping the process if there is too great a risk of influential stakeholders manipulating other parties. Positions 4 and 5 correspond to critical positions. There again, the objective of more equity is stipulated, but the resources used to achieve it differ. Indeed, these researchers call for procedural non-neutrality, considering that attempts should be made to counter power asymmetries, in particular, by empowering certain stakeholders. A distinction is made here between two possible critical positions that have been qualified as post-normal and strategic, respectively. According to Funtowicz and Ravetz (1994), a post-normal stance seeks to improve the quality of the interactions bringing a group to take a decision, more than the quality of the decision itself. Researchers adopting position 4 interpret this by considering that special attention should be paid to the quality of the communication arena, not only in terms of equity, but also in terms of legitimacy, this legitimacy stemming from the support of stakeholders for the proposed communication arena. Researchers adopting position 5 qualified as strategic aim to bring the group of stakeholders to take a decision. To achieve this, the support of all stakeholders is not necessarily deemed necessary. In particular, they make provision for the possibility of establishing mechanisms to exert outside pressure on certain recalcitrant groups of stakeholders whose presence is essential for formulating a viable agreement.

Axis 2: initial analysis of power games—not a priority or necessary?

The second axis is an indicator of the methodological choice made regarding the type of initial analysis of power games, reflecting how the researcher takes account of the socio-political context in which they operates (see Chap. 5). There is a consensus within the ComMod group on the fact that the ComMod process reveals power games (Daré 2005), and that it is important to give adaptive consideration to the power games thus highlighted. However, the issue of the importance of a preliminary analysis of power games is not the subject of a consensus. As pointed out in Chap. 5, a number of arguments can be made for a greatly reduced or no initial diagnosis. In addition to the advantage of lower costs, the designer with limited knowledge of the context in which they operate may highlight the fact that this could allow them to remain neutral and not influence the course of the process, leaving the stakeholders in control of the direction they take. Others put forward the pointlessness of an initial analysis of power games based on a survey lasting a few weeks or even months, as these intimate social interactions are difficult or even impossible for an outside observer to identify, and whose analysis is moreover necessarily subjective. Another key argument for a reduced initial analysis is that the implementation of the approach itself provides a better understanding of the socio-political context. However, does the designer of the ComMod process not then run the risk of understanding too late the importance of a given power game? As imperfect as it is, can a ComMod approach do without on an initial analysis of strategies and power games, as well as any existing power asymmetry? Table 6.2 describes the graduation of axis 2 relating to the importance attached to a given initial analysis of power games.

Table 6.2 Meaning of the graduation of axis 2 on the initial analysis of power games: non-priority or necessary?

Position	Definition
1	The initial analysis of power games is not a priority because it is pointless (it is impossible to understand power games in a society to which we do not belong) and unnecessary (the process itself reveals power games)
2	Although the initial analysis is not entirely worthless (in absolute terms, if we could understand power games), it is pointless (we cannot understand power games in a society to which we do not belong)
3	Based on pre-existing knowledge of the general context, the initial analysis is carried out on a group basis with local stakeholders selected and brought together during participatory workshops
4	An in-depth initial analysis with individual interviews is required as it helps to identify the most significant obstacles to the emergence of an equitable concerted process
5	An initial analysis of the anthropological type (by long-term immersion), providing intimate knowledge of the society is a prerequisite for implementing any participatory process

Table 6.3 Definitions of the graduations of axis 3 on the perception of the risk of the process implemented resulting in initial power asymmetries being reproduced and reinforced

Position	Definition
1	Low or negligible risk of the views of certain stakeholders in a weak position not being heard, dialogue considered as sufficient to allow all the stakeholders concerned to be heard
3	Non-zero risk of certain views not being heard, but power asymmetries do not necessarily dominate the process, the arguments put forward may prevail
5	High risk of processes resulting in the initial power asymmetries being merely reproduced in case no specific intervention is geared to empower the least influential stakeholders

Axis 3: perception of the risk of reinforcing power asymmetries in the initial context

While there is a consensus within the ComMod group on the existence of power asymmetries among stakeholders, there is no consensus as regards the risk of seeing these power asymmetries dominate the ComMod process to such an extent that the process would reproduce or even strengthen the initial power asymmetries. Axis 3 (Table 6.3) examines the way in which the designer sees the context in which they operate, and more specifically, their assessment of the risk of the process implemented resulting in initial power asymmetries being reproduced and reinforced (i.e., domination of the process by the most influential stakeholders and the views of the weakest stakeholders being unheard). The researcher's position on this axis thus depends on both the context itself and how they perceive it. In proposing this axis, we wanted to test the idea that the position adopted by the researcher depended in part on the context in which they found themself. We acknowledge, however, that given that this context may change during the process, the perception of risks and the position adopted are also likely to change.

Table 6.4 Meaning of the graduation of the subsidiary axis on the objective of the processes implemented

Position	Definition
1	Better understand the system (research objective)
2	Share with local stakeholders the researchers' view of the system (i.e., knowledge, perception) exchange of views between researchers and local stakeholders)
3	Facilitate an exchange of views, not only between researchers and other stakeholders, but also among these other stakeholders (facilitate a collective learning process)
4	Support a concerted process between local stakeholders ahead of decision-making.
5	Support local stakeholders up to decision-making on collective action

Subsidiary axis: objective of the process implemented

The fourth axis is about the objective of the implemented participatory process. It is slightly different from the other axis because it is not a direct indicator of the position adopted by the researcher with regard to power asymmetries. Moreover, the position on this axis is highly variable, since the objective of a participatory process often changes over time, especially in ComMod processes, which are inherently adaptive. As a result, this axis was not used to characterize the profiles describing various possible positions with regard to power asymmetries. However, we decided to keep this axis in the test, firstly, because it is important when discussing the legitimacy of a process to know its purpose (we will come back to this later), and secondly, because we wanted to test the idea that there is a correlation between the objective and the stance adopted with regard to power asymmetries (defined by the other three axes). For this purpose, we have defined five possible objectives for implementing a participatory process (Table 6.4).

Results: Contrasting Profiles

The test that 20 ComMod researchers agreed to take consisted of two parts. In the first part, they were asked to indicate their position on the four axes of analysis. Figure 6.1 shows the results for the first three axes, as the answers given for the fourth axis could not be included given the variety of objectives pursued jointly or successively by each ComMod researcher. This figure is a first illustration of the diversity of the positions in which ComMod researchers recognize themselves.

In the second part of this test, four profiles corresponding to four types of combinations of positions on the first three axes were briefly described. The researchers were asked to identify the profile they felt closest to or propose, where appropriate, a new profile better matching their position.

Figure 6.2 shows that most of the ComMod researchers who took the test (16 out of 20) correspond to profiles 3, 4 or in between the two.

As can be seen in the discussion, the results of Fig. 6.2 show a bias. They mask the fact that a researcher may adopt one position or another depending on the

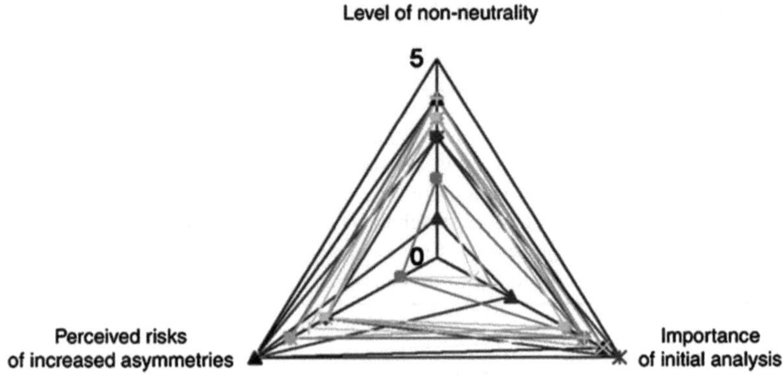

Fig. 6.1 Position of the 20 ComMod researchers who took the test on the three axes stipulating their position with regard to power asymmetries. On the first three lines, when a researcher said that they considered they were between two positions, we kept the intermediate value. While such an intermediate value made sense for the first three lines, this was not the case on the objectives line

Fig. 6.2 Distribution of the 20 ComMod researchers who took the test on the profiles identified

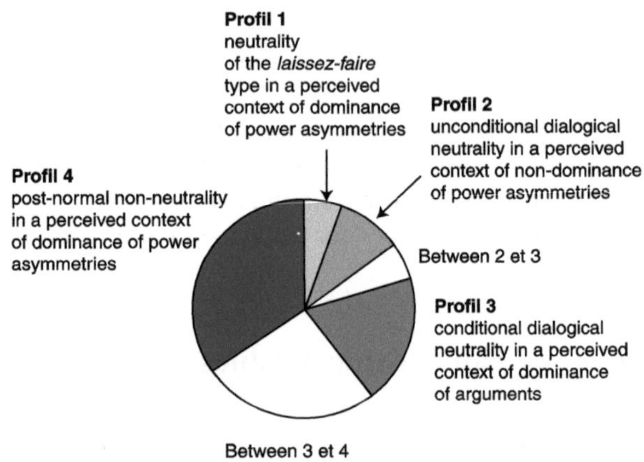

intervention context or even the time the process takes place. These figures are based on a dominant profile at a given moment in the life of a researcher. They are given here for illustration. What we are interested in is the identification of the different possible positions, each with their own consistency.

These profiles are presented below, starting with the brief description available in the test through which the respondents identified themselves (or not). This presentation was then refined through discussions arising out of the test and concrete examples illustrating them. The hybrid profiles proposed by six out of the 20 researchers who took the test are presented following the four types initially identified.

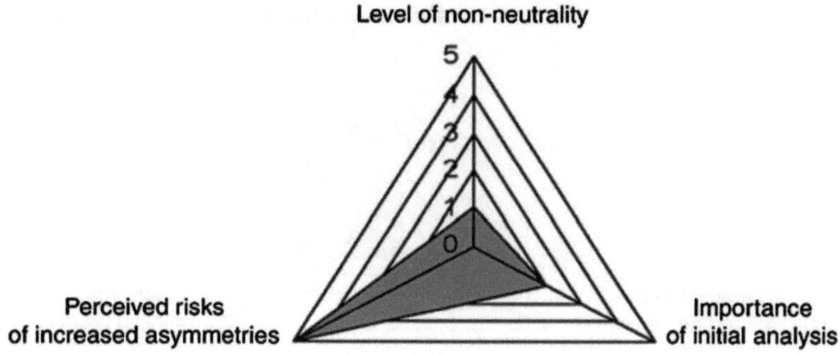

Fig. 6.3 Standard position of researchers in profile 1 on the three axes stipulating their position with regard to power asymmetries

Profile 1: Neutrality of the Laissez-Faire Type in a Perceived Context of Dominance of Power Asymmetries

The researchers corresponding to profile 1 (one researcher interviewed) (Fig. 6.3) consider that the power asymmetries inherent in the socio-political context are inevitably reproduced in the ComMod process, but they do not seek to counter them, as they consider that it is not their role to promote greater equity, and that it is pointless because the actions taken are generally not sufficient to change the existing power asymmetries, and what emerges is beyond their control. They also consider that while the initial analysis of power relationships is not necessarily pointless, it is not possible to have sufficient knowledge of the social mechanisms at work.

The researchers in this profile believe that participation is a means and not an end. It is not an end in that there is no underlying ambition to promote values such as participative democracy and social equity. The ComMod process is seen as a way to respond effectively to a request made explicitly by an external agent, such as, for example, to make a group of stakeholders aware of a given common problem to facilitate its resolution. The existence of such a mandate is a key explanatory factor in the coherence of this profile. Indeed, it is the existence of this mandate that legitimizes the ComMod process in the view of these researchers. They thus call for an absence of bias in favour of procedural equity when it is outside their mandate.

Another key explanatory factor in the coherence of this profile is the idea that the participatory processes implemented are not able to counter power asymmetries in the socio-political context. For example, in the Tarawa experiment in the Kiribati islands, the project team retrospectively identified various categories of stakeholders, that is, consensual players, pseudo-players and meta-players (Dray et al. 2007). Whereas the former were open to dialogue, the latter, who had strong interests to defend, dominated the discussions, holding on to their initial positions until the end of the workshop. The Tarawa assessment report showed that the

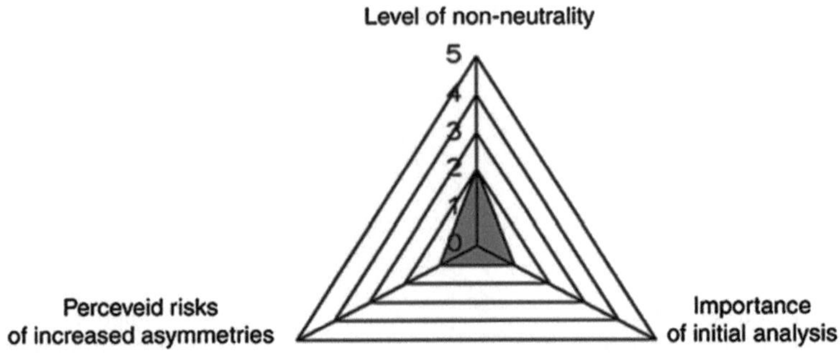

Fig. 6.4 Standard position of researchers in profile 2 on the three axes stipulating their position with regard to power asymmetries

efforts made by the project team were considered to be pointless in attempting to ensure that these pseudo-players took the view points of others into consideration. The meta-players for their part corresponded to stakeholders involved at higher institutional levels who were not invited to the role-playing game workshop and who, subsequently, imposed their own decisions without taking into account any of the proposals that emerged from the workshop. In retrospect, it appears that it might have been better to include these meta-players in the initial role-playing game workshop. Would a more thorough initial analysis of power relationships have helped to anticipate these constraints? One of the designers of this project, recognizing themself in profile 1, was doubtful. For such an initial analysis to make it possible to anticipate the complexity of power games, a thorough analysis of an anthropological type would be necessary. As this ideal is incompatible with the resources and time schedules of most research and development projects, the designer considered that a faster analysis carried out in a few weeks or months would have been insufficient and, therefore, pointless.

The next section shows that while the researchers in profile 2 also consider that an initial analysis of power games is not a priority, they think so for quite different reasons.

Profile 2: Unconditional Dialogue-Based Neutrality in a Perceived Context of Non-dominance of Power Asymmetries

The researchers corresponding to profile 2 (two researchers interviewed) (Fig. 6.4) consider that in a participatory process, sharing of knowledge and good quality dialogue are levers for a greater equity in the concerted process. They call for a neutral position with regard to power asymmetries in that they seek to give all stakeholders an equal say, without taking the side of a particular group, and without seeking to increase the voice of certain stakeholders. The initial analysis of power asymmetries is not a priority, as the ComMod process reveals them anyway, which logically leads to them being mitigated, at least to some extent.

In experiments carried out by researchers in this profile, the call for the unnecessary nature of the initial analysis may go further. In the Larzac experiment in southern France, for example, the designer of the process emphasized his quasi-absence of knowledge of the socio-political context was a way to ensure a lack of bias as a designer, which is a *sine qua non* condition for his legitimacy. However, this position raises the problem of the implicit knowledge of a context. Furthermore, in calling for a lack of bias, the designer runs the risk of imposing a bias unconsciously. They thus do not escape the need for a reflective analysis of his practices. In the assessment report on the Ouessant experiment in northern France, in the face of participants pointing out his strong powers of persuasion, the project designer was worried, explaining that the only thing of which he wanted to convince people was the merits of a method based on dialogue and discussion. He felt legitimate to convince them because of the existence of a mandate that explicitly called for the implementation of such a method. This mandate itself can be considered as a bias that should be made explicit. We will come back to this later.

Besides this question of legitimacy, the second explanatory factor determining this position is its fundamentally dialogue-orientated nature. Without denying the existence of power asymmetries between stakeholders, this position is based on the assumption that dialogue leading to a greater mutual understanding among them can give rise to the emergence of proposals taking the interests of all into account. 'When I place a sheep breeder and a forestry officer face to face, the power asymmetries are obvious and cannot be mitigated. I'm convinced, however, that the power of the most powerful will be better used because it takes more account of the interests of the least powerful' (M. Étienne, personal communication). This position is based on the capacity for empathy of the most powerful stakeholders and assumes that agreeing to participate in a discussion process such as ComMod is, in itself, a sign of such a capacity for empathy, and thus reduces the risk of manipulation. It should be noted here that the researchers who recognized themselves in profile 2 have only carried out ComMod processes in France, which might have influenced their position. Further analysis would, however, be required to provide conclusions about the existence of correlations between intervention countries and the positions adopted.

Profile 3: Conditional Dialogical Neutrality in a Perceived Context of Dominance of Arguments

In the contexts in which they operate, the researchers in profile 3 (four researchers interviewed) (Fig. 6.5) consider that the power asymmetries of the initial context do not necessarily dominate discussions, which may be based more on arguments. They call for a neutral position with regard to power asymmetries, not wanting to

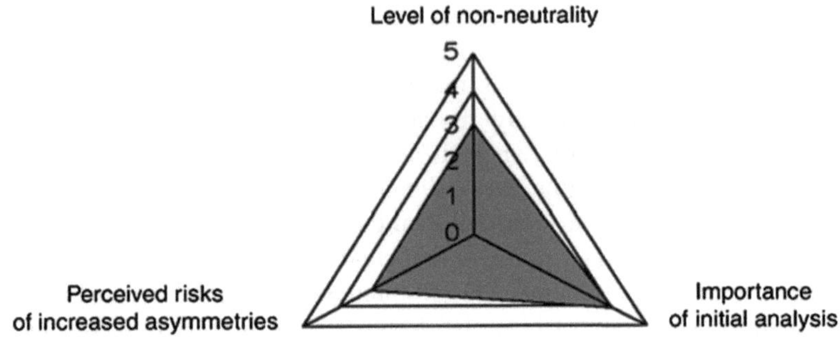

Fig. 6.5 Standard position of researchers in profile 3 on the three axes stipulating their position with regard to power asymmetries

take sides for one group of stakeholders as this could be detrimental to the legitimacy of the process. However, they assess the effects of the process they implement and may decide to stop it if they consider it potentially harmful to certain groups of stakeholders already in a weak position. This neutrality is thus only relative, or conditional, and is specified as such. Concerned to 'know where they tread', they attach importance to the implementation of an initial analysis of power relationships in the form of individual interviews.

This profile is intrinsically a profile between dialogue-orientated and critical stances. Researchers adopting it are dialogue-orientated like those of profile 2 as they consider that good quality dialogue should allow stakeholders to exchange their points of views through expression of argument rather than power relationships. In this respect, they call for certain neutrality and a low level of intervention in the facilitation of the process, so ensuring their legitimacy. They recognize, however, the limits of the dialogue-orientated approach in contexts of conflict or coercion, and adopt a cautious approach considering that, unlike profile 2, a thorough initial analysis of the socio-political context and careful monitoring of the power games at work are needed, at least as a precaution. While they are not very interventionist within the communication arena, they allow themselves to intervene in the process by stopping it if the risk of manipulation by those in positions of power becomes too great.

For example, in the SAGE Drôme experiment in southern France aimed at establishing dialogue and mutual understanding between farmers and water managers in a catchment, farm advisory professionals were invited to the test of a role-playing game. They viewed the role-playing game as a potential consulting tool to help them 'convince' farmers of the need for concerted water management. 'Such proselytizing use poses a problem and cannot be allowed by the game designers because the underlying model has not been validated for this type of objective and the risk of manipulation associated with it' (Barreteau 2007). The existence of this risk of manipulation partly influenced the decision of the research team not to seek to continue the ComMod process.

Profile 4: Post-Normal Non-neutrality in a Perceived Context of Dominance Power Asymmetries

Researchers in profile 4 (seven researchers interviewed) (Fig. 6.6) believe that because of the power asymmetries among stakeholders, there is a significant risk of the ComMod process reproducing and finally, increasing the initial inequalities. An in-depth initial analysis of power relationships is thus required to identify any potential obstacles to the emergence of an equitable concerted process, and then scale the ComMod process to try to overcome these obstacles in the proposed communication arena by strengthening the voice of the stakeholders in a weak position.[6] They thus stipulate their non-neutrality with regard to power asymmetries, a position that poses questions about its legitimacy. Within the context of a post-normal approach, their objective is to provide a communication arena as rich and equitable as possible. This position establishes its legitimacy by seeking the support of local stakeholders: their non-adhesion to this arena calls into question its legitimacy and makes the designer question the process they are implementing.

The researchers in this profile feel they have a certain responsibility towards the stakeholders who, with regard to the question raised by the ComMod process, have strong interests to defend but little capacity to do so. They fear that the process will harm them if they are not able to defend their interests in the communication arena. These 'weak' stakeholders may be very different in different contexts. For example, in a Senegalese experiment, the project team chose to empower local officials dealing with higher administrative levels to enable them to have more of a say on decisions regarding land use in their municipality. In the Thai Mae Salaep experiment, however, local officials were seen as dominant stakeholders who could mask the views of some of the marginalized individuals in the community if no precautions were taken. Furthermore, while in the ButorStar experiment in southern France, environmentalists were seen as stakeholders in a position of weakness in the negotiations with hunters and farmers, in another context, in Nan, in mountainous northern Thailand, the environmental stake hold by the national park was, on the contrary, in a position of power compared with the villagers that would be affected by the establishment of the park. The position of the researchers of this profile implies that they systematically stipulate which stakeholders they consider to be vulnerable and which should be given special attention and what capabilities of these stakeholders they want to reinforce. Then comes the question of how should ComMod researchers, while adopting a critical position, intervene in the communication arena to limit the risk of initial power inequalities being widened? To do this, they can intervene at several levels (this list is not all-inclusive).

- The selection and sequencing of the methods used (e.g., discussions in plenary sessions or in small groups, individual interviews, etc.): for

6 This profile corresponds to a 'critical companion stance', which was stated, tested and discussed in Barnaud (2008).

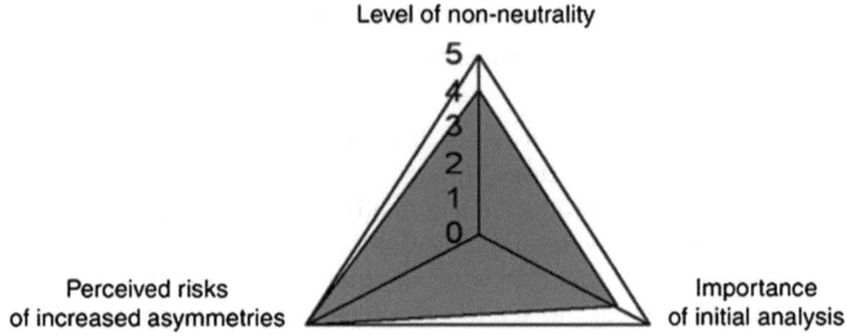

Fig. 6.6 Standard position of researchers of profile 4 on the three axes stipulating their position with regard to power asymmetries

example, in a Thai experiment, when facilitating discussions among farmers belonging to different ethnic groups, the ComMod process designers realized that the Hmong could not express themselves in plenary sessions. It was decided to hold separate workshops for the Hmong and Thais before returning to a common collective workshop (Becu et al. 2008).

- The choice of dynamics represented in the models: in the Mae Salaep Thai experiment, for example, the role-playing game was deliberately designed to highlight the problem of unequal access to irrigation water, a problem that wealthy farmers, benefiting from irrigation, had an interest in covering up.
- The selection of participants and the invitation procedures: the selection of participants is a powerful lever of action in the communication arena. The process designers may try to ensure that all the participants in attendance will be able to defend their interests. They may also influence the creation of possible coalitions and alliances that could strengthen the position of weak stakeholders. Still at Mae Salaep, the grouping of farmers without access to water around a charismatic religious leader was crucial in allowing them to defend their interests (Barnaud et al. 2008c). This leader corresponded to what some authors in the field of adaptive management called 'champions' (Gilmour et al. 1999). However, when the designers themselves select the participants, this can raise a question as to the representativeness[7] and legitimacy of these participants.

[7] The participants asked to participate in the participatory processes often 'represent' other stakeholders sharing common interests in relation to a given problem. However, these are not necessarily 'real representatives'. Indeed, to be truly representative of a group calls for this group to be recognized as such, which is different from a group regarded as such by researchers, which corresponds more to the concept of category. This also calls for members of this group to have appointed or elected this person to represent them. In most cases, a number of participants correspond to false representatives, who have no responsibility with regard to those belonging to the same category. In this case, if the process is designed to support a group decision process consideration needs to be given to widening discussions to all those 'represented' in this way (Barnaud 2008).

– The choice of animation procedures during workshops: for example, the animator of a plenary debate may choose not to push a group of participants too fast towards a consensus that might only reflect the opinion of a minority.

Profile 4 is the one that raises the most questions about its legitimacy as the choice to be non-neutral might be seen as a form of interference or intrusion. For example, when the process designer selects the participants, the composition of the group may seem arbitrary to local stakeholders who would have preferred to make the selection themselves. However, the power games at work in a process of self-selection may lead to the exclusion of certain categories of stakeholders. A designer adopting the position described in this profile, is torn between, on the one hand, a desire to let the local stakeholders guide the process, without intervening, at the risk of the process being manipulated by the most influential stakeholders, and on the other, the concern to structure the communication arena to allow the voice of the least influential stakeholders to be heard, at the risk of weakening the legitimacy of the process. The designer adopting a critical stance is in a situation of continuous instability, calling for him to question continually his practice (Barnaud 2008). The adoption of a critical stance also raised the question of the risk of creating conflicts where there were none. By highlighting power asymmetries and empowering certain stakeholders, the participatory process may upset the established order. For example, in the case of the Bhutanese Lingmuteychu experiment, while the creation of a village committee to manage the resources of the subcatchment area reduced a conflict over the sharing of irrigation water between two villages upstream, it in turn fuelled a similar conflict between two communities further downstream. Local people may prefer a situation, which while no doubt unequal, is at least peaceful. For these reasons, adopting a critical stance calls for the designer of the participatory process to give continuous critical consideration to their own legitimacy and the legitimacy of the process they are implementing in the eyes of the local stakeholders. Such legitimacy is acquired by systematically making explicit the intervention's assumptions, goals and sides taken (especially with regard to power games), so that local stakeholders can either reject them or accept them as legitimate. Systematically making explicit the assumptions underlying the participatory process is one of the ComMod approach's methodological pillars. We will discuss its limits in more detail in the last section of this chapter.

Hybrid Profiles

Out of the 20 researchers who took the test, five considered themselves to be somewhere between profiles 3 and 4. For example, the designer of the Kat Aware experiment (Farolfi et al. 2008) combined parts of the definitions of the initially proposed profiles to produce a hybrid profile better suited to him (the text in italics was changed from the definitions of the initial profile). 'Researchers in this profile consider that because of the power asymmetries among stakeholders, there is a significant risk of the ComMod process reproducing and finally, increasing the initial inequalities. An in-depth initial analysis of power relationships is thus required to identify any potential obstacles to the emergence of an equitable concerted process, and to scale the ComMod process to try to *understand* [instead of overcome in the initial definition] these obstacles in the proposed communication arena. They call for certain neutrality with regard to this balance of power, not wanting to take sides. However, they assess the effects of the process they implement and may decide to stop it if they consider it potentially harmful to certain groups of stakeholders already in a weak position. This neutrality is thus only relative, or conditional, and is specified as such. Concerned to "know where they tread", they attach importance to the implementation of an initial analysis of power games in the form of individual interviews'. They added: 'In other words, while aware of the risks, everything possible should be done to understand *ex ante* the power asymmetries in order to avoid macroscopic errors in the approach and its implementation. 'However, when animating the workshops, their position is not necessarily "interventionist" but more dialogue-orientated' (S. Farolfi, personal communication). While only four researchers considered their position to be between two profiles, the comments of a larger number of researchers point to such an intermediate position or an alternation between profiles 3 and 4 (and sometimes between profiles 2, 3 and 4), depending on the characteristics of the context of the process undertaken or how it changes over time. It would seem that this is a common feature to many ComMod researchers, as discussed in the next section.

Discussion: From Divergence to Concordance

We showed above the existence of contrasting positions with regard to power asymmetries within the ComMod group. However, our method aimed precisely at identifying these differences. We will now analyse what, over and above these differences, brings ComMod researchers together on this issue.

Adaptive use of the Complementarities Between Dialogue-Orientated and Critical Stances

Owing to our quantitative analysis method, the results presented above mask a more qualitative aspect that arose out of the responses to the test. It is the flexibility in the positioning of those researchers who, while fairly easily identified in a dominant profile, may in fact see themselves in a stance or another adaptive approach according to the intervention context or objective. Some researchers put forward, for example, the idea that in a context where they consider that power asymmetries are likely to predominate, a critical stance and a thorough initial analysis of the power games with individual interviews are required, while in a context where the power asymmetries seem less important, a more dialogue-orientated stance and a lighter analysis of the power games may be sufficient (these remarks highlight the importance of the context discussed in Chap. 5). Others think that when the process objective is to produce knowledge or facilitate a learning process, a dialogue-orientated stance is adequate, while when the goal sought is negotiation or decision-making, a critical stance is required. Still others suggest that the process objective should be scaled to the context. Among them, some believe that in the case of moderate power asymmetries, the goal is about collective learning, while in cases of more important power asymmetries, the target is more to facilitate a real negotiation process. Others suggest on the contrary that the more power relationships are unequal, the more targets should be modest (targets of mutual understanding and collective learning instead of negotiation and decision-making).

The comments of the researchers who took the test also show that the flexible, scalable adoption of one position or another may be sequential depending on successive phases in the process. Several researchers cite, for example, the possibility of adopting a critical initial stance to prepare a communication arena as balanced as possible (e.g., by organizing specific workshops to empower the stakeholders considered to be in a weak position), which then allows the adoption of a dialogue-orientated stance with an approach as hands-off as possible. Others instead suggest that in the initial stage of the process (exchanges of points of views, increased mutual understanding), a dialogue-based stance may be sufficient, while, when moving on to the scenario exploration stages towards a point where group decisions are involved, a more critical stance is necessary.

Broadly speaking, these comments show that ComMod researchers have a common awareness with regard to the issue of power games, which each researcher uses as a basis for establishing a personal representation of these power games in his intervention context, before choosing his position. Furthermore, a certain unity emerges from the ComMod group around the adaptive use of the complementarities between the dialogue-orientated and critical stances.

From a theoretical point of view, the ComMod groups is thus more in line with the ideas expressed by tenants of critical stance like Ulrich (2003) or Leeuwis (2000), who urged us to go beyond the conventional opposition between dialogue-orientated and critical approaches. These authors refer to the theory of communicative action of Habermas (1987) (which is traditionally used by dialogue-orientated authors) to demonstrate these complementarities. We saw above that Habermas identifies three main types of actions: the instrumental action, the strategic action and the communicative action. The strategic management of power asymmetries (recommended by critical authors) actually comes within Habermas' strategic action. Such strategic action attempts to remove obstacles to equity (obstacles that Habermas calls 'communication distortions') and to promote the emergence of collective learning based on communicative action (what Habermas calls an 'ideal speech situation'). Moreover, while dialogue-orientated stances are usually associated with social learning theories and critical stances with negotiating theories, Leeuwis highlights the sterile nature of this opposition. Indeed, researchers working in the field of negotiation usually make a distinction between distributive and integrative negotiation processes (Follett 1940; Leeuwis 2004; Carnevale 2006). In the first case, the stakeholders confront each other to decide how they could 'share the cake' in a zero-sum game. In an integrative process, the stakeholders instead reformulate the problem to 'make the cake bigger', paving the way for positive-sum games. Such a process is more demanding in terms of cognitive change as it calls for the stakeholders to consider the interests and values underlying their initial positions. Leeuwis (2004) argued that collective learning is needed to favour such integrative negotiation processes, once again stressing the complementarities between dialogue-orientated and critical stances.

Making One's Position Explicit to be Able to Question its Legitimacy

We saw above that there are several different ways to view the legitimacy of a participatory process. This is primarily linked to the objective of the process. When the objective is to produce knowledge, priority is given to determining its scientific legitimacy (does the scientific community see the ComMod process and the knowledge it generates as valid?), while when the objective is to support a negotiation or decision-making process, priority is more to determine its 'social' legitimacy (in the eyes of the stakeholders of the society in which it takes place). Furthermore, the need for 'social' legitimacy will differ according to the goals sought: it will be stronger, for example, in the case of a process aimed at supporting a decision-making process than in a process aimed simply at sharing knowledge.

Thus the need for legitimacy may change over a ComMod process according to the intervention's objectives, and conversely, objectives may change depending on the legitimacy of the process. It is not uncommon to see ComMod processes that begin with low social legitimacy and initially aiming at 'modest' knowledge production and sharing goals, move, at the request of stakeholders, towards decision-support objectives that need higher social legitimacy.

Moreover, we saw that different positions with regard to power asymmetries correspond to various ways of thinking about the legitimacy. First, the intervention's level of social legitimacy can be linked to the existence of a request or a mandate (see Chap. 5 for a more detailed discussion on this point); it is the main source of legitimacy in the case of profile 1 (*laisser-faire*). In dialogue-orientated profiles 2 and 3, legitimacy is determined by the designer not taking sides for any stakeholder or point of view. Finally, in the critical profile 4, the process's legitimacy is determined by the participants accepting the assumptions and objectives of the process as legitimate. This calls for the designer of the process to make these assumptions and objectives explicit. However, while it is evident in the case of profile 4 because of its asserted non-neutral position, it seems that whatever the profile, ComMod designers should make their assumptions explicit. It is indeed one of the ethical rules of the ComMod Charter, which is expressed as follows: 'make explicit any implicit assumptions in the experiment: this is an objective in itself and the process involves developing procedures for making explicit such assumptions' (Collectif ComMod 2005).

Thus, even in low-interventionist stances such as dialogue-orientated profiles 2 or 3, researchers should state their position to participants. Getting the stakeholders to sit around a table and put their knowledge on an equal footing is not a neutral stance. Participants should be aware of these assumptions so that they know into what they are getting involved. By the same token, in profile 1, the origin of the mandate that gave rise to the process should be systematically clarified to the participants.

Following on from the above results and discussion, the methodological and conceptual enhancement of companion modelling can be pursued on the 'power games and legitimacy' theme. Indeed, the methodology chosen by ComMod researchers to increase the process legitimacy, which involves systematically stating the process' goals and assumptions to enable their validation or invalidation by the participants, raises three questions: (i) could and should all assumptions really be made explicit? (ii) do the stakeholders really have the means to validate or invalidate these assumptions? (iii) which stakeholders are supposed to validate the assumptions for the intervention to be seen as legitimate? Do all stakeholders need to validate the assumptions? How to deal with situations where only certain groups of stakeholders refuse to participate in the process?

With regard to the first question, it should first be mentioned that in certain situations, it will not be in the designer's interest to reveal all his assumptions to all stakeholders, as this may go against the objectives he is pursuing. It is not necessarily appropriate, for example, to fuel a conflict by stating what each stakeholder said about the other. How can the assumptions that should necessarily be clarified be distinguished from the others? According to what criteria? Besides, it is unrealistic to think that everything can be made explicit. Indeed, these assumptions include factors such as the designer's values, culture and tacit knowledge. In selecting the assumptions to be clarified, there is still a bias, which should in theory itself be made explicit. However, even if the clarification of all assumptions seems unrealistic, attempts should be made to go as far as possible. For this purpose, tools such as logbooks (see Chap. 8), in which the choices made and the events transcribed as the process proceeds, may be useful. It is indeed also through their methodological choices that a designer reveals its assumptions. Finally, a ComMod process is often covered by an inter-disciplinary team. Such teamwork may stimulate discussions regarding the team members' different

[8] Daré et al. (2007) analysed the specific role of sociologists in a team designing a ComMod process. In addition to the knowledge they provided on the social context, they 'questioned the match between the model and its social uses' (p. 111). They also supported a particular position: 'we need to go beyond the debate between commitment and detachment to stake a claim to a scientific, humanist and pragmatic commitment' (p. 113).

positions and assumptions with regard to the socio-political context, and thus facilitate the process of clarification of these assumptions.[8]

These concepts, such as equity, are also generally subjective and plural and would ideallyrequire prior debate to recognize these differences and reach a shared definition. Moreover, even if significant clarification efforts are made at the start of the process, the participants may not have the means to really understand all the details of the process in which they are invited to take part, and will in part discover them as it proceeds. While it is essential to allow stakeholders gradually to acquire enough understanding of the process' assumptions and thus increase their abilities to alter its course (see Chap. 10), the difficulty of such an undertaking should not be underestimated. To offset these difficulties, the implementation of a rigorous protocol for monitoring/assessing the effects of the participatory process on the participants and their perception of this process is essential[9] (see Chap. 8). Any reluctance or deadlock situation should be considered as opportunities for the designer to question the legitimacy of the process.

Finally, the third issue requiring further attention is the identification of the stakeholders needed to validate the process for the intervention to be seen as legitimate. This calls for questioning, on the one hand, the legitimacy of the choice of stakeholders considered to be representative of a virtual entity (e.g., the community, farmers, politicians, civil society, etc.), and on the other hand, the most effective procedure for identifying the key stakeholders to be included in the process. In particular, this raises the problem of particularly conflicting or coercive contexts in which certain groups of stakeholders, even though concerned by the issues addressed by the ComMod process, refuse to join in the discussions so creating a deadlock situation (see Chap. 5). In such situations, some participatory approaches adopt a strategic stance involving the use of mechanisms that place outside pressure on such stakeholders. However, such practices appear to go against ComMod's ethical framework. Does this mean that ComMod is not applicable in such contexts?

Conclusion

Faced with the need for ComMod researchers to provide better formalization of the positions they adopt with regard to taking account of the socio-political context, in particular, with regard to the power asymmetries inherent in intervention contexts, this chapter provided a rapid method for clarifying these positions in the form of a short multiple-choice questionnaire. When applied to ComMod researchers, this method revealed both the existence of contrasting dialogue-orientated and critical profiles within the network, as well as two major points of agreement within the

[9] This calls for continuing the research efforts undertaken in the monitoring/assessment methods during the ADD-ComMod project (see Chap. 8).

group. First, most ComMod researchers call for open-ended, adaptive positions and recognize themselves occasionally in dialogue-orientated approaches and sometimes in critical approaches, depending on the objective pursued, the process phase, how they see the intervention context and any changes in the power games during this process. They thus share the views of authors who consider dialogue-orientated and critical stances to be more complementary than contradictory. Second, although the various positions in which the ComMod researchers recognize themselves correspond to various ways of considering the legitimacy of participatory processes, they share the same methodological procedure when questioning and seeking to reinforce this legitimacy. This legitimacy is based on the systematic clarification of their assumptions to allow stakeholders to validate or invalidate them. Our analysis enables us to refine and interrogate this major characteristic of the ComMod stance. It is indeed unrealistic to think that everything can and should be made explicit. Therefore, ComMod researchers now have to face the following questions. How to select the assumptions that should be made explicit? According to what criteria? How to select the stakeholders who are supposed to validate the assumptions for the process to be seen as legitimate? All this is likely to fuel a debate that could lead to the drafting of a new version of the ComMod Charter far more explicit regarding the way to take into account the socio-political context of the intervention.

Chapter 7
Assessment and Monitoring of the Effects of the ComMod Approach

Pascal Perez, Sigrid Aubert, William's Daré, Raphaèle Ducrot, Natalie Jones, Jérôme Queste, Guy Trébuil and AnneMarie Van Paassen

Justifications and Assumptions

Why Is an Assessment of ComMod Needed?

The assessment of the effects of companion modelling is currently still a theoretical and methodological field under investigation. However, neighbouring fields of research provide relevant elements of reflection. For example, research on

P. Perez (✉)
SMART, University of Wollongong, Wollongong, Australia
e-mail: pascal.perez@uow.edu.au

S. Aubert
UR GREEN, CIRAD, Antananarivo, Madagascar
e-mail: sigrid.aubert@cirad.fr

W. Daré · J. Queste · G. Trébuil
UR GREEN, CIRAD, Montpellier, France
e-mail: william's.dare@cirad.fr

J. Queste
e-mail: jerome.queste@cirad.fr

G. Trébuil
e-mail: guy.trebuil@cirad.fr

R. Ducrot
UMR G-EAU, CIRAD, Maputo, Mozambique
e-mail: raphaele.ducrot@cirad.fr

N. Jones
University of Queensland, Townsville, Australia
e-mail: n.jones3@uq.edu.au

A. Van Paassen
University of Wageningen, Wageningen, The Netherlands
e-mail: Annemarie.vanPaassen@wur.nl

M. Étienne (ed.), *Companion Modelling*, DOI: 10.1007/978-94-017-8557-0_7,

integrated assessments aims to provide public policy decision-makers with relevant information for decision-making. This information generally arises out of an interdisciplinary process for consolidating, interpreting and communicating knowledge from various sources (Hisschemöller et al. 2001). In this context, Siebenhüner and Barth (2005) tried to determine under what circumstances the use of a model would be consistent with the objectives of a participatory approach. It should be stressed, however, that there is a marked difference between an integrated assessment using certain forms of modelling and the ComMod approach itself. Hisschemöller et al. (2001) considered that the aim of integrated assessment is to inform the parties involved without seeking to promote convergence of opinion. As discussed in Chap. 2 and in more detail in Chap. 6, the ComMod approach, by contrast, seeks to support the sharing of views, the consensual exploration of scenarios and, where appropriate, group decision-making. From an approach assessment standpoint, it thus seems necessary to go beyond analysing knowledge creation and learning processes and also focus on interaction and mobilization processes, or even legitimacy and accountability principles to provide an understanding of 'what happens' during a ComMod approach and what are its consequences.

Trying to establish an assessment protocol, which is both scaled to the implementation contexts of the ComMod approach and sufficiently robust to cover all the aforementioned analytical dimensions, is quite a challenge. Expecting to create a perfect protocol at the first attempt is just wishful thinking. The protocol created for, and used during, the ADD-ComMod project corresponds to a first draft of a more advanced assessment methodology, the initial stages of which are presented at the end of this chapter. Within this project, the objective of the assessment process is initially two fold:

- to estimate the impact of the approach in terms of mobilization and involvement of local stakeholders in group learning and decision-making
- to improve the ComMod approach's underlying methodology and theory.

The second goal probably influenced the structure and content of the assessment protocol presented later in this chapter. As each structure has its own constraints, it is likely that the proposed format limited the assessors' ability to answer the questions raised by the first goal as much as they would have liked. However, the comparative nature of the ADD-ComMod project (27 case studies) imposed a methodological rigour called for by the diversity of the case studies and the assessors themselves.

Theoretical Foundations

This involves jointly mobilizing the theoretical foundations for using the models and the participatory approaches used to structure the assessment. The question is to identify how and at what stage the combination of modelling and participation is meaningful. To build an assessment framework, it thus seems essential to identify the specific methodologies and tools used during a project, and highlight their ability to fulfil the expected functions and to achieve the desired goals.

This assessment approach draws on the work of Webler (1999) who proposed a 'craft-theory-dialectic' as a means of advancing the field of public participation. According to Webler, what is required is to implement a practice that can learn from theory and a theory that draws on practice. Similarly, the ComMod approach is characterized by an empirical richness, which, on the basis of successive approximations, can bring about a more widely accepted solution. According to Webler (1999), to advance the field of public participation it is important to go beyond the answer to the question of 'what works?' and look at 'why does it work?' and 'and how could it work even better?' The mainstreaming of these two issues provides a link between theory and experience on the ground. Indeed, the case studies to be assessed differ according to a number of criteria, but share the same theoretical principle, that is, companion modelling. This theoretical foundation makes it possible to compare how these assumptions behave in practice as experienced by both the designers of the approach and the participants themselves.

Before describing the background of the assessment protocol proposed in the ADD-ComMod project, consideration should first be given to other theoretical foundations used by some assessors, in addition to the common protocol. These contributions have sometimes been deemed necessary to provide a better response to the first assessment objective concerning the mobilization and group decision. In general, these contributions stress the need to implement an assessment during the course of the process so as to adhere as closely as possible to the participatory process itself. The assessment then becomes a reflexive exercise with the project participants and designers relating to both approach procedures and its purposes. The 'most significant change' method, proposed by Davies (1998) and Dart (1999), is briefly presented in Box 1.

Without denying the obvious interest of this approach, it is useful to recall that almost 80 % of the case studies were assessed in the context of an *ex post*

Box 7.1: Most significant change.

The most significant change method calls for the players on the ground (including researchers) to indicate their experience or observations about a change (positive or negative) deemed significant in a specific area (e.g. knowledge acquisition, perception of the other stakeholders, changes in behaviour or practices, etc.) during the process undertaken and explain why. After indicating their experience, the participants in a plenary session choose the issues considered to be most relevant in each of the areas considered to draw the research team attention to this specific point. The research team is then required to take action on this point and scale its approach in the direction desired by the stakeholders on the ground. This type of monitoring/assessment of the effects of the process seems appropriate for the ComMod approach and should be carried out iteratively in support of the successive cycles of activity;

According to the changes observed and identified, such an approach may lead the team to re-consider its objectives. Above all, it provides them with a better understanding of the effects of the group process on the various stakeholders, and allows them to explore and share the values and preferences involved. Finally, this technique allows the group to understand what is or is not obtained by the process, to clarify what is really wanted and accordingly to define its next steps to move in the desired direction. The benefits of the technique are as follows (Davies 1998; Dart 1999).

- It is not based on 'objective indicators' that may control or ignore the diversity of knowledge, values and interests.
- It does not measure or list general experience but instead seeks to identify what is exceptional, surprising and a source of inspiration.
- It does not normally define indicators, as in-process learning is more deductive than inductive; it is a question of learning from recent events and keeping an open mind.
- The analysis is carried out and the indicators are defined by the participants rather than by the researchers.
- Dense contextual descriptions about complex and uncertain processes are more informative than statistics.

Conventional assessment methods stimulate the ritual performance of tasks, while the technique gives rise to new ideas for learning and tailoring in relation to new areas of interest and changing situations.

After Dart (1999)

procedure, sometimes 2 years after the project concerned. An example of the common protocol enhanced by some assessors is presented later in this chapter.

Background of the Assessment Protocol

The proposed assessment protocol is based in part on the 'theory-based evaluation' paradigm, which focuses specifically on the identification of a project's rationale. It provides a deeper understanding of the functioning of a project and allows priorities to be assigned to the issues requiring further attention (World Bank 2004). This paradigm implies that the vast majority of projects observed are based on an implicit or explicit theory explaining how and why such a project should succeed (Curnan 1998).

An examination of the various case studies within the ADD-ComMod project should show how the ComMod approach and its results vary in different contexts. This comparison should make it possible to identify patterns and critical factors in the group-learning and decision-making process. The application context, that is, the visibility or urgency of certain biophysical dynamics on the one hand, and socio-political arrangements on the other, significantly influences the ComMod approach. It is essential, therefore, for the assessment framework to consider all ComMod activities and their effects in their context.

The first step in the process for creating a multi-site analysis framework is to identify a comparison subject and measurement units. Several authors have focused on the problems associated with the conceptual boundaries of scientific projects on sustainable development because of the complexity of the socio-eco-systems considered (Cash et al. 2006; Reed 2008). The limits of these projects are often vague in terms of their scope and duration. It is often difficult to know when a project starts and stops. Assessors need to have specific knowledge of the time, space and subject limits of what they are supposed to observe. The importance of this initial focus increases when the diversity factors between the case studies of the ADD-ComMod project are considered:

- the diversity of participatory modelling techniques
- the various problems and controversies considered
- the interdisciplinary nature of the projects
- the use of assessors from several disciplines and cultures
- the stages of maturity of projects, with some still under construction, while others have been completed for several years.

To clarify the purpose of our study, a question should be asked: what are the common points between all these participatory modelling projects? A first common point is the scientific approach. The assumption made is that all the cases studied share a set of theoretical assumptions underlying their procedural rationale. These assumptions are based on concepts related to participation, the use of mediation tools, and group-learning and decision-making. Such assumptions are interpreted within social, institutional, political and environmental contexts. It follows that the way the process is structured to a certain extent reflects this underlying rationale. A project includes sequences during which tools and methods are implemented in a specific context. The proposed assessment protocol is thus based on:

- the project context (including the goals to be reached)
- the process (methods and tools used)
- the underlying theoretical foundations.

In this context, the assessment seeks to understand why the approach designers used certain methods and artefacts, and then analyse how these assumptions were interpreted and translated at the implementation level. Common assumptions can be analysed in different contexts by comparing the experiences of both designers

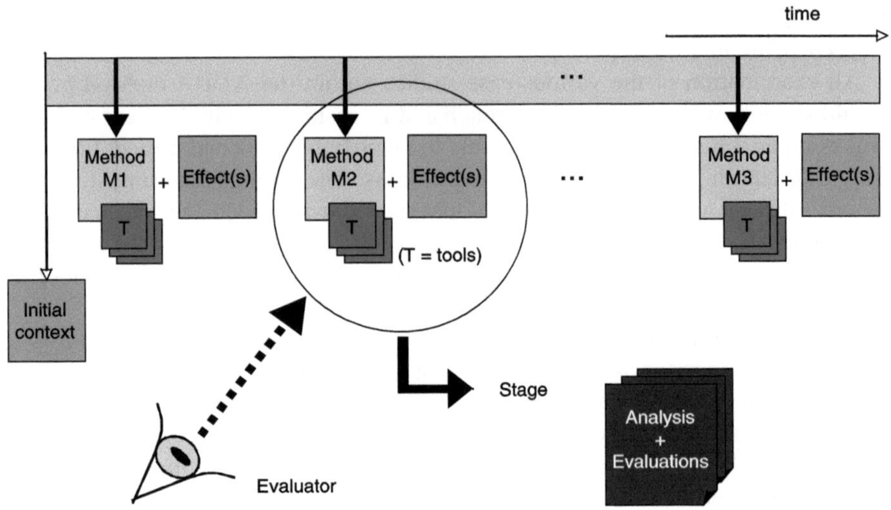

Fig. 7.1 The implementation of the ComMod approach

and participants. An assessment based on the theory is not limited to assessing a project's outcome and consequences. It also establishes a bridge between the two describing how in practice some activities contribute to certain achievements and impacts (Patton 1990). It is important to note here that the ComMod approach does not normally have an action plan. The action plan is actually designed and adapted along the way. This strategy allows certain unexpected side effects to be taken into account in the development of the process.

It seems difficult to establish a protocol that provides both an overview of many cases of highly diverse studies and the mainstreaming of contextual type adaptations. A theory-based assessment calls for the developing of a model—or an image—of the project rationale explaining how it works. Argyris, quoted in Patton (1990), refers to the 'professed theory' to describe what designers would like to see happen during the project, as opposed to the 'theory in practice', which describes what happens in reality. The proposed assessment protocol seeks to define the professed theory by working with designers and identifying the sequence of methods used and their expected effects (steps). The theory in practice is defined by questioning participants and designers to establish what actually happened in practice. One of the main objectives of assessment is thus to identify, characterize and explore these 'influential factors' to advance the field of companion modelling (Fig. 7.1).

Table 7.1 Methods for gathering data to define the Canberra protocol

Canberra protocol elements	Professed theory	Theory in practice
Designer questionnaire	Interviews with designers	Interviews with designers
Designer views	Project documents	Project documents
		Direct observations (ongoing projects)
		Previous assessments
Participant evaluation guide		Participant surveys
Participant views		Previous assessments
		Informal interviews (ongoing projects)

The Assessment Protocol

Description of the Assessment Protocol

The proposed assessment framework, known as the Canberra Protocol (CP), consists of two parts, the designer questionnaire (DQ) and the participant evaluation guide (PEG). The first part aims to capture the experience of the design team, including the theoretical rationale underlying the projet (see Box 2). The second part guides the information gathered on the experience of the participants. The theoretical rationale determines the professed theory as defined by Argyris (*in* Patton 1990). The experience of both the team of designers and the participants is used to rebuild the theory in practice. Table 7.1 summarizes the data-gathering methods that can be used to inform the CP.

The DQ defines the information to be gathered from the project team. It is entered by the assessor from the project documents and existing reports and completed during interviews. The project documents provide the assessor with a good overview and detailed information on the project context. Direct observations by the observer can also be used in the case of ongoing projects. The DQ includes two subcomponents: the 'context' and the 'process'. Figure 7.2 shows a diagram of this questionnaire.

Designer's Questionnaire

Context

Each project takes place in a social, political and economic context, which influences its design, functioning and results. Knowledge of these parameters is crucial to understanding the why and how of the project components. In the natural resources management field, the site's socio-political situation and physical configuration are key aspects of the decision-making process. The ComMod approach itself focuses on identifying the interactions between environmental and social dynamics. The DQs 'context' section looks at the relevance of the participatory modelling procedures with regard to certain socio-political and physical

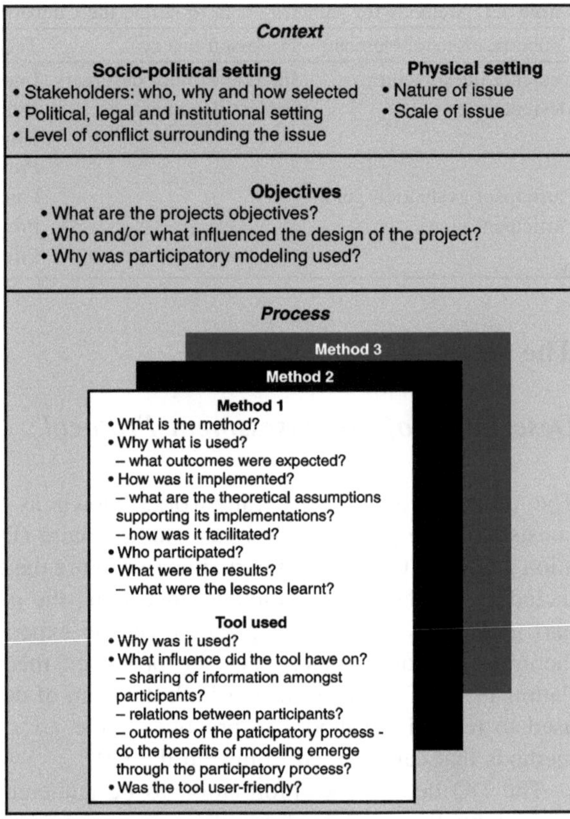

Fig. 7.2 Designers Questionnaire (DQ)

environment situations. The project's initial objectives are also important in informing the assessor of the project's initial intention. Finally, the assessment determines the relative influence of tools and methods to achieve these goals. To this end, the DQ also focuses on the project team's justification for the adoption of a participatory modelling approach. This contributes to validating the theoretical assumptions underlying the ComMod approach and the participatory model in general. The assessor may also make use of this justification in interviews with participants to update the relationships between professed theory and theory in practice.

Process

This part of the DQ provides a systematic record of the project's structure and sequencing of the methods used. The project team and assessor work together to restructure the succession of events as a series of methods (or steps). In some cases, this breakdown and the identification of a beginning and an end can be

Box 7.2: Applying the Canberra Protocol. Case study of Tarawa, Republic of Kiribati.

The project was developed on the Tarawa Atoll, capital of the Republic of Kiribati (South Pacific), in 2004–2005. The main goal was to resolve a conflict between the government and indigenous landowners over the management of groundwater stocks. These stocks are made up of small freshwater lenses and are very sensitive to any type of contamination. The project was initiated by a deep ethnographic study and progressed to the design of a computer role-playing game. Then, role-playing games involving landowner representatives and government agencies lead to collaborative scenarios of participatory management of water reserves. Nevertheless, the upheaval of contradictory government stands undermined the whole process.

The evaluation of this study case took place at the beginning of 2007, 2 years after the end of the project (*ex-post* evaluation). First, the evaluator that had been involved in two other evaluation processes, went through the documents and papers redacted by the leaders of the evaluation group. Then she adapted the DQ before interviewing the two main facilitators of the ComMod process. This preliminary study required 1 week of work, plus 5 h for the interviews. Sequencing the timetable of the project into steps (method + effects) permitted the definition of the collective key moments. With the agreement of the facilitators, the evaluator decided to focus her inquiry on the ethnographic interviews and the role-playing game. An open questionnaire was elaborated to cover the main topics of the DQ while taking into account the local socio-cultural context.

In February 2007, the evaluator spent a week in the Tarawa Atoll to interview 13 participants, with the help of the previous local responsible for the project. The landowners were interviewed at home and in their native language (I-Kiribati). Representatives of the government were interviewed at their office in English. Each interview lasted approximately 1 h. Despite the long delay after the end of the project, participants felt comfortable with the content of the enquiry, but the time span significantly influenced the accuracy of their judgements. The necessity of using an interpreter also limited the capacity of the evaluator to go deeper on specific key points.

The final evaluation report was published in September 2007, after several discussions with the facilitators of the ComMod process.

difficult. The project's rationale is based on this series of methods. The aim of this section is to identify the critical blocks that had a significant influence on the project workflow. Once these blocks are identified, their internal functioning can be focused on through global research issues. The blocks are also used to guide interviews with participants on the basis of the PEG (see below). Finally, any differences between activities carried out, theoretical assumptions and results (as perceived by participants and designers) can be identified and studied at block level.

The use of mediation tools (or models) plays a central role in a participatory modelling approach. This aspect of the CP is based on a similar initiative, known as 'harmonizing collaborative planning' (HarmoniCOP) (Mostert et al. 2007). An important part of this work includes the deployment of a framework for assessing the contribution of information and communication technologies as part of participatory processes. Some elements of the HarmoniCOP framework have been included in the CP because of common assessment objectives. The project team is thus required to explain why each method or tool was used and assess certain criteria arising out of the literature on HarmoniCOP (Maurel et al. 2007). These criteria are grouped into three topics as shown in Fig. 7.2.

The Participant Evaluation Guide

The PEG assists the assessor in establishing his investigation. It suggests what information should be obtained from the participants to understand their experience of the participatory modelling process. This PEG organizes the responses in the same structure as in the DQ, allowing the responses of participants to be compared with those of the designers. This allows the establishing of a consistent picture of how a participatory modelling initiative takes place from the standpoint of all participants. The survey conducted by the assessor, the formulation of questions and the conducting of interviews depends greatly on the social and cultural context and technical terms. The PEG thus only provides a guide for interviews and leaves it to the assessor's initiative to tailor the formulation of questions and the survey plan to the requirements on the ground (Fig. 7.3).

Due to its symmetry with the DQ, the PEG can provide information about what the participants think of the context surrounding the project: what were the aims of this project, who was involved, and why. In terms of procedure, the interview to be conducted is structured by the succession of blocks identified by the DQ. What is of primary interest here is to know what participants appreciated or not, and what benefit it was to them.

Description of the Assessment Guide

The assessor's guide (Aubert and Perez 2007) has three objectives:

- to provide assessors with the subjects and criteria to be used to guide their analysis of the documents and information gathered
- to ensure consistency between individual assessments and between the contractual requirements of the ADD-ComMod project

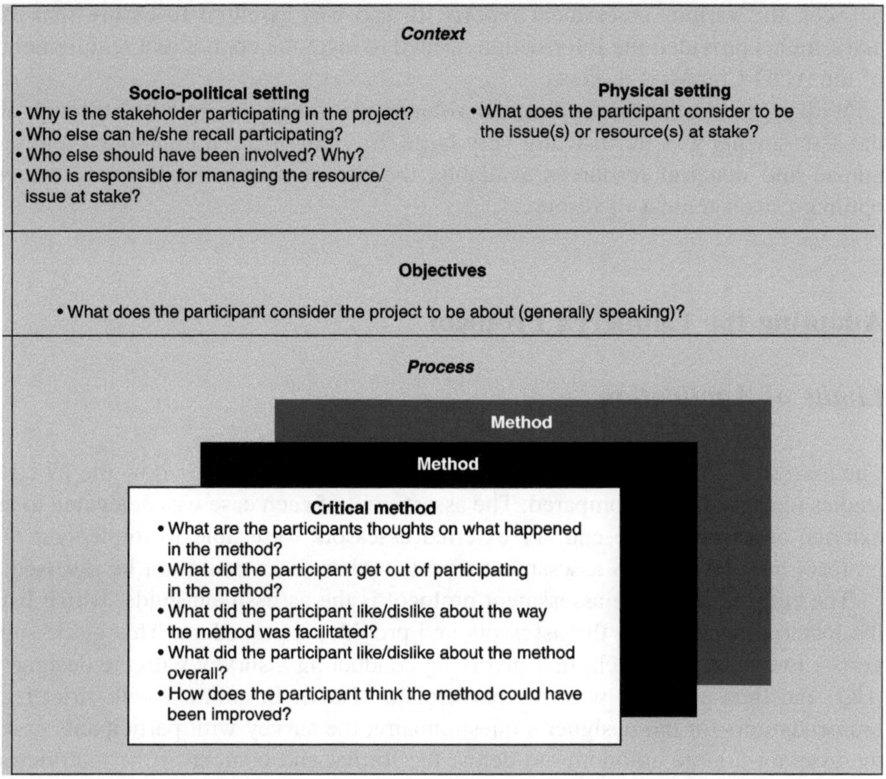

Fig. 7.3 Participant Evaluation Guide (PEG)

- to create a structure common to all the individual assessments.

With regard to the first point, it should be remembered that the CP, which acts as a methodological guide for implementing each assessment, does not contain any actual assessment criteria. The assessor's guide thus provides the key topics for using the information taken from the CP. This dichotomy between tools and criteria could frustrate the assessor if the wealth of material gathered could not all be put to good use at the level of the assessment subjects. However, it is important to note that all the material available for each project (CP, but also the description outline) can be reused later for further analysis.

With regard to the second point, one of the main challenges from the start of the ADD-ComMod project was to keep overall consistency as high as possible. Twenty seven projects dealing with various environmental issues in different countries and in various institutional contexts do not lend themselves easily to a unified analysis, especially as the assessments were divided among a dozen experts with very different profiles. Over and above a common methodology (CP), it, therefore, seemed essential to provide a focus document ensuring consistency between the various assessment reports. It also was essential to ensure that the assessments provided the information needed to meet the contractual requirements of the ADD-ComMod project.

With regard to the third point, the volume of information gathered through all the assessments was intrinsically very large. In view of the time allotted and the human and material resources available, the use of a common format ensured optimum management of resources.

Adapting the Canberra Protocol

Limits of Application

The assessment protocol (CP) provided a generic framework to allow the 27 case studies identified to be compared. The assessment of each case was delegated to an external assessor. In the end, 11 external assessors were able to implement the protocol and draw up 18 assessment reports. Other cases could not be assessed.

The entry point in the assessment protocol is the assessment guide, which lists the material available to the assessors and provides a procedure. This guide suggests a two-stage approach, first involving conducting a survey with the designers (DQ) and then a survey with the participants (PEG). In contrast with strict recommendations for the designer's questionnaire, the survey with participants gives the assessor a large autonomy to define the format and context of the interviews. With regard to the initial guidelines, the 11 assessors can be divided into three groups:

- three assessors followed the guidelines (seven assessments)
- four assessors modified the protocol and justified the changes made (five assessments)
- four assessors changed the protocol without justification (six assessments).

Finally, the assessor's guide requested from each assessor a critical feedback on the assessment protocol itself (Fig. 7.4).

An Example of Adaptation

One assessor (two assessments) considered that the aim of the ComMod approach should be taken into account when defining an appropriate assessment methodology. According to this assessor, the CP, based largely on integrated assessment methodologies, is not sufficiently tailored to assessing the ComMod approach as not enough importance is attached to measuring any improvement in communication and relationships between stakeholders and their capacity for collective

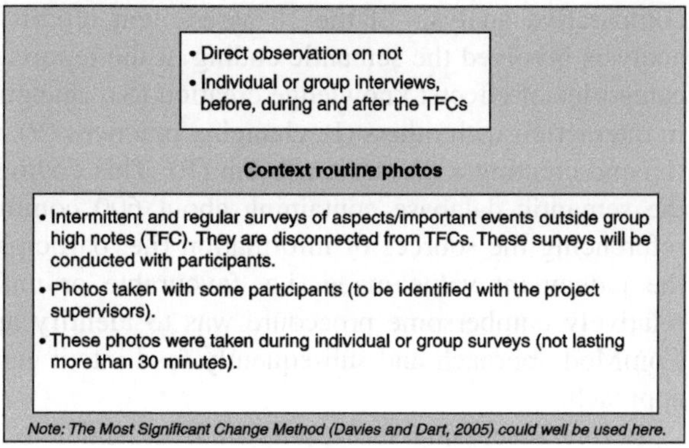

Fig. 7.4 Structure of a ComMod process monitoring/ assessment protocol

action. As a result, the assessor suggests using Leeuwin's (2004) learning and negotiation of innovations theory, and Vermunt's (1998) education theory, which outlines the cognitive, emotional and mega-cognitive dimensions, while incorporating those contained in the CP.

- Cognitive learning: learn about and discuss the aspects and dimensions of the problem examined, create a sense of urgency and identify possible solutions.
- The use of solidarity and group commitment principles: this requires a good understanding of the situation and individual values and interests, particularly by the facilitator, and the commitment to a group standpoint.
- The assessment of group legitimacy used to solve the problem in relation to other stakeholders, and through the use of the ComMod approach.

Results from Completed Assessments

Meta-Analysis Framework

From the 27 initial projects, 18 assessments only were completed. In accordance with the terminology used in the CP, the ComMod approach consists of a series of methods making use of possible artefacts. These methods sometimes correspond to group high points, the effects of which may be useful to assess. The set consisting of a method and its effects (positive or negative) is then called a milestone, as shown in Fig. 7.1.

The assessor's guide (Aubert and Perez 2007) thus asked assessors to identify these high points and their effects on the basis of the surveys conducted with designers and participants. Assessors were not asked to classify the risks and opportunities associated with these milestones on the basis of a pre-established interpretation framework. This operation was handled in the phase during which a comparative analysis of the 18 assessment reports was carried out. This meta-analysis involved the semantic coding of the reports based on the following seven categories of effects: knowledge creation (S), changing perceptions (V), assistance in interaction with others (I), changing practices (P), independence (A), legitimacy (L) and creating a discussion forum (E). This coding was carried out manually in the semantic database containing about 600 comments in six categories cross-referencing the sources of information (i.e. participant, designer or assessor) and the judgement value made (i.e. favourable or unfavourable). The aim of this relatively cumbersome procedure was to identify any potential invariants in the ComMod approach and subsequently feed into a generic conceptual model of the approach.

Before discussing the results, it is essential to note that the analysts had a crucial role to play in creating typologies. Indeed, as each assessor used their own nomenclature, it was essential to unify these different descriptions of the stages within a single typology. After some hesitation, seven categories were selected:

- identification workshop (AI)
- field surveys (ET)
- designing workshop (AC)
- modelling workshop (AM)
- model validation workshop (AVM)
- role-playing workshop (AJR)
- prospective workshop (AP)

Results from the Meta-Analysis

Assessment Approaches

This section provides an overview analysis of 18 assessments conducted during the ADD-ComMod project (Tables 7.2 and 7.3).

Of the 18 assessments implemented, most (13) were conducted after the project (*ex post*). Only four case studies were assessed along the way in real time. In half the cases, the assessment took place more than a year after the project. These two aspects limited both the quality of the information gathered from the participants and the ability of the CP to record detailed information about the process under way.

The assessments were conducted by 13 experts with contrasted profiles, six had fewer than 10 years of experience in their field and the other seven could be considered as senior experts. The experimental nature of the CP, together with the study contexts and their individual disciplinary benchmarks, called for some assessors to make significant changes to the proposed assessment protocol (10 cases) or to the structure of the report itself (five cases). Finally, it is worth

Table 7.2 Assessment method

Case study		4	1	3	24	16	17	25	26	27	6	7	31	32	8	9	10	18	19	Total
Type of assessment	Formative/ongoing	1				1	1		1									1		5
	Ex-post		1	1	1			1		1	1	1	1	1	1	1	1		1	13
Assessment method	Attending workshops	1				1	1		1									1		5
	Discussions with participants	1				1	1	1												4
	Using the designer's questionnaire	1	1	1	1	1	1	1	1	1	1	1	1	1	1	1	1	1	1	18
	Using previous assessment material												1	1			1		1	4
	Using a participant's questionnaire	1	1	1	1	1	1	1	1	1	1	1	1	1	1	1	1	1	1	18
Time between the end of the project and assessment	Real time	1					1		1									1		4
	<1 year		1			1		1					1	1						5
	1–2 years			1	1						1	1			1	1			1	7
	>2 years									1							1			2

Table 7.3 Assessors

Case study		4	1	3	24	16	17	25	26	27	6	7	31	32	8	9	10	18	19	Total
Experience	<10 years	1	1	1	1		1		1	1					1	1	1			10
	>10 years					1		1			1	1	1	1				1	1	8
Method in use	Compliance to the Canberra protocol	1	1		1	1			1	1								1	1	8
	Adaptation to the Canberra protocol			1			1	1			1	1	1	1	1	1	1			10
Assessor's report	Following guidelines	1	1	1	1	1			1	1	1	1	1	1				1	1	13
	Not following guidelines						1	1							1	1	1			5

7.4 Discussion forums

Case study		4	1	3	24	16	17	25	26	27	6	7	31	32	8	9	10	18	19	Total
Nature (highlights only)	Identification workshop	1	1	1					1		1	1	1	1	1	1	1			11
	Design workshop	1				1						1	1		1	1		1		7
	Modelling workshop		1	1		1	1	1					1					1		7
	Role-playing workshop		1		1	1	1		1	1	1		1	1	1	1	1	1	1	14
	Prospective workshop	1	1	1				1	1		1	1	1	1						9
	Model validation workshop			1				1							1	1		1		5
	Strategy workshop	1									1									2
	Field survey		1		1						1	1	1							5
Use of discussion forums	1–5 workshops	1			1					1							1	1	1	6
	5–10 workshops		1			1			1			1	1	1						6
	More than 10 workshops			1			1	1			1				1	1				6
	Participants withdrawing		1	1			1	1			1	1	1							7

recalling that six assessors had not signed the ComMod Charter, and could thus be considered as outside experts.

Creation of Legitimate Discussion Forums

The discussion forums created by the ComMod approach and to which the participants were committed are definitely the strengths of all the projects assessed (Table 7.4). The role-playing workshops appeared to be a particularly effective way of ensuring interaction between stakeholders. They also contributed to the mutual understanding of the issues and acceptable options for the management of natural resources.

The initial legitimacy of the key participants in nearly all the projects lies in their ability to inform other participants (16 cases), make use of resources (15 cases) or manage resources (16 cases) (Table 7.5). During the project, problems of legitimacy within the group appeared to be recurring but also limited in importance (11 cases). Put another way, the question of legitimacy seems to focus mainly on the absence of players considered, retrospectively, as being critical to the debate (17 cases).

The friendly (14 cases) and innovative (14 cases) nature of the methods used is a key factor in the cohesion of the groups created (Table 7.6). Conversely, the instrumentalization of the process (eight cases) or the limitation of technical options (eight cases) constitute barriers to cohesion.

In almost all cases (17), the methods proposed by the ComMod designer were considered as exogenous by participants and other co-designers (Table 7.7). The role-playing or simulation model creation phases were often seen by the commodian as periods during which the consolidation of the information is checked (10 cases). However, the innovative nature of the approach (10 cases) and the remarkable efforts made as regards transparency (12 cases) contributed to minimizing the negative impact caused by the two preceding points.

In short, the ComMod approach thus appears to be synonymous with quality in a participatory process allowing a group of stakeholders to formulate a shared vision of the issues and acceptable options for the management of natural resources.

Perceived Advantages and Limitations of the ComMod Approach

Table 7.8 summarizes the comments made by the assessors themselves and the comments gathered from designers and participants. The advantages and limitations of the ComMod approaches implemented were broken down into seven categories of results (see above). The measurement unit used was the total number of comments attributed to participants, designers or assessors during the meta-analysis of all the assessment reports (see below).

Table 7.5 Legitimacy of participants

Case study		4	1	3	24	16	17	25	26	27	6	7	31	32	8	9	10	18	19	Total
Type of legitimacy	Having useful information or expertise	1		1	1	1	1	1	1	1	1	1	1	1	1	1		1	1	16
	Belonging to the leading institution (client)	1	1			1	1	1			1	1		1				1		9
	Responsibility in resource exploitation	1	1		1	1	1		1	1	1	1	1	1	1		1	1	1	15
	Responsibility in resource management	1	1	1	1	1	1	1	1		1	1	1		1	1	1	1	1	16
	Representative mandate	1	1			1				1	1	1	1		1	1	1	1		11
	Involved in research-action initiative			1				1	1	1		1	1			1	1	1		9
	Respected personality			1	1															2
Questioning legitimacy of some participants	Frequent						1													1
	Rare	1		1	1	1		1	1	1	1	1	1		1					11
	Non-existent		1											1		1	1	1	1	6
Viewpoints of participants	At least on participant not considered legitimate			1	1		1													3
	Absence of participants considered legitimate	1	1	1	1	1	1	1	1	1	1	1	1	1	1		1	1	1	17
	Some participants are excluded	1					1	1												3

Table 7.6 Cohesion within discussion forums

Cas study		4	1	3	24	16	17	25	26	27	6	7	31	32	8	9	10	18	19	Total
Positive factors (according to participants)	User-friendly methods in use	1	1	1		1		1	1	1	1	1	1	1	1	1	1			14
	Neutral standpoint of facilitators					1				1				1						3
	Innovative and interactive methods in use					1	1	1	1	1	1	1	1	1	1	1	1	1	1	14
	Participants belong to pre-existing social networks					1					1	1			1			1	1	6
	Diversity of viewpoints						1	1			1	1	1			1		1		7
Limiting factors (according to participants)	Limited range of selected outcomes		1			1	1	1						1		1		1	1	8
	Intrumentalization by one participants		1		1		1	1		1			1	1			1			8
	Staff turn-over/renewal of participants			1							1									2
	Unclear objectives/ unexpected developments			1		1	1		1			1		1				1		7
	Inability to reach initial objectives										1							1		2
	Language, cultural or disciplinary barriers						1		1			1			1					4
	Short-lived forums										1		1		1				1	4

Table 7.7 Influence of project designers

Case study		4	1	3	24	16	17	25	26	27	6	7	31	32	8	9	10	18	19	Total
Interventions limiting participant's autonomy	Deciding on selection of participants	1						1			1		1							4
	Controling the collection of information	1					1	1	1	1	1			1	1					8
	Controlling the integration process	1				1		1	1	1		1		1	1		1		1	10
	Orientating the outcomes	1														1		1		3
	Imposing tools and methods	1	1	1		1	1	1	1	1	1	1	1	1	1	1	1	1	1	17
	Limiting training opportunities						1	1	1	1	1		1		1			1	1	9
Means used to mitigate designer's influence	Transparent processess to collective choice	1				1	1	1	1	1	1	1		1	1	1		1		12
	Innovative methods suited to all participants	1				1		1	1	1		1	1			1	1		1	10
	Clear definition of designer's roles and prerogatives			1																1
	Participants decide on the issues and outcomes to discuss					1	1	1				1	1	1	1		1			8

Table 7.8 Advantages and limitations

Case study		4	1	3	24	16	17	25	26	27	6	7	31	32	8	9	10	18	19	Total
Favourable comments	Creating knowledge (S)	5	8	3	5	3	4	3	3	2	9	13	5	5	6	4	3	5	2	88
	Modifying perceptions (V)	2	2	1	1	4	3	2	0	2	5	1	1	2	7	3	2	2	3	43
	Enhancing interactions (I)	6	6	4	0	5	1	4	4	2	4	1	2	2	2	1	4	1	1	50
	Changing practices (P)	3	2	0	3	3	0	1	0	1	0	3	3	0	3	5	4	2	2	35
	Securing independence (A)	0	0	0	0	3	2	2	0	0	0	2	2	0	5	2	3	5	3	29
	Enhancing legitimacy (L)	2	1	0	0	3	3	2	2	1	3	5	1	2	0	2	6	1	2	36
	Creating discussion forums (E)	6	1	0	1	7	7	6	4	0	12	6	6	6	2	0	4	2	1	71

(continued)

Table 7.8 (continued)

Case study		4	1	3	24	16	17	25	26	27	6	7	31	32	8	9	10	18	19	Total
Unfavourable comments	Creating knowledge (S)	3	3	6	0	3	5	6	2	0	2	4	1	1	3	0	0	2	0	41
	Modifying perceptions (V)	0	0	0	0	3	0	3	0	0	0	2	0	0	2	2	1	4	5	22
	Enhancing interactions (I)	0	6	6	0	1	1	0	1	0	0	1	0	1	3	1	3	3	2	29
	Changing practices (P)	0	1	0	2	1	0	1	1	2	1	1	1	2	1	0	0	2	1	17
	Securing independence (A)	0	0	0	0	2	1	3	1	1	0	1	0	2	4	3	1	4	1	24
	Enhancing legitimacy (L)	6	2	1	2	4	6	5	2	2	7	6	5	5	1	0	3	5	3	65
	Creating discussion forums (E)	1	2	2	1	4	6	3	1	2	3	11	2	5	1	0	1	1	3	49

Table 7.9 Products and outputs

Case study	4	1	3	24	16	17	25	26	27	6	7	31	32	8	9	10	18	19	Total
Computer model		1	1	1	1	1	1	1	1			1	1	1	1	1	1		14
Publication	1	1	1	1	1	1		1	1		1	1	1	1	1	1		1	15
Contractual or statutory agreement	1								1	1	1					1			5
Creation of institutions			1								1					1			3
New project/ training course		1		1	1					1			1			1	1		7

Logically enough, the creation of knowledge (88) and discussion forums (71) attracted a large number of favourable comments. Support for interaction with other participants (50) was also a strong point of the approach. Broadly speaking, the negative comments related to problems of legitimacy (65): legitimacy of the approach, tools, certain participants, the issue discussed or the solutions recommended. Paradoxically, the creation of the discussion forums (49) also drew criticism, often as a result of the frustration felt by some participants towards a participatory process that was limited both in time and in its ambitions.

It is sometimes difficult to establish with certainty the nature of the products generated through a ComMod-type iterative participatory approach (Table 7.9). Easily identifiable products, such as publications and computer applications (simulation models), may suggest that the ComMod approach is purely an academic exercise. In fact, the other three products identified (i.e. management agreement and formalizing a new group or project) concerned 11 out of the 18 case studies assessed.

In 14 cases, some participants expressed the desire to continue the ComMod approach on the same territory or in new situations (Table 7.10). However, the replication process was hampered by the major restriction of the ComMod approach, that is, the participants' lack of independence (16 cases). The specific issue of the contribution of the ComMod approach to sustainable development is discussed in Chap. 9.

Advantages and Limitations of the Participatory Methods Used

Table 7.11 shows the nomenclature used in the overview table describing the various methods used (workshops) and their effects.

This table provides a quantified (and thus attenuating) outline of the information contained in the meta-analysis semantic database. Indeed, it is essential to be able to return to the recorded comments and the value judgements made. Some of this material will be used in our analysis of the principal features of the overview table. The -measurement unit used corresponds to the number of recorded comments.

In fact, field surveys and modelling workshops are only regarded as high points in the approach by a small number of assessors. For this reason, we have not presented them in the overview table. An analysis of these results for the specific types of tools used (see Chap. 4 for their nomenclature) is presented in Chap. 8.

In view of the conventional tools and methods of the ComMod approach, it is not surprising to see that role-playing workshops (AJR) largely dominated the discussions (295 comments). This was a key point of most case studies during which the discussion forum was structured (E), interactions increased (I) and plural knowledge developed (S). This is above all the time during which atavisms are

Table 7.10 Process continuation

Case study		4	1	3	24	16	17	25	26	27	6	7	31	32	8	9	10	18	19	Total
Type of follow-up envisaged	Replicating on the same territory				1	1		1					1	1		1	1			7
	Replicating on an other territory	1										1			1			1		4
	Applying to other situations			1				1		1			1		1		1	1	1	8
Participants interested in follow-up	Local communities				1					1		1	1	1			1			6
	Institutions	1						1							1		1	1		5
	Experts			1		1										1			1	4
Participant's sense of independence	Comfortable with method and tools							1									1			2
	Not yet comfortable with methods and tools	1	1	1	1	1	1		1	1	1	1	1	1	1	1		1	1	16

Table 7.11 Nomenclature used to describe methods used and their effects

P/+: positive effect according to participant	P/−: effect to be improved according to participant
D/+: positive effect according to designer	D/−: effect to be improved according to designer
A/+: positive effect according to assessor	A/−: effect to be improved according to assessor

shaken and individual perceptions are broadened (V). Participants, designers and assessors confirmed the essential aspect: 'something' happens during the role-playing workshops. Paradoxically, some assessors seemed more critical than their sources of information and questioned the legitimacy (L) of the groups formed, most often regretting the absence of players regarded as legitimate, and, therefore, the quality of the discussion forum (E) created.

The second stage in order of importance (112 comments) was the design workshop (AC). This is an essential group phase during which the problem and system are considered jointly. Although this conceptual model can be implemented later in a simulation model or role-playing form, the essential aspect of the design workshop is to create this 'world view'. Each participant contributes to the common knowledge (S), the first interactions take place (I) and the discussion forum is established (E), sometimes with a few misunderstandings between participants and designers.

The prospective workshops (AP) formed a much more heterogeneous whole than the two previous cases. Indeed, they included both the exploration of management scenarios downstream of the two previous workshops, as well as the use of much more specific methods for case studies not normally called for in a ComMod approach. For example, the designers of case study 4 used a risk analysis matrix to explore as a group the implications of the various resource management modes. However, out of the 91 recorded comments, a balance can be established between favourable and unfavourable opinions, whatever the source of information. It seemed that the position of these workshops at the end of the project often drew criticism during an *ex post* assessment (13 case studies) from participants or designers frustrated at not being able to 'go further' in the process (Table 7.12).

The identification workshops (AI), although present in most projects, at least in the form of a launch workshop, were rarely considered by assessors as high points in the group approach (41 comments). However, the (rare) references underlined the importance of this initial phase in terms of the group's legitimacy (L) during its creation.

Overall, the same conclusions were reached in this workshop analysis as those drawn on the scale of the projects as a whole. The high points of the ComMod approach allow the creation of a discussion forum, facilitate interaction with other participants and generate group knowledge mostly in a relaxed atmosphere. However, various questions in connection with legitimacy were often raised in the assessments, which, it should once again be remembered, were largely conducted afterwards. The likely 'reconstruction' process, which often occurs in this kind of exercise, therefore, needs to be taken into account. However, the examination of

Table 7.12 Different workshops and their effects

	P/+	P/−	C/+	C/−	E/+	E/−	Total +	Total −	Total
Identification workshop									
Creating knowledge (S)	3		1	4	1	1	5	5	10
Enhancing interactions (I)	1		1		2		4	0	4
Changing practices (P)			1			1	1	1	2
Modifying perceptions (V)							0	0	0
Creating discussion forums (E)		1	4	2	1		5	3	8
Enhancing legitimacy (L)	5	3	3	1	1	2	9	6	15
Securing independence (A)						1	0	1	1
Total	9	4	10	7	5	5	24	16	40
Designing workshop									
Creating knowledge (S)	10	6	8	4	5	2	23	12	35
Enhancing interactions (I)	3	1	4		5		12	1	13
Changing practices (P)	2				3		5	0	5
Modifying perceptions (V)	8	1		1			8	2	10
Creating discussion forums (E)	8	5	3	7	4	2	15	14	29
Enhancing legitimacy (L)		6	5		1	3	6	9	15
Securing independence (A)	3			2			3	2	5
Total	34	19	20	14	18	7	72	40	112
Role-playing workshop									
Creating knowledge (S)	17	7	15	2	7	2	39	11	50
Enhancing interactions (I)	18	8	5	3	7	9	30	20	50
Changing practices (P)	10	5	3	2	5	11	18	18	36
Modifying perceptions (V)	16	5	6	1	3	1	25	7	32
Creating discussion forums (E)	23	7	10	4	6	12	39	23	62
Enhancing legitimacy (L)	3	8	2	7	4	15	9	30	39
Securing independence (A)	4	8	2	3	6	3	12	14	26
Total	91	48	43	22	38	53	172	123	295
Prospective workshop									
Creating knowledge (S)	3	4	3		3	3	9	7	16
Enhancing interactions (I)	3	1	2		3	1	8	2	10
Changing practices (P)	1	3		1		2	1	6	7
Modifying perceptions (V)	4	1	1				5	1	6
Creating discussion forums (E)	12	9		4	2	2	14	15	29
Enhancing legitimacy (L)	2	6	1	3		6	3	15	18
Securing independence (A)	1		1	1	1	1	3	2	5
Total	26	24	8	9	9	15	43	48	91

the four case studies assessed along the way did not indicate any significant difference on this point compared with the other case studies.

Finally, the limited number of comments on changes in practices (P) probably indicated the limits of the approach. Group decision-making requires more than just providing better information in a fairer way. This point is discussed further in the next section.

Improving the Current Assessment Protocol

Limitations of the Current Assessment Protocol

It should first be remembered that the assessment protocol had a two fold had to answer two questions:

- does the participation of stakeholders and researchers in a companion modelling process improve stakeholder involvement in group decisions concerning their management of renewable natural resources or not?
- if so, what are the components of the process (e.g. approach application contexts, combination of tools, etc.) that allow these results to be achieved? What improvements in the method are required to demonstrate its advantages and clarify its areas of validity?

Thus, the objective was not to assess the impact of ComMod approaches on the context, except from a sustainable development standpoint. The ADD-ComMod project was an exploratory phase in the assessment of a ComMod approach. The proposed assessment protocol (CP) is the result of a collective intention to understand what, in a ComMod project, produced a particular effect to improve and promote a given practice in support of sustainable development.

Two questionnaires were created: one for the project designers (DQ) and the other for the participants (PEG). Both questionnaires addressed social, cognitive, environmental and organizational issues. Any trends in the resource concerned or in the socio-political system were included in the participant forms. However, this information was not central to the processing of the resulting information. This last phase calls for a little more explanation. Why is this information not included?

The principle adopted in the protocol was to disaggregate the ComMod approach into meaningful sequences in terms of impact, and to compare the many methods and artefacts used in the 18 case studies. Applying ComMod principles, the CP is based on a simplified participatory approach model that defines the comparison framework: in a certain context, the designers organize, with participants, methods involving or establishing artefacts that promote interactions having effects on the participants and designers, which may even lead to a change of context (Fig. 7.1).

Such simplification was needed in order to compare a large number of case studies. This simplicity, however, has a cost. Such a model does not reflect important dimensions in a companion modelling process. In this model, socio-economic factors and power relationships were part of the context and considered to be passive. Only the methods and artefacts used by designers produced effects, with the context merely influencing these effects. In practice, these methods are organized in time and place. Hutchins (1995) considered that any action takes place in a historical and cultural context. In this context, role-playing, plenary sessions and other group exercises for building a shared representation of the situation were carried out in specific places, before and after certain events and in a cultural and historical context. The influence of these parameters may be difficult to mainstream in a comparison exercise but become relevant in the monitoring and accurate assessment of each case study.

Furthermore, the linearity of the sequences of the theoretical model used does not reflect the complicated, multi-faceted and unpredictable dialogue processes (Beuret 2006). In case study 10, for example, one series of dialogue workshops gave rise to several action offshoots: a complaint was filed in court and followed up by some stakeholders; another group tried to involve an influential personality from one of the villages concerned. Meanwhile, agricultural activities continued. Some of these action offshoots may or may not have contributed, later on, to solving the initial problem of irrigation management. The metaphor used by Beuret (2006) to describe this phenomenon is a tree offshoot. This image is appropriate except that the tree branches considered may sometimes join up and merge. In some cases, the iteration, sequence and/or combination of several trends and events, one of which may be linked to the ComMod approach, is more useful in explaining the effects observed.

Another limitation of the current protocol is the separation of those involved in these participatory processes into two main groups, the designers and the participants. Both groups were interviewed separately using different forms and the assessment focused on the distance between the intentions of the designers and the perception of the participants. Only the effects caused by the designers on the participants were registered. The effects caused by the designers because of their participation in the approach were not taken into account. In the CP, the designers are not considered like all the other participants because of their involvement in the design of the participatory process; this distinction challenges ComMod principles.

Improving the Assessment Protocol

The results of the assessments conducted and their comparisons in the ADD-ComMod project now allow us to propose some pathways to improving the protocol. First, it is essential not only to focus on the participatory modelling process itself, but also on the context in which it takes place, and the eventual influences of one on the other.

The wide variety of possible situations in which the companion modelling approach is used and the wide variety of objectives invalidates the proposal of an assessment protocol based on a linear model of such an approach. This proposal, which was relevant in a comparative and exploration perspective, isn't suitable when it comes to monitoring specific case studies. On the basis of the experience of the 18 case studies assessed, we now propose the establishing of an assessment process that can be combined with the ComMod approach itself.

Such a mechanism should thus combine several dimensions: individual and collective dimensions; discrete and continuous gathering of information;

procedural and contextual influences. It should include stakeholders in the design and implementation of the protocol and also be flexible and scalable over time according to changes in the social and environmental system concerned and the issues involved.

In several case studies, the assessors were asked to assess the effects of a series of events or interactions that took place several years ago. In practice, some tools have proven invaluable in analysing the dynamics involved and their satisfactory or unsatisfactory outcomes. In particular, keeping a project logbook was the most effective monitoring activity undertaken by some design teams (seven projects). This logbook took the form of a computer spreadsheet or just a simple notebook. This logbook should list all the events in the ComMod process. For each activity, the logbook should indicate the date, duration of the activity, organizer, participants, activity coordinator in the case of a group activity, type of activity, purpose, place, media used during the activity, sequence phase, and products of this activity (including the meeting reports, the intermediate objects obtained and the language used as the latter may significantly influence the involvement of a particular partner). For intermediate products and purposes, a physical link (in the case of a notebook, model, etc.) or a virtual link (for digital files) to the location of these products should be stipulated for ease of access.

However, keeping a logbook raises questions. Each of the seven logbooks produced and analysed during the ADD-ComMod project were kept by a single person, often the head of the design team. Many events in a ComMod process, however, happen in parallel or include only subgroups of stakeholders. To overcome this difficulty, an alternative system for recording events should supplement the logbook. Consideration could, for example, be given to a moral contract binding the stakeholders and project supervisor to setting up a simplified reporting system in the form of summary sheets, which would at least have the merit of briefly recording these aspects at regular intervals. This would provide regular 'snapshots' of the context and aspects and events that the participants considered important to record. The compilation of these snapshots taken from various viewpoints would provide a rich picture of events in and around the project, which can then be used in a more conventional *ex post* assessment.

In practice, the ComMod approach is structured around events bringing together participants, designers and observers. Role-playing, participatory modelling, discussions and other group activity sessions are organized during these events. It is essential to effectively assess these major events.

During the assessments carried out as part of the ADD-ComMod project, reports, minutes and verbatim accounts have proved invaluable. Written during, or immediately after events by a designer or an outside observer, these documents reflect the discussions, arguments, positions, alliances and power relationships between stakeholders, the technical choices made, the representation components considered, etc., or the minor events that punctuated the organization and the workflow of these major events. Several gathering levels and methods may be considered, but in all cases, the information must be gathered before, during and after the group interaction phase. Direct observation, whether by participants or otherwise (Arborio and Fournier 1999), is one of the preferred survey techniques. It requires observer training and the setting up of an explicit observation chart that should be the same throughout the process, even if the observer is changed. Semi-structured individual or group interviews should also be conducted with participants to complete the data. Such interviews can be used to assess the learning achieved and changing perceptions in the social and environmental system considered. On the basis of a sample of participants, we thus recommend that three series of interviews should be carried out; the first before a major event, the second just after and the third some time later. A period of 3 weeks is proposed to allow for events to be discussed calmly. The topics discussed during these interviews should first be tailored locally to make sure that they are suited to the social and cultural characteristics of participants.

(i) Key aspects of the event.
(ii) Technical learning about the issue considered and the socio-political context.
(iii) Learning about personal and other participants' opinions and values.
(iv) Ability to think and act as a group.
(v) Appraisal of the methods implemented.
(vi) Development of group skills.
(vii) Asset sustaining activities.
(viii) Actions initiated and considered when returning to actual situations.

The information gathered at the end of these interviews should also be archived in the logbook. The compilation of these items in the logbook may be made available to an assessor to allow him to carry out a more conventional *ex post* analysis. It is considered that this will attenuate any limits observed with the current version of the CP. A systematic comparison of the logbooks for the various case studies should also be carried out to continue to critically examine and capitalize on the approach. Furthermore, it also seems desirable to include the participants themselves in drawing up and updating these logbooks to tackle the principle of post-normal research as defined by Funtowicz and Ravetz (1994).

Questions About the Implementation and Use of the Assessment

Carrying out the assessments as part of the ADD-ComMod project has led to the formulation of new questions about conducting and using an assessment.

The first of these questions examines the relationship between the assessment and the ongoing participatory process. We believe that the results obtained during the interviews, the effects highlighted and the changing context can be mobilized as follows.

- to tailor the orientation and running of the process, in accordance with the needs and wishes of participants
- to improve current procedural aspects of the process (model, role-playing, logistics)
- to enrich the scientific understanding of the learning process (how to link the changes made to the project as a result of the results of ongoing assessments).

Another question raised during the project concerns the relationship between the designer of the activities, the project leader, the observer and the assessor. In practice, it is very difficult to coordinate the assessment of a major event directly with participant observation. While the conducting of surveys by the designer may solve certain obvious logistical problems, they may also give rise to a bias because of its 'identification' with workshops. The monitoring/assessment process that we have proposed must necessarily involve several people in a coordinated activity. How can the independence of observers be reconciled with the coordination of the designer's work? Should the non-neutrality called for in the ComMod Charter be applied to the assessor? What selection procedure should be used for an assessor who is both outside the project but has sufficient knowledge of the issues? What consideration should be given to the views of participants, which are sometimes contradictory to the ideas of the assessor and researchers?

Chapter 8
How do Participants View the Technologies Used in Companion Modelling?

Nicolas Becu, Pierre Bommel, Aurelie Botta, Christophe Le Page and Pascal Perez

The companion modelling approach mobilizes a number of tools to establish a representation of a given system. The most commonly used tools are simulation tools, principally agent-based computer simulation tools (virtual agent-based simulation model—VAM), role-playing games in their various forms (human agent-based simulation model—HAM and computer-assisted human agent-based simulation model—cHAM) and hybrid agent-based simulation model (HyAM) tools, which are hybrids of the two previous tools (Chap. 4 gives a detailed description of these four tools). However, other tools are also used to establish a view of the system and consider possible scenarios, such as spatial representations that may be based on various technologies (e.g. map, three-dimensional modelling, GIS, etc.) and knowledge representation diagrams (e.g. ARDI, systemic, UML diagrams, etc.). Some of these tools are based on computer technology and others only use paper and pencil, but all of them involve a set of standards and codes (known as the tool formalism) to represent a situation.

The participatory framework in which these technologies are used raises many questions. Is the tool formalism suited to the cognitive framework of users or

N. Becu (✉)
AGILE, Université de La Rochelle, La Rochelle, France
e-mail: nicolas.becu@univ-lr.fr

P. Bommel
UR GREEN, CIRAD, Brasilia, Brazil
e-mail: pierre.bommel@cirad.fr

A. Botta · C. Le Page
UR GREEN, CIRAD, Montpellier, France
e-mail: aurelie.botta@cirad.fr

C. Le Page
e-mail: christophe.le_page@cirad.fr

P. Perez
SMART, University of Wollongong, Wollongong, Australia
e-mail: pascal.perez@uow.edu.au

M. Étienne (ed.), *Companion Modelling*, DOI: 10.1007/978-94-017-8557-0_8,

participants? How do they interact with the tool? What are the effects of using the tool on the group of participants? Does the tool allow participants to deal with the question raised? Several chapters of this book provide answers on how to use these tools with stakeholders (Chap. 10) or explain the technical and methodological specificities of tools and combinations of tools (Chaps. 4 and 11). This chapter discusses the use of tools as seen by the participants, that is, the perception they have of how easy it is to use the tool (ergonomic design) in a given experiment, its advantage for the group and its ability to stimulate thinking about the system represented.

To answer these questions, our analysis is based primarily on the assessments carried out in the ADD-ComMod project (Chap. 7) of 18 case studies involving the ComMod approach, each using different tools according to the phases of the approach. However, as these assessments only partly answered the questions posed, we made use of other assessments to support our analysis.

The second section of this chapter describes the various tools used in the companion modelling approaches and the technological diversity on which they are based. The following section defines the framework used (i.e. ergonomics, effect of the tool and ability to encourage thinking) for the analysis of participants' perception of the tools used in participatory workshops. The following three sections present the results of this analysis for the three primary types of tools used, that is, simulation tools, diagram construction tools and spatial representation production tools. In the last section, we assess the strengths and weaknesses of various technologies and draw conclusions about their complementarities and the possible combinations of tools in view of the perception that participants have of them. In this section, we also examine how participants view their ability to be autonomous in using those tools.

Tools Used in the Participatory Mode

Before analysing how they are viewed by their users, we describe the type of tools involved and how frequently they are used in group situations in the ComMod approach.

Variety of Tools Representing Technological Diversity

Companion modelling approaches are based on a whole variety of tools to express and represent views of the system studied and its possible developments. The tools used with companion modelling stakeholders fall into three primary categories.

- The tools used to produce agent based simulations, including HAMs, which do not use any computer technology, cHAMs, which use computer technology for any function other than for specifying agent decisions, HyAMs, in which some agent decisions are specified by computer technology and VAMs, which are fully computerized.

- Tools for producing diagrams. The diagrams produced are generally based on the UML formalism, or on an entity/relationship-type formalism, such as the diagrams of the ARDI method, or on an *ad hoc* formalism based on a systemic approach. All diagrams are used to represent a system's entities, processes and interactions. The tools used to produce them are either manual (paper and pencil, marker board) or computerized (graphics software).
- Tools for producing spatial representations. These tools are used to locate in space the system entities, processes and system changes. There again, the variety of tools is based among other things on a diversity of technology. Some tools are based on computer technologies such as GIS, while others only use paper and pencil or a marker board to configure manually a spatial representation, and another type, known as a participatory 3D-modelling, is based on three-dimensional scaled relief model of a territory.

A last category of tools sometimes used in participatory mode is pseudocode a kind of programming tool. Pseudocode is a language close to natural language, which can be translated directly into computer code, but which calls for a stabilized lexicon.

The analysis in this chapter is primarily based on the assessments of participatory workshops held with the ADD-ComMod project. However, the assessment framework and the available results do not always distinguish the simulation tool from its media. Therefore, we were obliged to analyse the simulation tools independently of the spatial medium used, even though we realize that this introduces a bias. The tool, defined here as an object allowing a certain activity to be accomplished more effectively than without it, should thus be distinguished from the medium that is an object designed to impart information. Simulation model media and, in particular, spatial media, are covered in Chap. 4 and will not be discussed any further in this chapter. Let us just remember that 95 % of the time, simulation tools are based on spatial media that are also based on a technological diversity (e.g. game board, computerized cartographic interface, etc.), which affects the perception participants have of the use of the simulation tool.

Frequencies of Use of Tool

All companion modelling approaches with the stakeholders of the system represented take the form of a series of workshops each aiming to achieve a particular objective towards representation sharing. While some of these workshops do not involve any tools, most make use of one or more tools acting as intermediate objects for sharing representations between stakeholders. Here we focus on this second type of workshop.

For the purposes of analysis and because of the assessment system used in the ADD-ComMod project, two simplifications were made in dealing with these participatory workshops. First, in cases where several tools were used in one workshop, only the tool viewed as the primary tool by the assessor was considered. A workshop initially organized, for example, around the construction of diagrams and then on the basis of simulations carried out using a VAM, will be coded and analysed later in this chapter either as a diagram construction workshop or as a VAM simulation workshop, according to whether the assessor selected one or the other as a primary tool. The second simplification concerns the sample of workshops selected. The assessment system only included workshops considered by the assessor as collective key moments in the implementation of the approach. Some cases in which tools were used were not included in the analysis either because they were not regarded as collective key moments, or because they took place outside the group interaction.

On the basis of this analysis method, we reviewed the frequency of tool use in the cases in which companion modelling was implemented (Fig. 8.1). Role-playing games in its HAM or cHAM form alone represented half of the case studies. VAMs were used in one application out of five, while HyAM tools, which are hybrids between the two previous categories, were used in 5 % of cases. The other tools used during collective key moments were mainly group diagram production tools (19 %) and more rarely, spatial representation production tools (5 %) or pseudocode (2 %).

Due to the simplifications made by the assessment system outlined above, the statistics shown in Fig. 8.1 do not reflect the importance attached to spatial representation production tools in companion modelling approaches. Indeed, in many cases assessors did not include participatory mapping, 3D-modelling or participatory GIS workshops in the analyses of collective key moments. The trivialization of such workshops since the 1990s and which, as stressed by Chambers (2006), are now among the most common type used in participatory rural appraisals, partly explains the lesser attention given by assessors to these workshops. In addition, spatial representations were sometimes constructed at the beginning of a workshop where the main tool was a simulation tool,[1] or outside strictly group workshops. The frequency with which mapping is used in companion modelling approaches is thus actually much higher than that shown in Fig. 8.1.

The relative use of the various tools is not independent of the modelling phase in which the group is located, that is, design, modelling, validation or scenario-building. As shown in Table 8.1, the construction of diagrams was primarily used in design workshops. In one case study, however, the construction of diagrams was

[1] This is, for example, the case of role-playing game workshops where the game board is first jointly constructed with the participants.

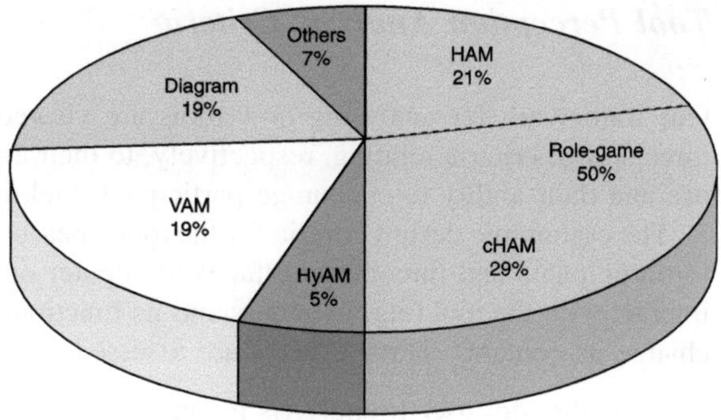

Fig. 8.1 Main tools used during collective key moments in case studies in the ADD-ComMod project

Table 8.1 Types of workshops during which tools were used (excluding spatial representation—production tools)

	Diagram	Pseudocode	HAM	cHAM	VAM	HyAM
Design (%)	88		11			
Modelling (%)		100			25	50
Model validation (%)				17	38	
Scenario-building	12		89	83	38	50

used to coordinate scenario-building workshops. Pseudocode, for its part, was exclusively reserved for modelling workshops. HAMs and cHAMs were primarily intended for scenario-building. In rare cases (three identified), these tools were used for model validation or design. VAMs for their part were primarily intended for scenario-building workshops. However, their mobilization in a participatory framework often involved the joint construction of the tool, which accounts for their prior use in modelling or model-validation workshops. Although the use of HyAMs was fairly similar to that of VAMs, the few workshops based on this type of tool do not allow such accurate statistics to be compiled.

For the reasons stated above, workshops involving spatial representation production tools could not be included in our statistics. However, the reports of individual cases of the ADD-ComMod project (i.e. outlines and logbooks) show that these tools were most often used in scenario-building or design workshops, particularly for building the spatial media on which the simulation tool was to be based.

Perception Analysis Framework and Available Data

This section mirrors the previous section on tools. It first describes the criteria for analysing how these tools are viewed by their users and then the information available for studying them.

Tool Perception Analysis Criteria

Our framework for analysing how tools are viewed by participants is based on three sets of criteria relating, respectively, to their ergonomics, the effects of their use and their ability to encourage participant thinking.

The ergonomic design criteria for the tools measure how accessible the tools are to participants and thus the interfaces (computer or otherwise) allowing users to interact with the tool (e.g. to understand its functioning and its results, and how to change its content). Three aspects are assessed.

- Is the tool user friendly (is it fun to use, what is its response time)?
- Is it easy to understand (is its formalism within the intellectual grasp of the stakeholders)?
- Can the participant handle it (change its contents)?

The second set of criteria assesses the effects on the group of using the tool as perceived by participants. For this purpose, five potential effects of use have been identified: (i) creation of knowledge; (ii) change in perceptions; (iii) help in interacting with others; (iv) change in practices; (v) creation of a forum for discussions between participants.

The latter criterion assesses the tool's ability to encourage participant thinking about the functioning of the system represented and its future. In particular, it assesses the link between the model and reality as perceived by participants as well as the tool's effectiveness in exploring changing trends in the reality on the ground (exploratory simulation).

It should be noted that the analysis criteria used are not independent of each other. For example, the tool's ease of understanding influences the creation of knowledge. Help in interacting with others and changing perceptions partly depends on the creation of a forum for discussions between participants. The analysis results clearly show these interdependencies.

Data Used for Analysing Tools

The data used for this analysis are the results of assessments of the ADD-ComMod project concerning 18 applications of the companion modelling approach conducted in various countries on five continents and in diverse socio-cultural contexts (Chap. 7). For each of these case studies, several workshops representing collective key moments were selected and assessed with participants. Thirty-three participatory workshops based on the participatory tools analysed in the context of the chapter were selected, thus allowing the participants' opinions of the tool used to be gathered. As indicated above, very little data were gathered on spatial representation production tools and for this reason these tools were analysed separately using other data.

Thus the available data essentially concern simulation and diagram construction tools. For both categories, more than 380 comments were gathered, which allowed us to analyse the ergonomic design of the tools, the effect of their use and their ability to encourage thinking. Although these data are qualitative, a meta-analysis carried out in the assessments of the ADD-ComMod project made it possible, in the case of tool effect criteria, to extract quantitative data as presented in the next section.

Having defined the materials and methods involved, let us now look at the analysis results on the basis of three main types of tool: simulation tools, diagrams and charts, and spatial representation tools.

Participants' Perception of Simulation Tools

Tool Ergonomic Design

Fun Side and Response Time

Participants in workshops, taking part in HAM-type role-playing games, viewed these tools as user friendly and fun to use; waiting times were generally short and were not attributable to the tool but rather to the workflow in the role-playing game. Indeed, these tools were generally designed with this in mind and, in particular, to ensure that participants were kept busy and active all the time. This was also true, but to a lesser extent, in the case of cHAM and HyAM tools, where participants often had to wait while data were entered or the computer performed other operations. The case of VAM tools was somewhat different because the participants could passively watch the simulation running on the computer screen. The tool itself did not encourage participant action and it was often up to the coordinator to ask participants to give their reactions either during the simulation or between two scenario simulations.

Ease of Understanding

Participants in workshops involving VAM tools were often held back by the tool's technicality, in that technical knowledge was required to interpret its results, and too much information had to be processed to understand its functioning and results.

In comparison, the HAM, cHAM and HyAM tools were easier to understand, as the participants were given a method for learning the rules and testing the model's operation, involving role-playing.

In addition, the computer interfaces used in all VAM tools, as well as in many cHAM and HyAM tools, could be difficult to read. In the case of the spatial representation interfaces that were most commonly used, the size of cells (in raster mode), the colour codes used as a caption and the changing states of cells after each simulation may be detrimental to some participants reading and accessing information. However, an *ex post* assessment of VAM workshops carried out with farmers in northern Thailand showed that three-quarters of the participants had a good understanding of the spatial interface and digital indicators, which could be read by nearly all participants after only two sessions (Becu et al. 2008).

Ease of Operation and Handling

This criterion corresponds to the participants' ability to operate the tool, that is, to explore and change its contents as part of the analysis. The ability to operate the tool provides participants with a better understanding of its functioning and results. Participants are then actively involved in operating the tool, unlike a passive use that only involves reading and interpreting simulation results proposed by a third person. Being able to use the tool autonomously is of course linked to its user friendliness, which also facilitates its transfer. This aspect is discussed in the conclusion. In this section, user autonomy is seen in the context of a participatory workshop and not with regard to any new project coordinator.

VAM tools were often seen by workshop participants as inflexible and only able to be operated and changed by the specialist. Some VAMs have user-friendly ergonomic interfaces, which allowed them to be operated by participants. Participants could, for example, start simulations by themselves, explore the various simulation results and even change the model's settings or rules if provided for; however, only the interfaced aspects of the model were accessible to the user. In other cases, changing the model required a programmer or a modeller to change the computer code.

HAM tools are just the opposite. The model's rules (at least the rules that relate to social entities and their processes) are oral rules given at the start of a role-playing game, but which can be changed at any time during the process. Moreover, workshop facilitators often want participants to change the rules as this indicates social adjustment. However, the possibility of changing the rules depends largely on the facilitator. The facilitator style will determine the extent to which the HAM tool can be operated by the participants.

Although these remarks are also true for the cHAM and HyAM tools, the increasing level of computerization of both these types of tools may, like their ease of understanding, make the tools less easy to handle and less user-friendly from the participants' standpoint. For instance, in a HyAM tool the participants cannot change by themselves the rules of the computerized social entities.

Effects of the Tool as Seen by Participants

The meta-analysis carried out in the assessments of the ADD-ComMod project made it possible to extract quantitative data on the effects of using the tools as seen by the participants. For this purpose, the responses obtained from the assessments were listed according to whether they expressed a strength or weakness of the tool as judged against one of the following five criteria: creation of knowledge (K), changing of perceptions (V), help in interacting with others (I), incentive to change practices (P) and creation of a discussion forum between participants (D). For example, a lack of knowledge creation or a reformulation of existing knowledge (without creating new knowledge) was coded as a weakness in criterion (K). A better understanding of the views of others and a change in attitudes towards others was coded as a strength in criterion (V). A tool that allowed the involvement and participation of all participants was considered a strong point of (D). In contrast, it was coded as a weakness of (D) when it restricted discussions because there were too many scenarios to be analysed or response times were too long during simulation phases between two discussions.

The percentage of the responses for this tool relating to a strength are shown in the light columns and those relating to a weakness are shown in the dark columns.

Figure 8.2 presents the results of this analysis and shows that the creation of knowledge and the creation of a forum for discussion between the participants were the two main effects of simulation tools according to participants (these criteria represent, respectively, 25 and 28 % of responses, all simulation tools considered). Help in interacting with others was in third position of the effects perceived by participants (21 %), while changing perceptions and changing practices represented, respectively, 12 and 14 % of the responses. However, these simulation tools have different characteristics.

Knowledge Creation

The HAM, cHAM and HyAM tools are well suited to the creation of knowledge. Participants saw them as tools, not necessarily for creating knowledge about a given system entity (although this is the case when the model illustrated a complex natural process, for example), but rather for highlighting knowledge, identifying constraints and changes, understanding interactions or behaviour and creating common knowledge. While VAMs were also recognized as good tools for creating knowledge, they had as many weaknesses as strengths in this area. Participants thus recognized that they acquired knowledge about the complex system (that they shared between themselves), but the difficulty of understanding the model and its rigidity (i.e. difficulty in using and changing the tool) were detrimental to the creation of knowledge.

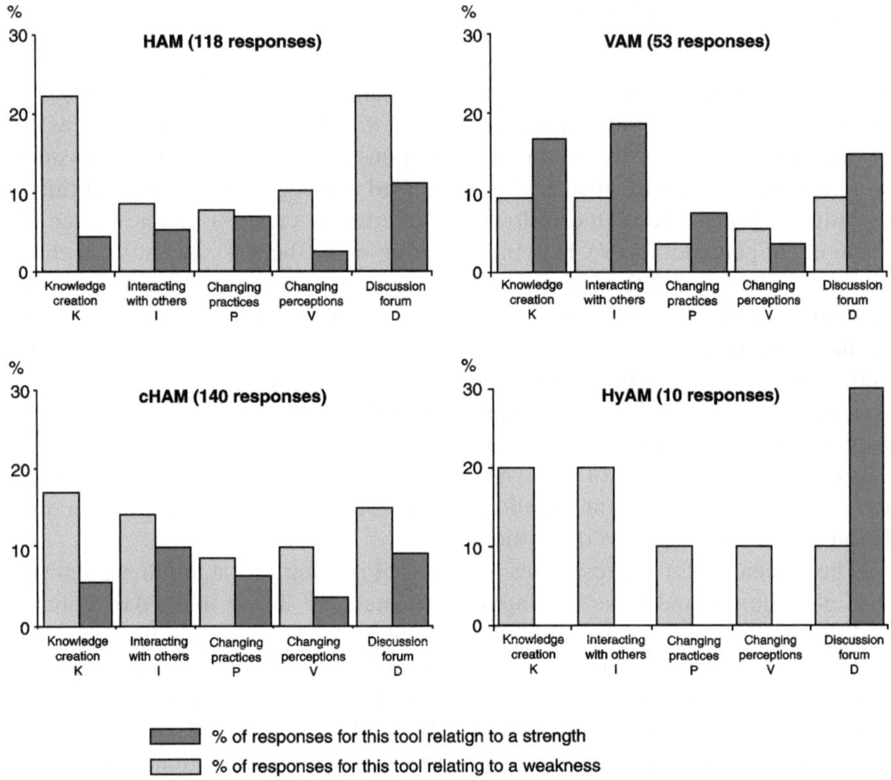

Fig. 8.2 The effects of simulation tools according to the workshop participants assessed by the ADD-ComMod project

Discussion Forum

The use of simulation tools in participatory workshops was always beneficial in creating a forum for discussion between participants. However, some tools were more beneficial than others. The results showed that the more tools are computerized (from HAMs to VAMs[2]), the more they had restrictions that were detrimental to the creation of a discussion forum. The HAM tool was clearly the tool best suited to discussing and following up cHAMs. The waiting time between two discussions (due to computer operations and the reading of interfaces) was the main restriction in workshops involving VAMs and more incidentally HyAMs and cHAMs. Similarly, the large number of parameters to be analysed was also a restriction in VAMs for this criterion.

[2] Due consideration should be given to the low number of responses in assessing the quantitative results in the case of HyAMs.

Help in Interacting with Others

The simulations tools each promoted interactions between participants (and indirectly with stakeholders outside the workshop) in their own way. They helped to build relationships with others (sometimes by reducing tensions between actors or by giving the weakest players a voice in discussions) and increased the opportunities for interacting. In the case of role-playing games and HyAMs, this involved gaming. In the case of VAMs, an explanation of the interactions between agents helped in establishing relationships between individuals. As role-playing games distanced participants from reality, it also allowed them to interact on aspects of the system that are difficult to state in public, such as corruption, for example. According to participants, the contribution made by simulation tools was to create a cooperative spirit within the group; they were seen as catalysts for triggering group decisions.

However, the use of VAMs and, to a lesser extent, computerized role-playing games, resulted in some participants withdrawing in certain cases due to the inflexibility of these models. These cases were particularly apparent when players disagreed with the simulation results or found they did not reflect their opinions (sometimes resulting in frustration, exclusion and even guilt). This could be avoided if it was possible for participants to change the game rules in an HAM tool. The lack of discussion time in the case of computer models was also cited as an obstacle to interaction.

Changing Perceptions

Although this was not the case most often quoted, the effect on changing perceptions was considered to be one of the strengths of all the simulation tools, whatever their implementation mode. It can be noted that what distinguished the HAM and cHAM tools was that participants mentioned fewer restrictions on change in perceptions than for the other two tools. In the case of VAMs, while spatial interfaces were generally welcomed, some participants did mention a certain mistrust with regard to the overly persuasive visual outputs, saying that they could arguably bias their judgement.

Changing Practices

Changing practices was the effect least often cited as a strength in simulation tools. In view of the results, it appeared that changes in practices were more often initiated in the case of workshops involving role-playing games. However, the little information available on this criterion was not sufficient to confirm this assumption. In addition, participants often stated their intention to change practices during meetings subsequent to simulation workshops (or repeated simulation workshops).

Ability to Encourage Thinking About the System and Its Future

Chapter 4 provides a reminder that, in the context of companion modelling, the validity of a model, as conventionally defined, is of little interest and it is better to assess its ability to stimulate user thinking and encourage creative action in relation to the issue involved. For this purpose, the simulation tool must be able to connect with reality; not by conforming reality but rather by producing results that make sense in relation to this reality. For this reason, the perception the participants have of the link between the simulation tool and reality is also taken into account in this section.

The assessments of workshops involving HAM or cHAM tools clearly indicated that these tools make participants think about the system represented. Similarly, the exploratory scenarios proposed by participants showed that the tool helped to stimulate their imagination and creative action. In addition, as the participants were generally able to identify the differences between the model and reality, this increased the model's legitimacy. However, some participants found the model too simple or thought that it ignored certain processes or options they considered to be important. These criticisms were encouraging to some extent because they showed that the participants were able to identify the limitations of the model. However, they also pointed to a weakness in such tools, namely that they do not always provide a grasp of the full complexity of a situation. Sometimes role-playing games may also not be well accepted by groups of participants. This may happen, for example, when hierarchic relationships exist between the different groups of participants. This dichotomy was clearly expressed in case study 24 (see Chap. 7), where some representatives of government agencies considered it was not normal to 'play' with serious issues, while the representatives of local communities found this quite natural.

HyAM tools have the same characteristics as the previous two tools, but differ somewhat with regard to their link with reality. These tools are based on agents, part of whose behaviour is controlled by the computer. Participants may sometimes place blind trust in the decisions of computerized agents without questioning their validity. This difficulty in challenging the actions carried out by the computer on agents is a recurring problem of VAMs. In addition, the understanding of difficulties related to these tools and their inflexibility may limit the model's legitimacy in the eyes of participants. In many case studies, the legitimacy of the VAM was thus acquired either by jointly building the model with participants (e.g. during workshops involving the ARDI method) or through a third party, who was neither a designer nor a stakeholder in the system represented, and who provided expertise in assessing the VAM.

Meanwhile, VAM tools showed a clear advantage over other simulation tools, namely the ability to provide a large number of exploratory scenarios easily in a short time (compared with other tools). This specificity, made possible by computerization, means that when the restrictions connected with its use are overcome, the VAM proved to be a particularly effective tool in encouraging participants to think about the functioning of the represented system and its future. For this reason, VAMs were used in many case studies following a role-playing game based on the same conceptual model. A role-playing game facilitated understanding of the conceptual model and the VAM tool allowed participants to explore a whole range of scenarios.

The Construction of Diagrams as Seen by Participants

According to the data gathered, the technological diversity found in diagram construction methods does not seem to influence the perception that participants have of these tools. This section thus presents the results for these tools as a whole, while making a distinction in some parts between the types of formalisms used.

Ergonomic Design

The construction of diagrams was not perceived as being much 'fun'. Its advantage, however, was that it did not keep participants waiting.

According to participants, the construction of diagrams did not really pose any problems of understanding, except that they had to learn the semantics involved. They may be of varying complexity depending on the formalism adopted. Participants in UML diagram construction workshops thus mentioned the high initial transaction cost due to the learning of the formalism. The graphic nature of the diagrams can facilitate understanding and there are many textbooks on learning formalisms aimed at audiences of different levels. In most applications, the workshop facilitators chose a diagram formalism whose complexity was appropriate for the audience.

The diagram construction tool was, in many cases, paper and pencil or a marker board on which participants draw by hand. Boxes, arrows or text then only had to be added or deleted to change the diagram contents. When the construction tool was a graphics software program, participants did not view it as being any less user friendly than the previous case.

Effects of the Construction of Diagrams as Perceived by Participants

Figure 8.3 shows in quantitative terms (using the same method as in Fig. 8.2) the analysis results of the ADD-ComMod project assessment on the diagram construction effects.

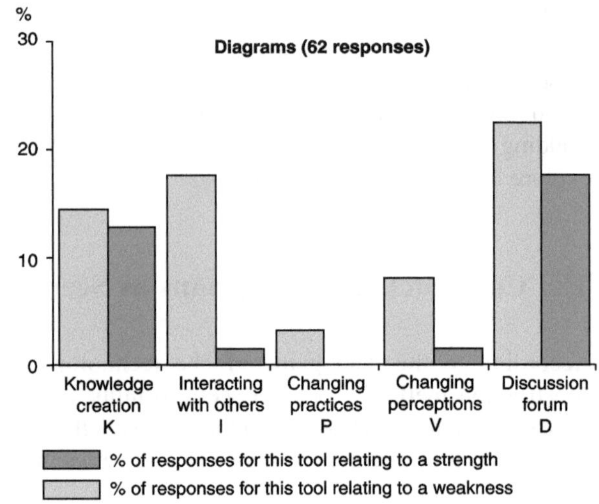

Fig. 8.3 Diagram construction effects according to the workshop participants assessed by the ADD-ComMod project

The workshops that involved the construction of diagrams were primarily seen as discussion forums between the participants (and across disciplines when specialists from various disciplines attended). This generated a group impetus that promotes interaction among participants. The explanation of the interactions between agents provided by the construction of diagrams helped in establishing relationships between individuals. However, the long and sometimes tedious design of the diagrams was felt by some participants as restricting communication between them (e.g. lack of time for discussion, difficulty of seeing things with hindsight).

The creation of knowledge is one of the strengths of constructing diagrams. Compared with other tools, it works differently, since it is a joint conception process for sharing, reorganizing and consolidating knowledge. However, the participants in these workshops deplored the inability of the tool to be interpreted for prospective purposes. In addition, the construction of diagrams, when carried out with subject specialists from different backgrounds, tended to change disciplinary perceptions. It was less clear when carried out with local stakeholders of the systems represented. Moreover, it did not show any effects on practices.

Ability to Encourage Thinking

The question of the tool's ability is somewhat different in the case of the construction of diagrams as it is often a design exercise carried out before the development of a simulation tool. The assessment criterion is then more about the tool's capacity to clarify the modelling assumptions, and diagrams do this effectively. In addition, the construction of diagrams can also prompt participants to think about the reality and stimulate creative action. However, it is a presentation mode more suited to thinking about the way the system works rather than about its future.

Tools for Producing Spatial Representations

Tools for producing spatial representations show a wide range of technological diversity and a wide variety of spatial representation formalisms. Those used in companion modelling are no exception to this rule. In this section, we take some of this diversity into account by adopting four main types of tools for the analysis (Fig. 8.4).

- Manual mapping of land use and land cover. This was achieved by the participants using paper and pencil or a marker board. It generally described land use and located the various system entities in space. In almost half the cases, the maps produced in companion modelling used a space grid that imitates the raster mode, which can be then imported in an agent-based spatial simulation or a GIS.
- Mapping of waterflows. The diagrams produced generally represented the hydrographic or hydraulic network of the territory studied in the form of links (river sections) and nodes (network confluences or inlets/outlets). It goes without saying that this type of tool was only used in applications where water is one of the system's important resources.
- Participatory 3D-modelling (which produces terrain mockups). This technique was usually used in case studies where the relief of the terrain affects the actions taken by the players. The production of the model with participants could then explain the relationships between relief and practices. In all the case studies listed here, the model produced was also then used as a game board for a role-playing game.
- GISs are the only spatial representation production tools directly involving the use of computers. Spatial representations are produced in raster or vector mode and are generally used to describe land use and land cover and to locate entities as in the case of manual mapping. In contrast to the latter, however, a GIS can overlap various layers of spatial information and is provided with sophisticated spatial analysis tools. The maps produced by GIS are often used later for producing computerized space media for simulation tools. Figure 8.4 shows that in almost half the cases, GIS was used. Yet, while the three other types of tools systematically involved participants in the production of spatial representation, the use of GIS was not always participatory. The dataset used did not allow distinguishing between participatory GIS and non-participatory GIS.

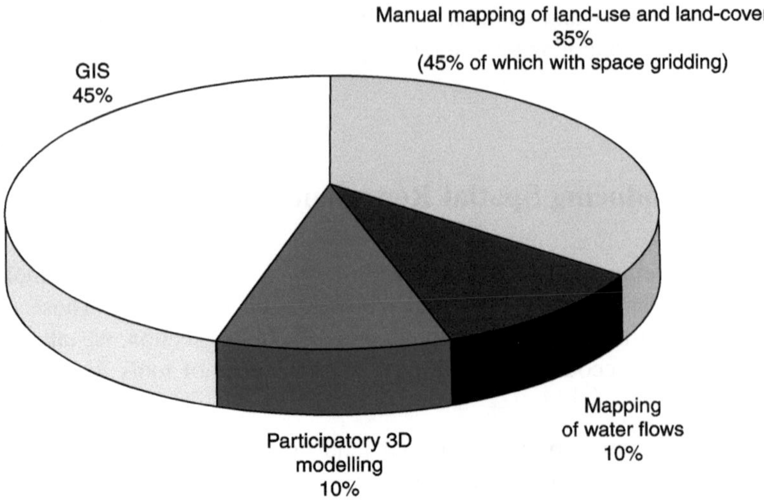

Fig. 8.4 Spatial representation production tools used during case studies in the ADD-ComMod project

As the assessments of the ADD-ComMod project did not produce a sufficient amount of data about these tools, their analysis was supplemented by information from the literature.

Tool Ergonomic Design

Manual mapping is relatively easy to understand and very user friendly. In this regard, Chambers (2006) wrote: 'the versatility and power of participatory mapping, the relative ease with which it can be facilitated, the fun, fulfilment and pride which people derive from it, and its multiple uses by so many stakeholders, have helped it to spread more than the others [other participatory rural appraisal methods]'.

He also indicates that 'flow diagramming' is just as ergonomic but more restricted in its use. Participants in a workshop assessed in the ADD-ComMod project indicated that the map was often interpreted differently from one participant to another, according to their own knowledge and experience.

In most cases (and unless prior training is given in such tools), the participatory use of GIS with local stakeholders calls for the presence of a facilitator who operates the computer software in response to requests from participants. As indicated by Chambers (2006), there is then a high risk of stakeholders being marginalized due to the attitude and behaviour of the facilitator who, through his skills in GIS technology, partially controls the production of the map. Participatory 3D-modelling is a relatively long exercise that can take several days. However, the elevation model, the high accuracy of these models and the ease in interpreting and understanding the mapping information they provide (Rambaldi and Callosa-Tarr 2000), are sometimes worth the time and effort involved.

Effects of Producing Spatial Representations

The primary effects of producing spatial representations are to allow the creation of a forum for discussions between participants and the creation of knowledge. The low-tech aspect of manual mapping and flow diagramming allows the greatest participation, as each participant can hold the pencil in turn. Producing 3D terrain models is more exclusive as it calls for proficiency in the process and the help of a facilitator. The omni-presence of the facilitator in participatory GISs means that these tools are the least suited to the creation of a discussion forum.

The creation of knowledge in the case of GIS is inherent in the process involved in the spatial representation joint construction and in reorganizing and consolidating the knowledge of participants. Although participatory GIS is the tool for gathering as much knowledge as possible, the quantity of information is sometimes detrimental to the creation of knowledge.

Link Between Spatial Representation and Reality

The ability of all these tools to provide a link between the spatial model produced and the actual space is undeniable. Spatial representations provide participants with outside world cues. The level of abstraction differs from one formalism to another, with chorems being the most abstract and 3D terrain models at the other end of the scale. Representing shapes in raster mode may seem more abstract than in vector mode and depends on the pixel size used. Yet, Barnaud and colleagues described three case studies involving the use of role-playing games where the game board had a different level of abstraction (Barnaud et al. 2006b). In all three cases, players could easily establish the link with reality and the abstraction level did not seem to have affected this criterion.

Implications for the ComMod Processes

The analysis, primarily focused on the results of the assessments of the ADD-ComMod project, enabled the participants' perception of the tools used in participatory workshops to be characterized. This section examines the implications of these results for companion modelling focusing in particular on the complementarity and the combination of tools and their possible transferability. We conclude by presenting analysis tools that are used more rarely in the companion modelling approach, such as the participatory video, 3D-visualization or pseudo-code language.

Complementarity and Combinations of Tools Based on Their Perception by Participants

It is clear from our analysis that VAMs should be intended more for exploring scenarios (in particular, because of their simulation execution speed), while role-playing games (HAMs and cHAMs) were perceived as more appropriate for creating a discussion forum. The later also provided useful opportunities for help in interacting with others. While intermediate HyAMs allowed interaction with others through role-playing games, waiting times were sometimes detrimental to the creation of a discussion forum. Their use, which is still rare, did not make it possible to go any further in terms of recommendations. Participants in workshops involving HAMs, cHAMs or HyAMs said that they had a good understanding of the tool. This played a positive role in creating knowledge through the use of these tools and for changing perceptions, which are both key criteria in the ComMod approach.

The main drawback of VAMs lies in their difficulty in being understood by participants, either because of the technical knowledge required or because there is too much information to be processed. This restriction, and the relatively long waiting times due to computer operations and, to a lesser extent, the tool's flexibility, is detrimental to the creation of knowledge and to facilitating discussions and interaction between participants. However, the case studies involving VAMs analysed in the ADD-ComMod project mostly included only one (or two) simulation workshops with the same participants for a given model.[3] A parallel experiment involved the setting up of a different protocol, whereby computer simulation workshops were repeated with the same participants using the same model being made somewhat more complex in each workshop (Becu et al. 2008). The results showed that after three sessions, most participants had acquired a sufficient understanding of the model to overcome any barriers to knowledge creation and the creation of a discussion forum.

It should also be noted that our analysis showed that computer interfaces (including spatial interfaces) seem to have a persuasive appeal to participants, which could be—detrimental to the objectivity of their judgements.

The specific features of simulation tools with, on the one hand, fully computerized tools that are difficult to understand but which have a high potential for exploring scenarios and, on the other hand, highly ergonomic role-playing games promoting discussions between participants, demonstrate the complementary of

[3] The workshops were repeated up to *n* times, but with different participants each time.

these tools and that their combined use can be particularly useful. Researchers who make use of companion modelling have understood this well. In many applications, VAMs are used following on from a role-playing game based on the same conceptual model. Role-playing games facilitate understanding of the conceptual model, and the VAM allows participants to explore a whole range of scenarios. If the exploratory simulation is an important aspect in the approach, such a combination should be promoted. If the objective is to facilitate discussions and interactions between stakeholders, however, role-playing games may be sufficient.

The construction of diagrams is clearly a design tool promoting the creation and sharing of knowledge. It is also relevant for facilitating discussions and establishing relationships between individuals, as it clarifies the interactions between agents. The relative complexity of the formalism of some diagrams (in particular, the UML formalism) may be an obstacle to participants' understanding of the tool; a formalism appropriate to the specific audience should then be chosen. Diagram construction workshops are thus highly complementary to simulation tools and may be usefully combined with VAMs to explain their contents and thus correct their defects.

Spatial representation production tools allow the creation of a forum for discussions between participants and the creation of knowledge. In companion modelling, however, the main advantage of these tools lies in the construction of a space medium used as a game board or computerized spatial interface for a simulation tool. The ergonomic design of the tool is largely due to the technology used. Chambers (2006) moreover indicated that the more the technologies that are difficult to understand and use are involved, the greater the likelihood of some participants being marginalized and other participants, who have a better grasp of the tool, tending to monopolize it and regard it as a source of pride and empowerment. This is also true for tools other than spatial representation production tools. In some socio-political contexts, the type of technology used to represent space is an important parameter to be taken into account according to the type of stakeholders involved. So in an application in a catchment area in northern Thailand, Promburom and Bousquet (2008) used GIS with government institutions and policy-makers, since such actors were more receptive to this type of technology. On the other hand, when workshop participants were farmers from the catchment area, the use of two-dimensional game boards was preferred.

Transferability of Tools According to Participants

In the previous chapter, it was seen that in 14 cases, some participants expressed the desire to continue the ComMod approach in the same territory or in new situations. In most ComMod approaches, however, the tools are not transferred. In the approach cases carried out, this request could have had a wide range of objectives such as, for example:

- to broaden the participant audience by repeating the approach, which often calls for knowledge of how to tailor the tools accordingly
- to broaden the feedback audience in handling not only the results but also the tools; this is particularly the case when these tools are used for teaching purposes
- to continue the approach without the logistic support of the initial modeller(s).

Depending on the objective sought, the methodological and ethical issues raised by such a transfer, vary. What is transferred exactly, the tools, the ability to build them or the ability to use them in a given companion approach? Should tools be transferred without the approach? Who should they be transferred to and with what consequences? Which resources are provided for this transfer (these transfers)? These issues in part give rise to problems in learning (Chap. 10), involvement in local power games (Chap. 6) and participant 'outscaling' (discussed in Chap. 11).

Although few participants raised this issue in the assessments of the ADD-ComMod project, their remarks showed that they do not feel they have the ability to be autonomous in coordinating a workshop based on the same tools. The more complex the formalisms used, the more conclusive this was, as seen in the case of cHAMs and even more so for VAMs. However, specific coordination capabilities also were often mentioned in the case of VAMs. Past experiments also showed that if tools are transferred, such transfers should be considered as far upstream as possible of the project, firstly, to allow go-between individuals to be trained in the process,[4] and secondly, to assess the potential impact of such go-betweens in better preparing for the transition with all the participants in the approach (this aspect comes under the general principle of transparency of project supervisors with regard to their approach). In addition, computer-based tools call for constant updating to ensure compatibility with operating systems and constantly changing modelling software (Meadows 2001).

Other Tools Used in Companion Modelling

Computer programming tools are also sometimes used directly with the actors. The primary tool in this case is pseudocode. Although it is written in terms based on natural language, pseudocode often seems somewhat obscure to participants and a certain learning curve is involved. When compared with graphic diagrams, its written form does not facilitate its reading. Due to its lack of user friendliness, pseudocode is generally not intended to be handled directly by participants. Workshops using this tool generally include a modeller or computer specialist whose role is to translate natural language (participant proposals) into pseudocode.

[4] The training issue is, however, hampered because stakeholders usually do not have the time to learn a formalism (Meadows 2001).

It happens that in some applications, stakeholders ask for highly realistic spatial representation as a basis for discussion, without being either too technologically sophisticated or too costly in terms of development time. 3D-visualization can be a good compromise. This tool is complementary to simulation tools and comes after the definition of prospective scenarios on resources and landscape evolution. When a scenario is established with the participants, the future state of resources and landscape is imported in a GIS, which is then used to produce a 3D-cinematography. Endusers can then use the 3D-visualization tool to fly over the virtual landscape and zoom to specific areas that they need to analyse in detail for the consequences of the various scenarios. This tool is especially efficient in creating a discussion forum between participants about natural resources and landscape evolution issues in mountainous areas (Gibon et al. 2004).

Other approaches using participatory video involve supporting local actors in the joint construction of audiovisual tools designed to model a problem affecting them. The making of videos, used as intermediate objects in the participant network, is a pretext for promoting discussion and moving forward as a group towards a shared representation of the situation. There is a wide range of participation procedures that involve the distribution of production functions (Colin and Petit 2008; Shaw and Robertson 1997).

The descriptive strength of the images illustrating or complementing the audio content explains why videos are often considered by participants as providing objective proof. In addition, the oral nature of the message (accessibility) and the reproducibility of the medium (mass distribution) make it a powerful mouthpiece (Lunch and Lunch 2006). Finally, the use of interviews can contribute to public legitimization of the views expressed (and thus alter the relative weights of the stakeholders in the local system), reinforce the value given to the video content and encourage the involvement of participants in its design.

While participatory videos do not provide for prospective simulation, their viewing does, however, promote the implementation of thought-provoking work (mirror video), or introspective work (Langlois 1995). Within the network, it thus makes it easier to view with hindsight the rhetoric of individuals and the interactions of players. This seems essential for the group recognition of the coexistence of different action rationales.

The participatory video can also create new discussion forums, both real and virtual (Colin and Petit 2008), giving rise to interactions between stakeholders when previously any communication seemed impossible for social reasons or geographical remoteness. It thus facilitates reconciliation and broadening of the network of participants (Snowden 1984).

It should be finally noted that the fun side of this approach is not unrelated to the success of participatory video approaches on the ground. Indeed, in areas where there is an overabundance of research or development aid work, the implementation of these innovative solutions often provides new impetus for actions with local stakeholders (Colin and Petit 2008; Dagron 2001).

Chapter 9
ComMod: Engaged Research's Contribution to Sustainable Development

Sigrid Aubert, Christine Fourage, AnneMarie Van Paassen, Pascal Perez, Raphaël Mathevet, Cécile Barnaud and Martine Antona

A Stance Justified by the Needs of Sustainable Development

The Emergence of the Sustainable Development Concept

The question of how to conciliate ecology, economy and society was first raised in the 1970s with UNESCO's Man and the Biosphere (MAB) programme. The term, 'sustainable development', a product of the growing awareness that ensued, was defined in 1987 as: 'development that responds to the needs of the present without compromising the ability of future generations to meet their own needs' (Brundtland 1988).

S. Aubert (✉)
UR GREEN, CIRAD, Antananarivo, Madagascar
e-mail: sigrid.aubert@cirad.fr

C. Fourage
UMR ESO-CARTA, Université Catholique, Angers, France
e-mail: cfourage@uco.fr

A. Van Paassen
University of Wageningen, Wageningen, The Netherlands
e-mail: Annemarie.vanPaassen@wur.nl

P. Perez
SMART, University of Wollongong, Wollongong, Australia
e-mail: pascal.perez@uow.edu.au

R. Mathevet
CEFE-CNRS, Montpellier, France
e-mail: raphael.mathevet@cefe.cnrs.fr

C. Barnaud
UMR DYNAFOR, INRA, Toulouse, France
e-mail: cecile.barnaud@toulouse.inra.fr

M. Antona
UR GREEN, CIRAD, Montpellier, France
e-mail: martine.antona@cirad.fr

M. Étienne (ed.), *Companion Modelling*, DOI: 10.1007/978-94-017-8557-0_9,

Sustainable development asserted itself as a response to a preoccupying global situation, notably the fragility of ecosystems and natural resources and the imbalance between extreme poverty and wealth. The search for solutions to these problems was part of a perspective allying 'sustainable resource management', 'social cohesion' and 'economic rebalancing'.

At the 1992 Earth Summit, Agenda 21, a set of action goals[1] and principles,[2] was established[3] to give form to sustainable development. Agenda 21 proposed an overall action plan to governments, development institutions, United Nations bodies, and independent groups active in all of the fields in which human activity influences the environment. Since then, the programme has been continuously discussed and refined within the framework of various international, national, regional and local agreements.

Consequently, the goals of sustainable development are relatively explicit and there is a consensus within the development and scientific community of United Nations member countries.[4]

- Social, economic, and environmental policies must be addressed in a spirit of synergy and with a long-term perspective. 'If threats to future quality of life are not anticipated, they will have irreversible consequences and will lead to a sharp increase in costs for society.' A message was thus sent to public policy-makers to assume responsibility for choices facing society here and now and for their future consequences.
- Environmental concerns must be taken into account in daily life as in public policy.
- Better sharing of material and immaterial resources must be foreseen. This concerns both natural resources as well as access to goods and services, the reduction of inequalities, and the development of individual and collective 'to do and to be' (Sen 1982) capacities. This assumes a patrimonial development approach.[5]

---

[1] Return human beings to the centre of sustainable development concerns because people have the right to a healthy and productive life in harmony with nature, notably through a fight against poverty respectful of present and future generations. Preserve global balances and environmental resources for long-term development by altering development modes and eliminating unsustainable production and consumption modes in favour of sustainable ones.

[2] The principle of precaution, the principle of integrating the environment into the development process, the principles of responsibility and international solidarity, the principle of paying for pollution and the principle of participation for a new governance.

[3] Declaration of principles concerning forests and the Rio Declaration on the environment and development.

[4] At the 1992 Rio Summit, 150 countries committed themselves to linking the challenges of economic and social development with those of the environment to create a more unified world preserving resources and natural environments. This commitment was reconfirmed at the sustainable development summit in Johannesburg held 26 August to 4 September 2002.

[5] According to de Bradt and Gadret (1998) a communal heritage is constituted by: 'the ensemble of objects and products (including the natural environments that were or were not exploited by man and therefore a kind of "product" for him) to which this collective group or a sufficient

- Tensions produced by unbalanced development must be reduced: sustainable development concerns social equity between generations and within the same generation, the fight against social vulnerability, and an improvement in access to goods and services.
- Sustainable development thus generally is considered as an 'operating concept that should enable the evaluation of risks, inform opinions and guide public action' (Aubertin and Vivien 2006). However, as long as sustainability remains difficult to define and imagine from an ecological, social or economic perspective, the means to achieve sustainability will be controversial and uncertain. In principle, the use of the concept assumes the existence of norms enabling the evaluation of sustainability or managing contradictions between the three pillars of sustainable development (ecological, social and economic): 'Yet these norms, particularly in the field of territorial development, cannot be defined scientifically or abstractly: while one may know approximately what the sustainable management of resources might be, one is incapable, for example, of defining an appropriate threshold in terms of an environment's "carrying capacity"' (Theys 2002).

The definition of sustainable development requires the examination of 'the effects of economic and ecological decisions on the social dimension, but also the effects of decisions taken within the social sphere' (Dubois and Mathieu 2002). This returns to the notion that the three pillars of sustainable development rest on an ethical (fairness) and political (geopolitical stability) foundation, which presumes individuals' access to the ensemble of goods and services, the reinforcement of people's personal and collective capacities, and fairness given available and transferable resources. The social sustainability of development, approached either from the angle of risk of social dysfunctions[6] within any society or from that of preventive solutions developed by social stakeholders to address such risks, is determined by the elaboration of 'principles of social precaution enabling the anticipation of risks and guiding social stakeholders in their undertakings'

(Footnote 5 continued)
portion of its members attach a value because what is involved are realities that testify to the identity of this collective group by establishing a temporal link between its past and present (witness to the past), and/or between its present and what it imagines for its future (witness to plans)'. Torrès (2002), specifies: 'the expression "ensemble of objects and products" is very broad: it encompasses objects and natural environments but also architectural monuments, works of art, sites and landscapes, as well as information that can exist in different forms. Local social networks, customs, oral traditions, significant collective memories… can furthermore be considered as non-material components of heritage. The patrimonial goods require an economic treatment that is different (existence value, analysis in terms of "support and services") than that applied to standard goods'.

[6] 'Biodiversity conservation favours a collective good, in principle open to all humanity, but the social cost and restrictions associated with it are not borne equally. A way of softening this negative effect consists of seriously considering instituting social compensation mechanisms that will compensate the regional society for the advantages conceded to the collective. The definition of the compensation for the losses borne shall be the focus of extensive negotiation between public bodies and the affected population' (Lima 2002).

(Dubois and Mathieu 2002). These principles of precaution or social prudence are understood within an ethic of 'responsibility' in relation to others whose goal is to guide and evaluate public policy decisions implemented within the framework of development strategies, and which aims to reduce inequalities in access to goods and services through redistribution policies.

The Social Legitimacy of Research Decisions

The emergence of the concept of sustainable development not only brought 'development' under debate but also generated new societal expectations from science. Science has been at once called into question and strongly solicited, notably in order to provide a rational foundation for this new vision of development.

In this context, the social legitimacy of research choices must be reinforced. 'Research efforts must take stakeholders' expectations into better account, develop surveillance methods and link these with procedures to define priorities and, above all, be more transparent.' It is necessary to 'move beyond the decision model or the alliance of enlightened technocrats with competent scientists imposing choices on the rest of society'. What is needed is not more science but a science anchored in democracy (Guesnerie 2003).

In response to this injunction, the ComMod approach places particular emphasis on taking into account stakeholders' expectations at various points in its application. The demand for companion modelling may or may not be explicit (Chap. 5). In either case, it is up to the research scientist to formalize the demand and facilitate its evolution so that it can both be understood through theoretical and methodological mechanisms and enable the involvement of the ensemble of stakeholders concerned by the social change.

As the demand generally emanates from particular operators, the formalization of the initial social demand is not always sufficient to take into account the expectations of everyone who may potentially be interested in participating in the ComMod approach, or of those who may question the work undertaken by others actively involved. It is necessary, therefore, to identify and appeal to stakeholders directly or indirectly involved in the management of the ecological or social system targeted and in the associated decision-making processes. This collective exercise is assisted by the commodian filling the role of facilitator. Given the importance of this stage, and of its determining impact on the rest of the process, the commodian should fully appreciate the opportunities and risks of the application of the ComMod approach and discuss them with participating stakeholders.

Once this stage is completed, the expectations of different stakeholders will be clarified and debated through the expression of their respective points of view. These points of view, rooted in different personal paths through life, express different value systems that sometimes may be incompatible with each other, yet all are considered to be legitimate. Companion modelling must organize these convergences, contradictions and incompatibilities to feed the dialogue between stakeholders and define future courses of action.

Companion modelling is not used to constitute *ex nihilo* a monitoring-evaluation system, although this may be one of the elements of a joint construction effort to be undertaken with participants in the approach. The approach is put at the service of sustainable development to further, within a given context, the emergence of values and priority actions expressed by social actors. These actions must enable the organization of the ecological and social system to evolve towards a configuration collectively defined as desirable, most often at the scale of a territory.

Transcending Traditional Divisions Within the Organization of Science

To achieve the objectives described above, the ComMod approach tries to go beyond the divisions and hierarchy induced by traditional scientific classification between cognitive (fundamental) and applied research. Each case study provides an opportunity both to increase knowledge and respond to a social demand. The ComMod group mobilizes an international community of research scientists from diverse disciplines who share the precepts stated in the ComMod Charter and who work in specific fields in which they are expected to seek recognition from their peers. In addition, the application of the process aims to obtain practical results which, in the context of the problems as they are understood, is rather ambitious.

The ComMod approach meets the requirements of sustainable development research: positioned at the intersection between research and action, it encourages the development of a systemic dimension, mobilizes and crosses competencies from natural and social sciences and introduces the notion of risk and uncertainty. The approach aims to increase the individual and collective capacities of the stakeholders participating in the process while promoting the participation of citizens in public policy decision-making.

The approach proposes one method to discuss how the modalities by which societies make collective choices and are governed highlight the challenges of sustainable development, and a second method to facilitate reflection and transversal action across academic disciplines as well as across envisaged social and environmental issues. The approach is open to improvement, and its application remains heavily dependent on the involvement of the stakeholders who mobilize it, their access to cooperation arenas and their capacity to position themselves within such arenas.

A Way to Apply the Principle of Participation

The ComMod group is dedicated to proposing to stakeholders of localized ecological and social systems a process that permits the principle of participation to be applied. To do so, it institutes a mediation process to help stakeholders handle uncertainty in decision-making processes. It also facilitates the stakeholders' understanding of the targeted ecological and social system by applying the principle of participation in different ways.

Different Modalities for Applying the Principle of Participation

The principle of participation is one of the fundamental principles of sustainable development. The first definition was provided in the Rio Declaration (Principle 10[7]) and was later elaborated by international law, notably in the framework of the Aarhus Convention that was signed in Denmark on 25 June 1998 by 39 countries.[8]

The principles of public information and participation in terms of the environment involve transparency in policy development (in terms of the direction as well as the consequences of choices made) and the empowerment of stakeholders. The information principle aims to provide stakeholders with data on their objectives and the measures and expected impacts of public intervention. Technically, the application of the principle of participation involves consultation, which seeks to collect the opinions or data needed to evaluate public intervention and participation, which should identify alternative or compensatory measures proposed and jointly studied by stakeholders.

The implementation of the principles of information and participation attempts to make economic agents and citizens assume responsibility for the impact of their actions, and to inform them of the risks or hazards to which they may be exposed. The principles underlie the transparency of policy development and the reduction of risks linked to the application of policies to better meet sustainable development goals. They also allow discussion on the social consensus needed to achieve these

---

[7] 'Environmental issues are best handled with the participation of all concerned citizens, at the relevant level. At the national level, each individual shall have appropriate access to information concerning the environment that is held by public authorities, including information on hazardous materials and activities in their communities, and the opportunity to participate in decision-making processes. States shall facilitate and encourage public awareness and participation by making information widely available. Effective access to judicial and administrative proceedings, including redress and remedy, shall be provided.'

[8] 'The Aahrus Convention concerns access to information, public participation in decision-making processes, and access to justice concerning the environment. It focuses on the following three themes: developing citizens' access to information held by public authorities, notably by ensuring the transparent and accessible diffusion of essential information; encouraging public participation in decision-making with environmental ramifications notably, encouraging public participation from the start of a development process "in other words, when all of the options and solutions remain possible and the public can have a real influence" the outcome of an individual's participation must be taken into account in the final decision, which must also be openly communicated; broaden access to justice concerning environmental laws and access to information.' URL: http://www.ecologie.gouv.fr/Communication-la-convention-d.html.

goals by ensuring that no stakeholder is omitted from the consultation and that everyone's interests are considered.

While these principles are constantly reaffirmed in political discourse, their mode of implementation mostly remains vague. In practice, the supply of information (the communication of meaningful data), and the effective participation of populations in public decisions, and therefore the empowerment of stakeholders, are far from systematic.

The ComMod process presents itself as a means of applying the principle of participation by being an explicit approach enabling, within a given context, stakeholders to reflect on their respective roles in terms of information and participation, and on the consequences of their investment (or lack of investment) in consultations.

The ComMod approach may contribute to discussions in existing consultation arenas, generally within a functional framework defined by the law or the prescriptions accompanying land planning and management. It may also involve the creation of new discussion arenas. In the first case, the discussion arena's pre-existing functional role will facilitate the insertion of companion modelling into the decision-making process. In the second case, this arena will need to be co-constructed through discussions taking into account how it will be integrated into the existing decision-making process and collective action. Procedures must be developed to determine the conditions under which other institutions will recognize the legitimacy of this new group to involve itself in the decision-making process.

In most cases, commodians wish to pursue all three application modes of the principle of participation (i.e. information, consultation and participation). However, given that it is heavily dependent on the intervention context, the ComMod approach might be limited to information, consultation or participation of a panel of selected stakeholders. This prior choice reflects a conscious decision taken by the ensemble of participating parties in a context marked by uncertainty.

Instituting a Mediation Process for Stakeholders to Assume Responsibility for Uncertainty in Decision Situations

Commodians understand the decision-making process as being 'the result of a process of interaction between individual stakeholders and/or groups holding different representations and weights in the negotiation' (Weber 1995a). In adherence to the provisions of the charter, they focus on a mediation process. 'As an alternative form of regulating conflicts, inserted to a greater or lesser degree into the judicial framework, mediation has become a fundamental concept (in fact a notion) to ensuring good governance at the national and international scale, thus contributing to globalization with greater facility as the English notion of mediation has experienced analogous changes' (Marshall 1984).

Various authors who have studied the concept in depth consider mediation to be a learning process that restores to 'mediation participants' responsibility for their actions (Guillaume-Hoffnung 2005), or a 'personal and social creative space, citizenship at work' (Six 1998). All agree on three essential points:

- mediation is an optional procedure that requires the free and conscious agreement of the people involved to engage in an action (the 'mediation') with the help of an independent third party (the 'mediator')
- mediation cannot be imposed; it is accepted, defined and carried out by an ensemble of actors
- for each party, the acceptance of mediation means committing oneself in good faith to seeking, with the assistance of the mediator, whatever is needed to establish a new balance in their relationship.

The 'third party' question has been, however, the subject of heated debate due to the stakes at play in the 'mediator' function. Commodians, who are intensely conscious of their own influence during the implementation of a ComMod process, do not accord themselves the role of 'mediator'. They believe the model fills this role. While this position distances them from the set of considerations related to the training and activity of mediators, issues related to the conduct of mediation feed into discussions on how to implement a ComMod approach.

The model becomes a 'mediator' to assist stakeholders in their search for viable and acceptable solutions concerning the management of the targeted ecological and social system. As is true in any mediation process, the implementation of the ComMod approach involves several stages: the presentation of facts, or '*theoria*', the confrontation of points of view, '*crisis*', and the development of a consensus, '*catharcis*' (Morineau 1998). The process also involves participants projecting themselves into the future. In a given conflict situation, effectively it is easier to find a compromise on a vision of future social, economic and ecological arrangements (generally on the scale of the next generation) than on current uses where conflicts of interest are often violent. Once the convergence of points of view is formalized, the way is open to reconsider the elements of the present that will enable participating parties to continue towards the envisaged future(s).

Nevertheless, no theoretical element can ensure that a mediation process will result in consensus. This will depend on the engagement and will of the stakeholders to invest themselves in the search for, and implementation of, a compromise. It will also depend on socio-political interactions existing prior to the process, and the possible rearrangements that the process might provoke (Lesage 2007). Such uncertainty often does not meet political timeframes or satisfy demands for rapid results.

Public decision-making processes consequently have a long way to go before they are based systematically on mediation. Autocratic, oligarchic, rational or irrational decisions may be involved, which may occur before, during or after a certain number of stakeholders have invested in a participatory process. The ComMod approach is subject to this state of affairs.

Although a number of research scientists in the ComMod group would like to accompany the decision process up to the effective implementation of the solutions devised by stakeholders, this in fact is not always possible because 'the accompaniment is situated upstream of technical decisions'. In reality 'the ComMod group lays claim to a modelling process enshrined in the decision process that is not the consultation process in and of itself' (Chap. 2). Companion modelling also emphasizes the importance of recognizing 'the uncertainty of decision situations in renewable resource and environmental management' (Chap. 3), and aims to help stakeholders handle this uncertainty while proposing a negotiated framework for social action.

Facilitating the Understanding of Ecological and Social System Operations

For individuals to be informed, express an opinion or participate in a mediation process in a complex situation, they must be able to understand and make themselves understood. To do so, we believe they must be placed first in a learning situation, be provided with information and organize this information into a meaningful whole.

By adhering to the charter, and by accepting the refutation of explicit hypotheses as an essential means of increasing knowledge, commodians place themselves in a learning situation. Furthermore, as is true for all mediation processes, the ensemble of partici-pating parties, not only the research scientists, should place themselves in a learning situation in relation to other people's points of view during the ComMod process.

The information allowing understanding of interactions between ecological and social systems is often disperse, incompatible, incomplete, redundant, unreliable or contradictory. Commodians try to get around this situation by soliciting not only the knowledge of the research scientist but also the layperson, the technician, the institution and the student (Chap. 3). Irrespective of how these different forms of knowledge are made explicit, their consideration is conditioned by their integration, at the source, in a co-constructed, conceptual model. What is important is to understand and make understood the organization of the ecological and social systems of the intervention through the expression of different points of view.

The modelling process, applied to transformations in ecological and social systems, their resilience and the search for the conditions of their viability, helps to overcome data inadequacies and barriers to experimentation through simulations. Modelling neither aims for nor is based on an exhaustive dataset; rather, its objective is to render data coherent through a simplification that makes sense to the stakeholders consulted and involved.

This is the basis that makes interactive forecasting possible, not to 'foresee the future' but to discuss 'envisaged futures'. The expression of different viewpoints allows the organization of the targeted ecological and social system to be envisaged. The commodian's role is then to 'to encourage the system of interactions presiding over change, to continuously monitor and make explicit changes in the system in order to be able to propose adaptations and to continually learn by observing the effects of these adaptations' (see Introduction).

This role involves making choices that are not neutral (Daré et al. 2007), because the mere presence of scientific observers shapes the representations that each person makes of the system. However, neutrality is not the objective of the ComMod research scientist as he is a stakeholder among other stakeholders. He is committed to ensuring the transparency of hypotheses and procedures and to maintaining the possibility of refuting them.

The Impact of the Implementation of the Principle of Participation

It is difficult to evaluate the impact of the implementation of the ComMod approach (Chap. 8). However, in terms of promoting sustainable development, we can affirm that the group's research scientists contribute to the clarification of dynamics regulating ecological and social systems, the development of tools and methods promoting stakeholder involvement for sustainable development, and the precision and operational potential of the concept.

Clarification of the Dynamics Regulating Ecological and Social Systems

'The application of the approach allows an increase in knowledge about the targeted ecological and social system through learning about the existence of different viewpoints and the consequences of their diversity on the way the system functions' (Introduction). This knowledge, related essentially to the management context, highlights and clarifies interactions between stakeholders and the consequences of these interactions on the dynamic of the targeted ecological and social system. Exchanges orchestrated through companion modelling around a contextualized set of problems allows participants to ask themselves questions regarding the resilience of ecological and social systems and, continuing in this direction, on the living conditions of present and future generations. This questioning enables individual and collective learning adapted to the promotion of sustainable development.

The confrontation of scientific and social viewpoints around and about the sustainability of a socio-ecological structure, and ensuing systematic discussions, contribute to the emergence of shared knowledge produced by the group and reintroduced to be validated and evaluated by each of the disciplinary fields mobilized. Beyond case studies and monographs, the implementation of the ComMod process authorizes the development of a constructivist approach concerning the operations of ecological and social systems. The relevance of the theoretical models developed by the 'thematic specialists' is often discussed during the co-construction of multi-actor conceptual models and is questioned in interactive simulations. Whatever the school of thought or discipline of the research scientist[9] involved in the ComMod process, these doubts and implications for the evolution of knowledge in their field often leads to the publication of scientific articles.

The scientific knowledge produced, whether contextualized or theoretical, enriches the concept of sustainable development and increases its operational potential in so far as the management of renewable natural resources (a favourite commodian domain) is a field where interactions taking place between stakeholders at different territorial—organizational levels strongly influence the well-being of future generations.

The Development of Tools and Methods Promoting Stakeholder Involvement

Along with other actors, commodians also generate knowledge about companion processes. Methods and tools are developed, adapted and improved during a dynamic research process aiming for ever greater efficiency and effectiveness. The value of these methods and tools appreciates, among other elements, proportionally to the involvement of stakeholders in favour of sustainable development.

In the fields of cognitive and computer science, commodians help to develop methods allowing the formalization of complex systems (e.g. the ARDI method, UML diagrams, ontologies, etc.), the placement of stakeholders in contextual situations (e.g. role-playing games, participatory mapping, focus groups, etc.), and the simulation of the co-evolution of ecological and social systems' (e.g. the development of simulation platforms such as Cormas and Mimosa, enriching knowledge about multi-agent systems, etc.).

The participation of stakeholders, envisaged through the formalization of complex systems, practical case studies and/or interactive forecasting[10] helps them to refine their knowledge of interactions between society and the environment, the impact of changes in systems influenced by humans and their possible evolution over time.

---

[9] In terms of economics, Kat Aware relates to an attempt to internalize externalities, Tarawa to economic ecology, Domino Réunion may be associated with questioning initiated by de-growth theories given the issues linked to maintaining agriculture in La Réunion in a context of strong population growth.

[10] These different forms of participation may be considered separately or simultaneously depending on the objective pursued and the resources available.

Furthermore, the development of a shared vision of a complex system enables participating parties to understand better the issues confronting the socio-economic structure in which they will insert themselves and the role they will play there. Companion modelling facilitates the expression of different viewpoints about a targeted ecological and social system, and the confrontation of these different viewpoints enables stakeholders to develop arguments that enrich democratic debate. Once thus articulated, the different viewpoints can be presented to territorial managers and decision-makers. In this way, the participating parties influence public decision-making while taking into account the necessary collaboration between institutional stakeholders. What is involved here is a contribution to governance because it helps to 'clarify the links that sustainable development institutes between stakeholders and land by proposing a decision-making rationale better adapted to assuming responsibility for the environment' (Laganier et al. 2002).

Interactions developed within the framework of the ComMod process thus a priori induce within participants 'a modification of knowledge concerning their interactions with the dynamic of renewable natural resources, a modification of power relations, a modification of their capacity to plan the collective use of resources, a possibility of transferring new knowledge to actors who have participated in only one of these processes' (Chap. 2). Potentially, these interactions reinforce the stakeholders' capacity to control the social changes induced by projects trying to stabilize social, economic and ecological organization, and reduce the vulnerability of some of them in the face of these changes.

The Contribution of Research to the Precision of the Concept and the Operational Potential of Sustainable Development

The implementation of the ComMod approach in several case studies stimulated numerous exchanges between research and development, particularly regarding how to reformulate the stakes of sustainable development and how to assess the risks and opportunities related to the management of renewable natural resources that present themselves to citizens, managers and decision-makers.

Given the extent of the hazards threatening the ecological systems on which people depend for their survival, the international community is imposing models aiming for sustainable development. The ComMod approach contributes to this effort by applying a model on a territorial scale. This is the scale best suited to development projects (Torres 2002)[11] We work to increase the resilience of

[11] 'Local stakeholders need to produce norms concerning their territory. The correct mechanism thus is the following: within a procedural, overall framework, one sets the overarching agenda (descending logic of the production of norms) and the local stakeholders produce norms to apply this agenda with their own perceptions and arbitrage (ascending logic). Local stakeholders cannot imagine a precise objective of optimal sustainability, which is necessarily a complex concept, but they can act on the basis of the overall agenda for which there are strong reasons to believe they

societies or, in other words, their capacity to adapt to change within an emerging socio-ecological system.

Sustainable development effectively involves a radical change in the structural organization of society. It means moving from a 'modern' social system focused on increasing productivity to exploit natural elements to satisfy human needs to another, post-modern system based on the sustainability of ecosystem management and the exploitation of renewable natural resources to satisfy the needs of both present and future generations. This change involves a significant capacity to adapt, which we propose to accompany. In this context, social resilience (USAID 2006) is understood by commodians to be the society's capacity to absorb the full amplitude of the perturbations revealed by this new vision of development, that is sustainable development.

We take a territorial-based ecological and social system, we construct a representation with stakeholders and, through the learning enabled by interactive role-play, and we reinforce the social resilience of the group of participants. Their capacity to adapt to change (e.g. in practices, representations, discourses, etc.) increases thanks to the appropriation of the concept and stakes of sustainable development and to the interactions that develop within the project group. The process improves the capacity of the group to identify and address perturbations (e.g. predator behaviour, inertia of unsustainable exploitation systems, etc.) that might destabilize the structure and organization of the socio-ecological system revealed by the co-construction. This has, furthermore, consequences for reducing the vulnerability of the most fragile stakeholders.

We thus adhere to the view of the Resilience Alliance (www.resalliance.org) that a social system's resilience aims to increase the capacity of stakeholders to anticipate and plan their future. Applied to socio-ecological systems, resilience is characterized by the amount of change a system can integrate while remaining functional and conserving control over its structure, the system's capacity to reorganize, and the degree to which it is involved in building and increasing its capacity to learn and adapt.

This being the case, companion modelling also reveals the importance of taking into consideration different organizational levels in the development of a decision process concerning an ecological and social system (Chap. 11). Up until now, our case studies generally focused on the local level, even if this level was linked to other organizational levels. At this scale, the analysis of social relations and the context of collective action enabled public decisions to be sensitized to social equity requirements that are integral to sustainable development.

Changing scales, however, is now one of our research priorities, first to expand the reach of the outcomes obtained at a local scale, but also because we know that

(Footnote 11 continued)
will converge on the set of objectives considered as simply appropriate. The production of local norms will therefore take place through empirical trial and error, step by step, in a context of daily grappling with the area's environmental problems and in relation to the overarching agenda articulated at a global level' (Torres 2002).

all change at the territorial scale can induce damaging effects of inequity, inequality or social exclusion on neighbouring territories. The model of the socio-ecological system that is co-constructed in the ComMod process is in reality subject to numerous other perturbations than those imagined in the model. These perturbations come notably from the reintegration of the socio-ecological system in dimensions involving multiple spatial and temporal scales. Beyond mobilizing the ComMod process to research management options for renewable natural resources at the local scale, it is appropriate to ask whether these options may be acceptable to, and accepted by, other organizational levels in the territory.

This leads us to take an introspective look at the nature of engaged research's contribution to sustainable development. The question is whether the ComMod approach should contribute to the development of participative democracy by seeking the—involvement of other organizational levels from the start of the process, or should it limit its role to uniformly reinforcing the capacities of all stakeholders in order to give them the argumentative tools for negotiations that they must undertake on their own (Chap. 6)?

Transparency on the Empowerment of Stakeholders

The application of the ComMod process comes up against a number of difficulties that are shared by all participatory approaches. The number of individuals willing to participate is necessarily limited as it is impossible to involve the entire population potentially concerned by the problem raised in the participatory process. In numerous cases, the legitimacy of participants is debated and their representative capacity questioned. Some therefore abandon the process in midstream just as others are joining. Thus there is a turnover, the importance of which is generally correlated to how long the participatory process lasts. Participants rarely are autonomous at the end of the implementation of a participatory process, one that remains long and expensive and which generally does not lend itself to the rapid implementation of concrete action. Finally, the involvement and sensitization of concerned managerial decision-makers in a territory is indispensable because they will be the supervisors of the concrete actions decided by the collective formed during the process. However, a hegemonic vision of political and institutional rationality is sometimes difficult to surmount.

According to some authors, these difficulties tend to weaken the relevance of participatory democracy as an engine of sustainable development. In our view, the main goal of sustainable development is to promote the involvement of all the stakeholders concerned by a renewable natural resource management issue through a process of information, decision and collective action. Ideally what we need to envisage, through the establishment of consultation arenas, are social relations fostering sustainable development.

While sustainable development is a process of change that concerns everyone, it is not possible to place *de facto* the entire population into a dynamic of thinking and acting differently and adopting another way of understanding development. The population cannot uniformly and instantly divest itself of the positivist, universal and linear vision of mankind's future that was developed over time to justify 'modernity'.

By considering sustainable development as a process induced by a complex and dynamic socio-ecological system that generates strong uncertainties, commodians assume the position of companions to change. They recognize that the impact of the ComMod approach is necessarily contingent and that they do not have at their disposal the means to guarantee perfect control over a sustainable development process whose implementation relies on considerations outside the scientific domain. In other words, ComMod research scientists place themselves in a post-normal framework. According to Funtowicz and Ravetz (1994), when dealing with decision processes in the highly uncertain situations lying at the heart of important social challenges, it is not the quality of the decision that needs to be improved, but the quality of the process that leads to making the decision. While the quality of the decision-making process may not necessarily guarantee the quality of the decision, the post-normal stance is based on the hypothesis that the former greatly influences the latter. It consequently is important to associate within decision-making processes diverse stakeholders holding divergent values and interests in relation to the stated problem.

Under these conditions, the objective is to mobilize oneself, according to the progressive articulation of stakeholder demands, to improve the quality of decision-making processes related to the conception, elaboration, implementation and evaluation of public decisions. To do so, one of the fundamental elements in the application of the ComMod process is seeking the maximum transparency possible to empower stakeholders.

The concern for transparency is founded on a culture of evaluation. In our view, this is part of a 'quality approach' in which the contractual terms established with partner stakeholders may be constantly redefined. Furthermore, by relying significantly on peer review, such evaluation aims for recognition from the scientific community regarding its approach and originality, and thus enabling its evolution and diffusion.

The accompaniment of socio-ecological organizational change applied to renewable natural resource management commits research scientists to constructing possible 'solutions' and their modes of implementation. While a research scientist normally is much less subject to socio-affective manifestations related to the identification and sense of belonging to a territory than members of civil society, he nevertheless holds, consciously or not, values that he draws on in such a social project (if only in referring to the framework of discussion and action of sustainable development). The research scientist must take care that his scientific status, which naturally gives him the legitimacy to set an agenda, is not experienced or seen as a problematic imposition laid down with a symbolic violence to which one agrees without actually believing. However, this must not lead him to remain removed from the practical solutions that the participants develop rather, the research scientist should continue to analyse them critically and transparently so that governance rules may come to light.

More transparent research for the increased empowerment of stakeholders may be included in a quality approach framework applied to the formalization of objectives and to the conduct and consequences of implementing a ComMod process.

Regarding Objectives

Before initiating a ComMod process, it is important to define clearly the objectives of the research with all of the participating parties. This process must be transparent in so far as it is the foundation for the legitimacy of the research activity. The Association française de normalisation (AFNOR) proposes three basic criteria to construct the quality process during this first stage: (i) relevance, or novelty and originality; (ii) opportunity; (iii) feasibility.

(i) The question of the relevance, or novelty and originality, of a new application of the ComMod process can be assessed by looking at the work already carried out by the group. This analysis is facilitated by systematizing the way different case studies are presented through the frameworks and various typologies proposed in this book. These also make it possible to assess whether the field of application, problem set, implementation and final goal of the planned project correspond to, and/or are likely to enrich, the group's practices. Furthermore, when the approach involves several commodians, or when the case studies are debated during annual meetings, discussions between members of the network help refine the evaluation of the project's objectives in regards to the state of the art.

(ii) Given its nature, the potential of a ComMod approach must be assessed on a case by case basis with the ensemble of stakeholders. The evaluation of objectives, therefore, will be undertaken in relation to the strategic directions and needs identified by each as a function of the situation. In this context, it is appropriate to assess the modalities and end goals of the application of the principle of participation as envisaged by each stakeholder. Is the question one of informing or training, evaluating, having pre-constructed decisions accepted or applied, or helping co-constructed decisions emerge and be formalized? The definition of companion modelling objectives, determined within each project by sustainable development stakeholders, inherits the conceptual uncertainty linked to the application of the principle of participation. Yet the effects of companion modelling are heavily dependent on the strategies employed by the concerned stakeholders with a view to the application of the principle of participation, and the approach relies on the active investment of these stakeholders. This investment is motivated and limited by the end goal they initially attribute to the principle of participation. The risk of manipulation exists. The commodian, who is aware of this situation, must make his position to pre-existing power relations explicit while helping everyone else involved do the same (Chap. 6).

(iii) The feasibility of the research consequently is heavily dependent on the context of intervention. In fact, the ComMod approach is not a theoretical, experimental tool seeking to find an optimal organizational solution for a given ecological and social system. It aims to accompany a political process anchored in the social, economic and ecological systems of a territory. It is, therefore the responsibility of the commodian to understand this context. In which political processes are potential ComMod participants involved or wish to be involved? How is this process integrated into the socio-political environment (e.g. organizational level of public decision-making, NGO intervention, etc.)? Is the commodian able to facilitate the emergence of shared knowledge, promote collective action, call into question prefabricated solutions (including his own) for the problem raised and work over the long term to reinforce stakeholders' capacities and defend their interests? These questions may be addressed through preliminary studies and surveys. However, the interactive understanding of the context also constitutes an opportunity for the commodian to clarify his own position and the nature of the ComMod approach, and to ensure that his point of view is understood and accepted by various partners. It is important to make explicit and to specify the objectives each stakeholder assigns to a ComMod approach as far up the organizational hierarchy as possible. Each stakeholder will formulate specific expectations, ones based on his own status, history and responsibilities, and in defence of his own interests, that deserve to be made explicit to all participating parties (Chap. 5).

The research scientist, who is evaluated by his publications, will use the information produced by the application of the ComMod approach to enrich scientific knowledge in his field of expertise. This objective must be explicit because it might conflict with the ethics of a mediation procedure, if this was what motivated the appeal to the ComMod approach, or with a desire for confidentiality on the part of those involved. Politicians, while they are often subject to an ethical and even legal obligation to communicate openly to the public the objectives and modalities of their intervention, may be led to invoke the need for secrecy, or to abuse their power when confronted with a sensitive, controversial or conflictual issue. For their part, participating citizens must establish their legitimacy, both in the eyes of the group they are supposed to represent and in those of their counterparts. In such a context, they must manage multiple contradictions while defending their personal and community interests.

The identification and formalization of political stakes and power relationships by the commodian in each case study and preliminary discussions with stakeholders concerning these issues are difficult, but contribute to the assessment and sharing of risks linked to the implementation of the process. On this basis, and as far as possible, the attempt to be transparent and formalize relations between stakeholders will improve the social utility and overall quality of the research work by anticipating the difficulties linked to possible manipulation, concealment of information or refusal to participate. Lastly, the planning of the implementation of the process assumes that there will be a match between available resources (e.g. technical and financial) and expected results. The formalization of this match, even if it is repeatedly modified afterwards, is a means of soliciting the involvement of the stakeholders approached. The precision and shared formalization of objectives incontestably guarantee a better appropriation of the context by all stakeholders, greater legitimacy for the implementation of the approach, and a better assessment of the promise of the solutions that may be developed. If the initial idea of turning to modelling is maintained, this exercise will help to verify whether the process meets the expectations of the research and the society, and if such is not the case facilitate stakeholder ability to reorientate the project.

Regarding the Conduct of the ComMod Approach

The implementation of the ComMod approach is subject to numerous uncertainties that are further amplified by its iterative character. Despite efforts undertaken during the collaborative definition of objectives, it may be difficult to envisage a priori a rigorous description of the research process. 'On the other hand, it is always possible, realistic and often useful to specify a posteriori the intellectual, technical and scientific path followed by the research team once they achieve their goal or when they choose to take a break in order to discuss their work. In this case, the quality approach favours knowledge and skill-learning mechanisms and a return to experience and making the most of knowledge' (AFNOR 2001).

AFNOR proposes four basic criteria for the construction of a quality approach in this second stage: transparency regarding the realization of objectives; respect for the time allocated to research; control of resources mobilized and used; seeking alternatives when initial avenues reach an impasse.

The logbook was developed to meet the objective of transparency (Chap. 2). It is kept by the promoters of the approach as work is underway to increase the transparency of the process and ensure the traceability of activities and results. This tool makes it possible to inventory intermediate results and place them in their production context, which allows, when need be, debate over whether the objectives or resources initially allocated should be revised. Respecting the time allocated for the implementation of the approach depends greatly on the availability and involvement of partners. It is here that animators play a fundamental role. They are responsible for organizing and circulating among the diverse partners the information exchanged or developed during collective key moments and to note actions undertaken between these periods. When required an animator must also, in association with the promoters of the approach, ensure that 'the necessary -transparency and foreseeable nature of the process does not result in a blockage of initiatives and projects, notably due to the time and resources required for consultations'.

In companion modelling, the most important technical and methodological points to master are—beyond a required command of social science data collection tools (e.g. interviews and participatory observation)—collective key moments and, when the need arises, the development of models. Concerning the first point, transparency and formalization falls once again to the animator who must be sufficiently experienced to be able to play the role of facilitator, mediator or guarantor of group unity depending on the situation. Throughout the process, these roles, made explicit by the work of the ComMod group, are updated and recognized by the different stakeholders involved in the process. Concerning the second point, it has been noted that participants do not always feel capable of contesting a model (notably when it has not been accompanied by role-play facilitating discussion and its appropriation by stakeholders). Their ability to refute a model depends on their capacity to grasp the level of complexity that the modeller chooses to represent, and on their will to invest themselves in a co-construction effort that would enable the more or less precise decoding of the model's black box. It is up to the modeller to minimize the difficulties that stakeholders encounter in questioning the model. The modeller must seek to differentiate between commonly accepted ideas about 'experts' models' (from which companion models strongly differ) and to diminish the risk of continuing confusion. 'The commodian, therefore, will consecrate part of his energy to clearly framing the quantitative or spatial validity of data produced, or to impeding the aesthetical or visual aspects of proposed model outputs from obscuring the relevance of discussions.'

The control of material and financial resources generally is imposed on ComMod research scientists by their parent institutions responsible for the administrative management and financial accounting of projects. These institutions, subject to budget controls, are developing tools that are increasingly powerful (and time consuming) within the framework of their own quality processes. Lastly, the search for possible alternatives to problems arising from the realization of the objectives that were set initially lies in the iterative character of the process. Companion modelling can lead to unexpected but interesting results, and orientate itself towards undertaking a 'new loop'. If, on the other hand, the problem is caused by certain partners' attempts to manipulate, conceal information or refuse participation, alternative solutions must be developed, as upstream as possible, with the remaining partners with the intention of continuing the research effort if it is still deemed relevant.

Regarding the Valorization of Results

'Research produces a set of often complex elements that cannot always be precisely defined and specified in advance. This set is notably made up of new information and knowledge, arguments regarding the limits and field of validity of this knowledge in addition to new, clearly expressed questions. Outcomes that do not correspond to the initial objective, as well as research avenues that have been dropped, can be of great value' (AFNOR 2001).

AFNOR proposes two basic criteria for the construction of a quality approach during this third stage, which consists of the validation of outcomes and their transfer. This must be carried out first with the partners of the research, in other words, the stake-holders involved in the ComMod approach. They are the first to validate the significance of outcomes and the envisaged use of knowledge and uncertainties attached to them. During each collective key moment, debriefing stages play this role. This validation is documented by the writing and validation of reports on the proceedings. It is, however, more difficult to validate results linked to participant learning, unless specific questionnaires are distributed, ideally given to participants not only during collective key moments, but also on a regular basis.

The transfer to the scientific community of the outcomes of a ComMod process can be envisaged, within the limits of possible confidentiality clauses, in various forms (e.g. articles in scientific journals, congresses, colloquia, expertise, patents, etc.). However, this transfer must also involve managers, decision-makers and citizens. The ComMod approach presents itself as a means to promote sustainable development in so far as it allows stakeholders who so wish (and who have a stake in so doing) to be involved in the design, development and evaluation of public decisions. These stakeholders' expectations consequently often go beyond having the outcomes simply validated by the people directly involved in the ComMod process. This is the context that the scope—and thus the modalities of transmission—of outcomes produced through the application of a ComMod process is placed.

Commodians, however, cannot assume that the consultations initiated will necessarily result in the taking or implementation of a public decision. These dimensions are the responsibility of stakeholders who are directly and indirectly concerned for two essential reasons. Beyond the identification of norms likely to establish an environmental policy, their formalization is an autonomous process (writing, legislation); once the environmental policies are formulated, their implementation concerns another type of dynamic (i.e. the effectiveness and efficiency of environmental policies given the continuously evolving social and environmental context).

By eliciting the participation of stakeholders in the analysis of socio-ecological dynamics and the potential impact of their actions on the systems studied, the ComMod approach creates a sense of responsibility among stakeholders directly involved in the application of the process. However, the transfer of outcomes is not always enough to elicit the same sense in stakeholders who remain outside the process yet who nevertheless play a crucial role in making and/or implementing public decisions.

The ComMod research scientist, who is aware of this situation, helps to reinforce the conditions that allow participants to influence public decisions and their implementation as far as his resources permit. This contribution can take the form of an adapted and detailed formalization of the outcomes obtained, or even the development of recommendations relative to the appropriation and communication of the outcomes of the process in order to fuel the decision process to come. In terms of sustainable development, it appears important to transfer the outcomes of engaged research to citizens, notably through training. The case studies may thus enrich the examples of training given by members of the group (Chap. 12).

The work undertaken under the ADD-ComMod project undeniably has helped to clarify a formal framework for the quality approach. This approach is considered as a tool to be used and adapted on a case by case basis with project sponsors (e.g. identification of criteria and indicators, definition of resource obligations, specification of expected products, means of communication outcomes, etc.). It ideally is complemented by a peer evaluation using a grid proposed by the ComMod group (Chap. 7), which enables the different points addressed above to be finely formalized through the aid of an external perspective. The quality approach and evaluation methods are now systematically planned by commodians, who are convinced that these contribute to the increased satisfaction of partners, enriching the knowledge produced by companion modelling, and the recognition and development of the group.

Conclusion

The ComMod approach is firmly part of the sustainable development process. It contributes to promoting engaged research aimed at the empowerment of people in the management of renewable natural resources and the fight against inequality and social vulnerability. The network of experts mobilized use improvable tools and methods to further scientific knowledge about understanding and managing the complexity of socio-ecological systems and contribute to a more masterful application of the principle of participation. In a context in which the end goals of the application of this principle are uncertain, supporters of the approach wish to give themselves the means to be transparent, as much regarding the objectives and conduct of the approach as the scope, through the control of risks and opportunities induced by the methods mobilized. The quality approach is one means to achieve this objective of transparency.

The ComMod approach hopes to contribute, through interactive research on modes of governance adapted to concrete situations, to the empowerment of a growing number of stakeholders faced with the challenges and means of sustainable development. Nonetheless, the ideas put forth within the framework of a ComMod approach do not lead systematically to collective action, which involves other considerations that cannot be controlled with certainty.

This does not mean that commodians disengage themselves from the use made of the process once their involvement is over. It is clearly up to them to help define the modalities by which participants may critically appropriate the solutions and actions defined during the process. This means that they are not in a position to decide public action, but they can think about and perhaps orientate it.

Chapter 10
Learning About Interdependencies and Dynamics

William's Daré, AnneMarie Van Paassen, Raphaèle Ducrot, Raphaël Mathevet, Jérôme Queste, Guy Trébuil, Cécile Barnaud and Erwann Lagabrielle

As mentioned in previous chapters, the companion modelling approach is based on principles laid down in the ComMod Charter (Collectif ComMod 2005). In this founding document, two fields of application were identified: to produce knowledge on the social and ecological systems under study and to facilitate cooperation between different stakeholders involved in a participatory process.

The process invariably involves a group, one that may or may not be constituted specifically for the companion modelling experiment. This group is composed of both scientists and individuals representing social groups and their interests. These

W. Daré (✉) · G. Trébuil
UR GREEN, CIRAD, Montpellier, France
e-mail: william's.dare@cirad.fr

G. Trébuil
e-mail: guy.trebuil@cirad.fr

A. Van Paassen
University of Wageningen, Wageningen, The Netherlands
e-mail: Annemarie.vanPaassen@wur.nl

R. Ducrot
UMR G-EAU, CIRAD, Maputo, Mozambique
e-mail: raphaele.ducrot@cirad.fr

C. Barnaud
UMR DYNAFOR, INRA, Toulouse, France
e-mail: cecile.barnaud@toulouse.inra.fr

E. Lagabrielle
University of Port-Elizabeth, Port Elizabeth, South Africa
e-mail: erwann.lagabrielle@gmail.com

J. Queste
UR GREEN, CIRAD, La Réunion, France
e-mail: jerome.queste@cirad.fr

R. Mathevet
CEFE-CNRS, Montpellier, France
e-mail: raphael.mathevet@cefe.cnrs.fr

M. Étienne (ed.), *Companion Modelling*, DOI: 10.1007/978-94-017-8557-0_10,

stakeholders formulate different and even contradictory viewpoints about the complex social and ecological system related to the issue raised. It is assumed that the viewpoints of all participants are legitimate, even if the foundations for this legitimacy may differ. The knowledge, on which they are based, whether empirical or scientific, is consequently considered to be relevant. The facilitator of a ComMod process will seek to draw out each participant's knowledge and opinions to clarify the hypotheses on which their arguments are founded in order to share them within the group. Thus, they may be understood and internalized by each member (Chap. 2).

Commodians assume that the involvement of participants in a companion modelling process contributes to modifying their viewpoints, opinions and representations. Through their interactions the ComMod participants learn about themselves, others, their relationships and interactions. During the whole process, from the acquisition of knowledge on the context (Chaps. 4 and 5) up to possible agreement for action, the facilitator combines modes of information production and exchange between individual participants and with the commodians. The ComMod approach is sequential, adaptive and iterative. As shown in Chaps. 1 and 2, workshops in which stakeholders and commodians interact play a key role in the dynamic of the process.

The objective of this chapter is to elaborate the issue of learning and answer the question: whether and how does the ComMod approach enable participants to learn for collective renewable resource management? More precisely, the objective is to gain a better understanding of how participation in a ComMod process enables participants to learn collectively about complex socio-ecological systems, multi-level interaction, their dynamics and interdependencies. The second objective is to describe the types of learning achieved and the dynamic of this learning. In other words, we call into question the postulate that the activities in ComMod workshops play a fundamental role in the learning process. Indeed, individual participants are involved in collective key moments.[1] Each of them has their own representation of the issue. The modelling approach aims to create a shared representation of a complex system, multi-level interactions and emerging overall dynamics. At the workshops, participants interact, express and discuss their viewpoints, but from what moment does real learning take place? How is the move made from the expression of multiple perceptions to a shared representation that is legitimate in the eyes of all ComMod participants?

In the next sections, we first review different learning theories and build a theoretical framework about learning. This framework enables us to analyse the learning triggered by the ComMod processes and to illustrate it through various examples. The last section discusses some elements likely to enable the consolidation of learning in ComMod processes.

[1] We will not address here the issue of the representation of social groups that was discussed in Chap. 5.

Learning Theories Related to a Companion Modelling Approach

We define learning as the acquisition of knowledge for effective action in the domain of existence (Maturana and Varela 1992; Röling 2002). Now, according to the ComMod Charter what is particularly involved is questioning how a sum of individual learning may lead to the emergence of collective learning. 'Such a family of models is a genuine knowledge-based system allowing interacting researchers and stakeholders to increase their personal and common knowledge of the system, current processes, and the situation of each actor-observer in such processes' (ComMod Group 2003). What is the relation between individual and collective learning?

Towards a Cognitive Theory of Individual Learning

A historical review of academic thought about learning shows that until recently, three main schools of thought, that is, behaviourism, gestalt theory and cognitivism, have dominated debates over learning revolving around the central question: 'how does someone learn and retain what he learns?'.

In the early twentieth-century, behaviourists such as Watson, Skinner and more recently, Deutsch, Krauss and Fischer focused on response stimulus behaviour. In their view, learning occurs through action, through a process of trial and error. Therewith, they rejected any explanation of observed behaviour by mental processes. Tolman was the first who considered contiguity and reinforcement of events as the principle explanatory variables of learning, but acknowledged that mental processes also play a role (Lecocq 2007).

In the Second World War, the Berlin school under the impetus of Wertheimer, Koffka and Köhler developed the gestalt theory. They disagreed with the behaviourist vision of learning because it does not include 'perceptive gestalts'. Proponents of the gestalt theory noted that we perceive objects in a global manner. Therefore, numerous components of reality are perceived simultaneously. Sensory fields order and give rise to segregations, articulations and regroupings (Dubé 1990). This represented a real change in thinking about learning. As with gestalt theory, the whole gives sense to the parts, contrasting sharply with the atomist vision of behaviourists based on the principle of contiguity between elements (stimuli and response). In the view of cognitivists, behaviourists pay too much attention to isolated events, stimuli and visible behaviour without tackling the set of mental processes in which they are inserted. While behaviourists address the issue of learning through a relationship to the environment, cognitivists are more interested in perceptions, and in the learner's representations. They consider them as elements of a pattern, of a whole brought into play during the learning process. In their view, learning is more a modification of knowledge than a pure modification of behaviour. The cognitivist approach draws from the group dynamics psychology work of Lewin (1946) and on criticism voiced by linguists of the Chomski school (Lewin 1946). The psychology works of Piaget and Vytgotski on the cognitive development of children have influenced greatly cognitivists (Goupil and Lusignan 1993). Piaget criticized the analysis of the fundamental processes of knowledge acquisition. He showed that learning constructs itself due to processes of balancing cognitive structures in response to environmental stimuli and constraints. Vytgotski proposed that the process of knowledge acquisition starts from the social (interpersonal knowledge) towards the individual (intrapersonal knowledge).

Over the last two decades, recent theoretical developments within the cognitivist field enabled the learning observed within ComMod processes to be analysed. These developments are positioned within a constructivist perspective, which considers social reality as a permanent construction process. There are multiple perceptions of social reality. Knowledge of this reality is distributed among each of us, with each individual able to understand only a part. Experiential learning theory is based on the four principal stages of cognitive development identified by Piaget: sensorimotor, perceptive, representative and operational. In his work, the French psychologist insists on the necessity of using different experiences in learning. During these experiments, the learner (for him, a child) is led to manipulate objects and proceed to concrete tests. In doing so, he is lead to reflect on the results of, or the questions raised by, his experiences (Goupil and Lusignan 1993). Following Piaget's proposition Bruner (1960) explored the links between mental and learning processes, especially those associated with methods of discovery and exploration. He showed the strength and durability of learning thus achieved. Kolb (1984) clarified the mechanisms. He considered that individuals learn through a cycle that alternates between stages of theoretical exploration and experimental practice (referring to Piaget's sensorimotor stage), observation (Piaget's perceptive), reflection (Piaget's representative) and action (Piaget's operational) (Piaget and Inhelder 1984). This learning cycle, known as 'Kolb's learning cycle', serves as a theoretical reference, especially for interactive training approaches. Within the cycle (Fig. 10.1), an individual undertakes an action without necessarily thinking about it. Then, he observes and reflects on his action and experience. Next, he interprets the facts and events and integrates them into a theoretical framework. Finally, he uses what he understood to try to predict what will happen next. In experiential learning theory, Kolb considers both the cognitive and the subjective or emotional dimensions of learners, but he does not analyse the influence of the latter on the learning process. The learning cycle then was discussed by Honey and Mumford (1992). They showed that each stage of the cycle mobilizes different conducts and attitudes, different observation and communication aptitudes, different values and beliefs, and all condition the success of learning.

The ComMod process is in keeping with the Kolb cycle. It allows participants to experiment with these four experiential learning stages, especially during the conceptual co-construction phase and simulation exercises with role-playing-

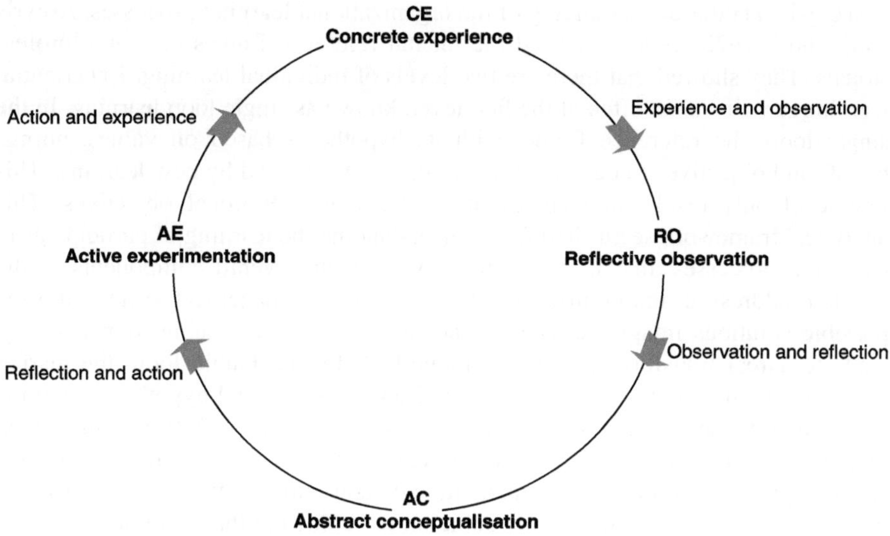

Fig. 10.1 Kolb's experiential learning cycle

games (Mathevet et al. 2007). For instance, during a role-playing game session, stakeholders are led to play the game and live a concrete experience related to their daily life (CE), to observe the behaviour of others (RO), to discuss and understand the dynamics and effects in order to induce desired and feasible actions (AC) and then to test them (AE). Moreover, simulations allow virtual experimentation to observe and understand better the long-term dynamics.

Memory plays a fundamental role in the learning process. Not only the immediate memory that passively registers, but the memory that contributes to the phenomena of recognition (allowing one to recognize a previously encountered object) and evocation (evoking an absent, previously encountered object through the remembered image). However, as long as recognition is individual and unconscious, evocation is constructive as this requires the construction of mental structures amongst which are mental models. For Piaget, our permanent and transitory mental models are necessary to recognize the world. They are also 'polarizing filters' of our perception of reality. Operational learning is committed to memory through procedural habits, or the accumulation of routines, whereas conceptual learning mobilizes knowledge frameworks built with concepts, laws of causality, and semantic and semiotic systems that condition a true 'intelligence' of the situation (Kim 1993). In ComMod processes, the co-conception and simulation stages are both based on the exchange of viewpoints. They should state clearly the 'polarizing filters' of each participant. Moreover, these stages allow the alternation between conceptual learning (i.e. on the issues under consideration) and more operational learning (i.e. technical or interpersonal skills).

To gain a better understanding of the organizational learning processes, Argyris and Schön (1978, 1996) analysed the mental reference frames of young professionals. They showed that there are two levels of individual learning. Experiential learning remains most often at the first level, known as single loop learning. In the single loop, the reference frame, with its hypotheses based on values, norms, beliefs and objectives to describe the world, is not modified by new learning. This first level only results in changes in practices or operational objectives. This analytical framework seems helpful in understanding the learning of participants in ComMod processes, that is, the issues involved, the overall components of the problem addressed, the complexity of the resource management system and the possible solutions imagined. The second level, known as double loop learning, leads to a real transformation of an individual. In the double loop, the learner questions the foundations of his reference frames, beliefs and hypotheses, and the norms and values he previously held (Argyris and Schön 2002). Single loop learning generates small operational and cumulative changes, while double loop learning produces more fundamental, strategic and radical changes (e.g. a change in reasoning where justifications become more ecological than economic).

Major learning theories only consider individual learning in a static manner. During the presentation of more recent theories rooted in cognitive theory, we first looked into the scale of individual learning. This first proposition allowed us to show the relevance of an analytical frame for the ComMod learning process. However, the question remains of how one passes from individual to collective learning.

Dependence Between Individual and Collective Learning?

In the ComMod approach, we assume that the interactions between participants with intermediate objects produce meaning and modify behaviours, perceptions and mental models of participants both as individuals and as a group. Hence, the learning realized must also be considered at the collective level. We also seek, therefore, to obtain a better grasp of the nature of collective or organizational learning. When can we consider that learning is no longer exclusively individual but concerns the group as a whole? How does individual learning evolve into collective or organizational learning? What conditions facilitate collective learning?

In the 1970s and 1980s, sociologists and psychologists showed that knowledge is a social construct, historically and culturally enshrined in social settings (Knorr-Cetina 1981, 1984; Latour 1987). Recent work by Lave and Wenger (1999), in addition to previous works (Brown et al. 1989; Salomon 1993) further developed this theme. Firstly, they showed that each individual's cognitive structures are not mobilized outside a precise spatial-temporal and social context. Secondly, they demonstrated that learning is distributed among several people through language, artefacts, and more broadly, the environment. There is thus a close relationship between individuals' situated learning and collective learning (Brown et al. 1989; Lave and Wenger 1999). Individuals are not passive receptors. They enter a learning situation loaded with their experience and life history and thus, they actively contribute to the re-construction of collective knowledge.

Now that we have established that individual and collective learning influence each other, we tried to understand better the complex retrospections between these two types of learning. Learning mechanisms have been analysed in numerous collective situations, that is, organizations, practice groups and multi-stakeholder platforms. An organization is a hierarchical group of people with a mission, precise objectives, and coordination and communication procedures to guide its members in their production activities (Weick 1995; Boudon and Bourricaud 2002). Practice groups (Lave and Wenger 1999; Wenger 1998) are constituted by individuals collectively engaged in the same type of activity. These individuals may work in different organizations but they share a history, knowledge and procedures. They share the same experience, the same perception of a problem and communicate to resolve it. Here, there is no hierarchy or formal coordination between individuals. Multi-stakeholder platforms are constituted by stakeholders who represent categories, positions, perceptions, values and different interests. Therefore, they do not *a priori* share objectives or interests (Aarts 1998).

Weick (1995, 2001) studied the construction of meaning in organizations. He found that, like individuals, organizations have objectives, rules, procedures and routines that help them work within their field of specialization. In the enactment theory, he shows that with the routines (tacit knowledge) collectively produced, members of an organization craft their environment and collective frame of reference by selecting, rejecting and interpreting information. This frame provides routines informing the behaviour of members of the organization. Individuals thereby develop skills that favour the reproduction of the system and its stability over time. This frame endows an individual's actions with meaning, builds itself through interactions between members, and enables known problems to be solved. When the organizational environment changes and the routines no longer bring the desired results, organization members commit themselves to a process of destabilization followed by reconstruction. In unknown situations, the construction of meaning becomes intersubjective, drawing on individuals' behavioural and cognitive capacities and allows improvization. Then everybody can observe the effects and assess the advantages of adopting the behaviours and new cognitive schema produced. The change is a result of an experiential learning process (or learning by doing) as described by Kolb (1984). Argyris and Schön (1978, 1996) found complex retrospection mechanisms between individual and organizational learning. The changes in individuals' mental models to construct shared mental models of an organization modify the perception of the organization and transform organizational values and paradigms. Consequently, this modifies the environment of the individuals and affects their own mental models. In their studies of organizational learning, they note that organizations usually engage in first level experiential learning (single loop) to resolve short-term operational problems. At times, radical changes in the environment can lead organizations to enter second level learning (double loop). A fundamental reorientation of reference frames follows. This requires a redefinition of the organization's rationale and a new look at tacit knowledge, theories in use and espoused theories to analyse whether these are in tune with the new environmental situation.

Lave and Wenger (1999) show that the learning dynamics in communities of practice resemble those of organizational learning. However, in this case all of the members carry out the same tasks and have identical levels of experience. In addition, there is no central decision-making power such as that found in organizations where a leader plays a defining role in learning processes and group actions (Schein 1985, 1988). The concept of multi-stakeholder platforms is interesting in the field of natural resources where interests are heterogeneous and indeed, competing. Aarts (1998) showed that in such situations, stakeholders tend to adopt negotiation positions that are strategic or conflictual rather than cooperative.

Chapter 4 showed that stakeholder groups are systematically constituted when a ComMod process is established. According to the definitions we just proposed, such groups are neither organizations nor practice groups but rather multi-stakeholder platforms. In ComMod platforms, the intention of the action, to draw participants into a learning situation, is clearly assumed and explained. By agreeing to take part in this type of process, participants show their willingness to interact, whether in a cooperative or conflictual manner. The conditions allowing an exchange between the mental models of individuals and the group, therefore, a priori are present. Consequently, can we apply the theories developed in the framework of structured organizations to ComMod platforms in order to explain individual and collective learning processes? To circumvent this methodological difficulty, we rely more on the concept of social learning derived from that of organizational learning.

Social or collaborative learning refers to the learning process of a set of people seeking to improve a situation through collective action. These groups may change and do not necessarily constitute organizations. The different definitions of social learning are drawn on the theory of communicative action (Habermas 1984). They emphasize the role of dialogue and intercommunication between members of a group to facilitate the perception of different representations, development of collective reasoning and action. At first, this theory only referred to individual learning resulting from social interaction within a group and founded on the observation of the other (Bandura 1977). It was then enriched by contributions on learning by Argyris and Schön (1978, 1996, 2002).

The concept of social learning was the basis for numerous approaches in the field of ecosystem governance aiming to increase a group's capacity to carry out joint activities linked to natural resource management (Daniels and Walter 1996; Ison et al. 2007; Pahl-Wostl et al. 2008; Rist et al. 2006). In these approaches, learning is based on interaction between numerous processes: the construction of networks and social activities, dialogue and communication around joint activities, the organization and management of knowledge. This learning is always situated. It links knowledge and relational practices to enable reciprocal interaction and discussion of these interactions. The urgency of the 'problem' (e.g. environmental, social, economic, etc.), and the interdependence between the stakeholders and their individual and collective stakes are the two main conditions that motivate stakeholders to involve themselves in social learning processes and collective action.

Numerous approaches rely on the development of various intermediary objects (Vinck 1999) that have both a nominal function (because they serve to support the activity, give it meaning and promote exchanges of knowledge between participants), and an interpersonal function (they support communication, collective action and the construction of networks). In the ComMod approach, these two functions often are assumed by simulations in role-playing games or through computerized models. These intermediary objects promote dialogue and understanding of other people's viewpoints. They help stakeholders to explore different scenarios of the future and to compare the costs and benefits of various management options.

Finally, the theories of experiential learning, organizational learning and its connection with social learning form a theoretical corpus. We have shown that this corpus allows interactions and mutual links between individual and collective learning to be taken into account. This corpus of knowledge thereby enables us to think about the learning of individuals and groups taking part in a ComMod process.

The Dynamics of Learning: a Process or a Result of Change?

Now that we have described the link between individual and collective learning, we must examine learning modes. Is the concept of a learning dynamic relevant to ComMod? In other words, how does learning take place? In the literature, there is tension between two poles of thought: is learning the process that enables learning, the dynamic that enables the acquisition of knowledge, or is it a result of this process?

During the 1960s and 1970s, learning was defined in psychology as a change in behaviour, thus the result of a process. As such, it is tangible, visible, palpable and recognizable. This perception of learning is closely linked to the development of experimental approaches in psychology. The interest of this perception of learning is to show the characteristics of this result and thereby, to illustrate its relationship to the change. However, it does not take into account the factors that allow this behavioural change. Other authors consider learning more as a process, focusing on what happens when one learns. Maples and Webster (1980) considered learning to be 'a process that induces a change in behaviour following an experience'. The question then becomes whether the individual or organization is conscious of becoming engaged in a learning process. If so, what are the consequences of this state of consciousness on the learning process? We subscribe to the definition of learning given by Maples and Webster (1980), considering learning to be a conscious process. However, there is still the question about the conditions of learning within a ComMod approach. Placed in a given situation, individuals and groups acquire information that can lead them to change their behaviour and even their mental model.

Ramsden (1992) identified two types of learning. The first is outside the learner as it is provided by a third party, the teacher. This type is assimilated with the additional knowledge thus acquired. This knowledge is stored in a person's memory and is likely to guide their actions. It is translated into skills or methods so it can be called upon at any time. The second is interior, personal to the learner. It plays a role in his relationship with others and the world. It helps the learner to interpret and understand reality and to find meaning (Ramsden 1992). A commodian does not focus his intervention on the teaching of knowledge but acts to promote exchanges of opinions and knowledge with local stakeholders. In doing so, he hopes to participate in the modification of the mental models of the participants in the ComMod process and also, through retrospection, of the groups to which they belong.

We should note some authors have emphasized that the advantage of working on learning processes is lost if it is not linked to an objective of action (Edelenbos 2005; Röling 2002). According to Röling (2002), collective cognition and distributed cognition are two paths to achieving an action-oriented objective. 'Collective cognition emphasizes shared attributes, that is, shared myths or theories, shared values and collective action. Distributed cognition emphasizes different but complementary contributions that allow concerted action, for example, the operation of the market and legal frame of policies.'

This theoretical analysis indicates that individuals in organizations and less formal groups learn better in action situations, when they are confronted by a shared problem that they try to solve. The complexity, openness and uncertainty of the social and ecological systems studied do not allow an ideal solution to a given problem to be achieved. According to the theoretical analysis, the development of routines, the production of rules and the emergence of new interactions between group members appear primordial from the viewpoint of the learning expected from a ComMod process. To show the effectiveness of learning in a companion modelling approach, we focused our analysis on collective situations of interaction that occurred during a companion process. Learning will be understood at two levels of interaction, that is, that of the learner and that of the group. Based on the theoretical corpus made (with social learning, organizational learning and experiential learning theories), we aimed for a better identification of the types of learning and their associated dynamics in the different case studies.

Analysis of Learning in ComMod Experiments

We will now look at the material on which our analysis is based and the analytical grid we produced for a better understanding of learning in ComMod processes.

Contributions of the Previous Theoretical Corpus to the Construction of an Analytical Grid of Learning in ComMod Case Studies

During a ComMod process a multi-stakeholder platform is built, and the theoretical corpus discussed previously allowed us to show that learning, produced through experience, is situated. Knowledge is a social construct, historically and culturally situated, and is ever-changing. Learning, individual and collective, is interdependent and dynamic, feeding each other through retrospection. There are two types of learning: single loop, where only the practices of individuals or groups are affected; and double loop, where individuals of a group recognize their own frames of reference and those of their interlocutors and bring them into question, which allows a more radical change. Lastly, the stakeholders involved recognize the urgency of the situation and their interdependence in order to be able jointly to address the question raised.

Recognition of multiple interdependencies occurs through processes of co-learning and interaction between stakeholders. These processes allow stakeholders to identify themselves and their interests and perspectives in relation to the situation studied, and give them certain legitimacy. These processes also define the frames of reference called upon and the conditions of their emergence. This work of mutual knowledge is helpful in understanding better oneself and others. It allows the broadening of one's vision of the world and the exercise (Aarts and Woerkum 2002). Furthermore, learning about the relationship between individual and collective interests allows them to imagine together the desirable options by recognizing the costs and benefits that these solutions may have at the individual level (Checkland and Scholes 1990).

Learning in ComMod thus must be understood at two interacting levels, that is, the learner and the group. This is because companion modelling approaches rally stakeholders who have heterogeneous frames of reference, interests and objectives to engage in a collaborative process on a natural resource management issue. Given the complexity, openness and uncertainty of the social and ecological systems studied, there are no ideal solutions. The resolution process puts into play strategic negotiation processes, including compromise and integrative negotiation. The latter require a true, mutual understanding and/or a creative process that promotes the redefinition of the objectives of the exchange.

The ComMod process systematically puts experiential learning to work. Companion modelling, particularly in role-playing games, places stakeholders in situations in which they may test, observe, represent their deductions and make operational. It follows, therefore, the Kolb cycle described above. The shared experience, based on principles of participation (i.e. universal right to speak, listen, interact, etc.) brings together stakeholders who would never or rarely meet or interact with each other, or who may even ignore the existence of their interactions and respective impacts on a resource that they nevertheless share. As the process proceeds, a feeling of belonging to a particular group is established that is based on these moments of exchange. The reflection organized at the end of the collective key moments (i.e. role-play debriefing, discussion-synthesis workshop) contributes to linking the experience, the opportunity to gain perspective on the actions and reactions of each, in the context of the real world. These processes of intersubjective communication take the form here of an exchange of experiences, arguments and clarification. It precedes the action, deliberate or improvised, that enables new management situations to be tested. To explain the reality of this learning, we have worked in particular on situations of collective interaction put into place during a companion process.

Beyond experiential learning, a transversal process in the ComMod approach, three other main types of learning are in play. These consist of learning about the content of the situation studied (nominal learning), learning about the participants and their interdependencies, and communicational learning. Given the wealth of material available, we chose a slightly more detailed analytical framework in which the first two points correspond to nominal learning.

- Learning related to the issue under consideration: general knowledge about the dynamics of the socio-ecological system, and the conditions in which the question addressed emerged.
- Learning knowledge and techniques that enable a better understanding of the technical options and their consequence on the socio-ecological system and its dynamics, and in this way think about the possible options that would allow the system to attain a desired state. This learning might involve the stakeholders as well as the research scientists involved in the process.
- Learning about others: on the one hand, this concerns knowledge of each actor's interests, skills and stakes; on the other, knowledge of the beliefs, viewpoints, norms and values of each actor.
- Communicational learning involves acquiring a mode of social interaction that permits the sharing of knowledge, learning and decision-making though experimentation with new means of communication. This point aims at the social learning of groups for collective decision-making, the mobilization of the stakeholders concerned, even the most marginalized, the mobilization of key actors, and the creation of alliances to help the process advance.
- Organizational learning involves the acquisition of knowledge about the stakeholders' organizational options and their consequences on the system in order to select the organization that is most suited to achieving the desired system state. Selecting the organization means identifying the joint objective of all the members, defining the rules of existence, establishing routines and describing its hierarchy. This allows one to verify whether the ComMod process permits a multi-stakeholder platform to evolve into a true organization.

To correctly explain the evolution of learning during the companion modelling process, the learning dynamic must be broken down into three stages: (i) initialization and creation of a stakeholder group (creation of social ties and mutual confidence); (ii) dialogue and learning about the social and ecological system dynamic, the problems encountered and objectives sought, ways of resolving a problem and achieving an objective; (iii) the organization of stakeholders in order to achieve the objectives in the field. These different stages highlight the fact that the learning dynamic is part of a broader group dynamic in which the creation of confidence and the involvement of different stakeholders are essential.

Materials and Method

The analysis is based on 14 evaluation reports produced through the ADD-ComMod project. The significant diversity of the case studies and the implementation modes of the evaluation protocol (Chap. 6) rendered the task delicate, particularly due to the absence of a firm definition of learning. Nevertheless, the wealth of material allowed a cross analysis of the evaluations. This was undertaken by comparing the views of two readers with those of the designers of the approach using the proposed analytical framework.

The analysis was complemented by monitoring collective key moments in each case study. There were different kinds of monitoring. Beyond monitoring certain collective key moments themselves, notably the game sessions by recording the actions and discussions of participants during the game and in the debriefing, some approaches specifically monitored learning. Methods varied between approaches; for example, there were individual, post-simulation surveys for analysing and understanding the actions and evolution of representations (Mae Salaep, Lingmuteychu), short pre- and post-game questionnaires (Camargue, AguAloca, Ter'aguas), and socio-anthropological monitoring (Njoobaari). These elements are particularly important in reporting learning dynamics.

Each type of learning is present in the quasi-totality of the case studies. Their combinations vary chronologically, qualitatively and quantitatively. We propose to to illustrate by concentrating on each type using one or several case studies that benefited from close monitoring at the time of implementation.

Learning in ComMod Experiments

Learning About the Issue Addressed

All of the evaluation reports mentioned that whatever a participant's involvement in the companion modelling approach may be, the participant leaves with a greater understanding of the complexity of the stakes and the dynamics of the issue addressed. One witnessed during the process an accumulation, interaction, production of knowledge, information and multiple and varied data contributed by the ensemble of participants in the experiment. This more or less powerful awakening to the complexity of the system studied (some may have already been sensitized before the process, others not at all) was demonstrated by the specification of a wide variety of practices, representations and perceptions of the goal and subject of the study.

The AguAloca study case (see the Appendix) is particularly illustrative of this type of learning. It aimed to facilitate consultation processes regarding the multi-use management[2] of water resources in a watershed committee in the metropolitan Sao Paolo region in Brazil (Clavel et al. 2008). After a series of thematic studies aimed at better understanding the dynamics of the watershed, a computerized role-playing game called AguAloca was developed in a companion modelling process mobilizing a multi-disciplinary team and a small group of managers. The game itself was played twice, the first time with engineers from different institutions (i.e. potable water supply company, water management department) and representatives of several towns, and the second time with several members of the watershed agency of which the committee was part. Each game session was monitored by: (i) two people who observed the development of the game and the individual and collective behaviour of players; (ii) two short questionnaires completed by players before and after the game to analyse their expectations, feelings about a session, and the evolution of representations of the management issues. An evaluation was carried out through semi-structured interviews nearly 8 months after the last game session with the players and actors who participated in the design.

Participants emphasized the contribution of the ComMod process in terms of understanding the overall issues, notably the significance of the terms 'integrated or shared water management' and 'collective action for water management', which were the two main principles on which the committee's work theoretically was based. Participants emphasized the highlighting of underlying interactions between different activities and processes on the watershed, as well as the interdependencies between actors, and between decisions, resources and actors. One player noted an accelerated learning about watershed management issues, learning usually acquired only after participating for 2 years in committee meetings—this, when the mandate for representation only lasts 2 years. The 'local' perspective of each stakeholder was articulated to produce a shared representation of the issue addressed and the stakes involved. One person mentioned that the game served as a reference framework to understand and analyse committee discussions. The testing of the model allowed the issue to be appropriated, which resulted in the transformation of how participants grasped questions. For example, technicians who initially were principally concerned by water quantities began to think about the qualitative processes highlighted by the game experiments. Specialists were called into explain their research results and a workshop was organized by the committee on the subject.

[2] Drinking water for the agglomeration, industrial use, agricultural use, dilution of sewage, flood protection and recreational use of reservoirs.

Learning About Others

All of the evaluations emphasized the stakeholders' learning about others. This learning focused on both a better understanding of each actor's interests, skills and stakes and on building an awareness of their beliefs, viewpoints, norms and values. The interviews undertaken in the AguAloca case study highlighted, beyond learning about the issue discussed above, the better understanding of the various watershed stakeholders' interests, practices and their impacts on the resource. Participants realized that they were all legitimate, concerned and dependent on each other, and that they had to work together and 'accommodate each other'. Participants particularly appreciated the possibility of testing other stakeholders' management difficulties, leading some to develop a greater capacity to listen and take into account the contributions of others during committee debates.

The Lingmuteychu case study (see the Appendix) illustrates another aspect of this type of learning. On the Lingmuteychu watershed in western Bhutan, seven villages share water from different rivers for domestic use and agriculture irrigation. The principal crop is high altitude (over 2000 m) rice grown on rain-fed terraces. The availability of water during the brief transplanting period is the main factor limiting yields. Terraced fields must be completely flooded sufficiently early in the year for transplantation to avoid a failed crop because rice flowers when temperatures are low. Numerous conflicts had arisen between the communities on the watershed over this issue. In 2002, there was a dispute between two villages, one downstream of the other, over the opening date of the principal floodgate of the downstream village's irrigation system located on the territory of the upstream village.

In this context, a scientist from the Bhutanese Bajo research centre began a ComMod process with participants from the two villages that took the form of three, three-day workshops in 2002 and 2005. Each workshop was organized around a role-playing game session in which the relationship between irrigation and cropping was acted out. These games did not refer to either the floodgate or the opening date but emphasized communication between and within communities. At the end of the second workshop, it was decided to include all seven villages on the watershed in the game. The evaluation of the ComMod process was based on three series of semi-structured individual interviews and on the participatory observation of a training session on collective action that the Bajo research centre organized on behalf of the watershed natural resource management committee. Of the 11 participants from the downstream villages that were interviewed, only one claimed to have understood something about the conflict in which he was involved. In contrast, six of them claimed to have learned about a conflict between two other villages on the watershed in which they were not involved. These six participants could explain the conflict and formulate advice for its resolution. Observations of other participants, discussions and debates that took place during the 3 days of the final workshop enabled them to acquire knowledge of the issues, positions defended, power relations and obstacle points. Based on this observation, the formulation of advice reflected the reflexive character of the exercise and a conceptualization of the information collected. The interviews, however, did not clarify how the learning process continued after the workshop, notably whether it was individual or collective.

Learning about others was particularly strong in the companion process that took place in the Vendres tidal area (see the Camargue case study description in the Appendix). This humid, 1600 hectare area situated in southern France includes nearly 900 hectares of reedbeds. The area has a high heritage value, notably for waterfowl, and is the object of multiple uses, that is, grazing, hunting, fishing and tourism. In response to the generalized degradation of the environment due to numerous conflicts of interest, a coordinated management plan was implemented starting in 2003 by the Syndicat mixte de la basse vallée de l'Aude (SMBVA). ButorStar (Mathevet et al. 2007), a role-playing game, served as a mediator in user discussions on the collective management of the tidal area. Two role-playing sessions were organized in 2006 by SMBVA with 12 users (Mathevet et al. 2008). The experiment was expected to enable discussion of the technical, social, economic and environmental stakes involved in the management of a humid zone. The evaluation was carried out based on individual questionnaires before and after the role-playing game, followed by a telephone interview 3–5 weeks later. The evaluation was completed 1 year later by a series of semi-structured interviews with players and organizers. The users' understanding of the effects of water management clearly had improved, as well as the impact of human activities, particularly those related to cutting and grazing, on the ecological evolution of reedbeds and on avifauna (Mathevet et al. 2008). Two-thirds of the players said they had learned a considerable amount about the effects of user practices on other users, and particularly about the needs of other actors. All of the participants emphasized the importance of sharing knowledge. For two-thirds of them, the experience had not significantly modified the way they viewed how the tidal area functioned. However, they said it had helped them discover the importance of considering the impact the environment had on human relationships and vice versa. After 3 weeks, they unanimously considered the experience to have improved their capacity to participate in group projects on the management and development of nature areas. 1 year later, the survey through semi-structured interviews revealed that the lasting value of some of this learning had diminished. Notwithstanding, the survey showed that all of the users were interested in renewing the experience.

Learning Technical Knowledge

In the majority of the evaluations, the stakeholders who participated in the companion modelling process declared that they had acquired technical knowledge about the structure, dynamics and functioning of the system studied. The diversity of knowledge acquired obviously is linked to the situations dealt with by the projects. This ranges from knowledge acquired through specific training, particularly in mapping, to a broader understanding of farming strategies, agricultural, forestry and pastoral dynamics, interactions between societies and resources, the role of certain economic regulatory tools (e.g. water management fees), interactions between agricultural and urban activities (e.g. risk of fires, urban sprawl on to farmland), and even procedures to implement new management mechanisms and their potential consequences on agricultural practices.

In the Lingmuteychu case study, several inhabitants of the downstream village began agronomic experiments following the first workshop, introducing a second crop (carrots, potatoes, turnips) before the rice crop in their cropping systems. This initiative was referred to several times as a result of the workshop. Asked about their experiments, the farmers explained that they decided to carry them out when they noticed during the role-playing game that such activities had an impact on the income of people practising it in the upstream village. Discussions took place between participants from the two villages that enabled a transfer of technical information. The diffusion of this technical innovation is a horizontal knowledge transfer, from one participant to another. The workshop served to stimulate this learning without actually supplying the corresponding technical information. Once the learning cycle was initiated, the interested participants collected technical information directly from the pioneering farmers. This learning was made possible by the discovery of the principles and results of double cropping practised in the upstream village. It illustrated learning about farming activities in other villages. The habitants of Dompola already knew that their neighbours had recently begun to cultivate potatoes before the rice crop but they had not copied this potentially lucrative practice. Role-playing games, particularly when the roles between the two villages were inversed, contributed to increasing their knowledge of the others' activities and economic results.

The Lam Dome Yai case study (see the Appendix) is a good illustration of another type of technical learning. This work aimed to deepen knowledge about the interaction between water use, management of labour and land and migratory flows in three different types of non-irrigated, family rice farms. At a time when the Thai government was planning a new wave of sophisticated and expensive hydraulic installations, the plan was to test a hypothesis that a greater availability of agricultural water would limit the extent of worker migration, the condition required for the sizeable planned investments to be profitable. Between 2006 and 2008, a series of round trips between the field and research laboratory enabled the co-construction of a MAS simulator. This work was punctuated by five workshops based first on role-playing games, and later on participatory computer simulations. These workshops grouped together students and teachers, agricultural workers, and rice farmers and their families working 11 different kinds of farms. The sustained monitoring and evaluation of the effects of this process were undertaken through participatory observations, systematic individual interviews following each collective key moment and a recording of life histories. While the types of learning are again diverse, this evaluation demonstrated particularly interesting results regarding technical learning. The rice farmers declared that they realized that they needed to be better organized in order to manage better the risk of drought on their farms. On the one hand, the experiment had led them to envisage previously unimagined situations and, on the other, to gain a better understanding of the relationship between water availability, migration and a lack of manual labour on small and medium-size farms. The arrival of an irrigation canal or a large community basin would increase agricultural incomes and allow the introduction of 'integrated systems' (i.e. planting several types of crop around an aquaculture basin also serving cattle livestock and adjacent rice fields). The improvement of knowledge concerning the relationship between the distribution of rainfall and the rice farming calendar was reflected particularly in a better understanding of risk-avoidance strategies for transplanting rainfed rice. In this way, numerous small and medium-size rice farmers declared that following their participation in the workshops they had changed the way they decided their agricultural calendar, varietal choices, use of water from their individual basins, and allocation of manual labour on their farms. One of them had undertaken work meant to supply him with more agricultural water. Lastly, the participants considered that they had acquired new knowledge about how to manage a community basin.

Communication Learning

One of the fundamental principles of the companion modelling approach is to consider and facilitate the expression of different viewpoints about the social and ecological systems studied. The participants gathered together consequently are heterogeneous. This heterogeneity is of various orders (e.g. social, economic, disciplinary), and regroups people of different status (e.g. individuals, representatives of one or several groups, scientists, members of the civil society or simply an inhabitant and citizen). Given this diversity, communication learning is thus a major challenge in the participatory approach. In addition, the analysis of evaluation reports showed that communication learning facilitates the autonomy of participants in terms of their participation in various group processes, from exchanging information to consultation and indeed negotiation. Participants' increased autonomy also relates to a change in the relationship between stakeholders and research scientists. Numerous research scientists recognized that their attentiveness to others was modified through their direct interaction with stakeholders. Numerous case studies showed that participants not only acquired an assurance that enabled them to speak more freely and express their viewpoints, they also learned to accept both that other people may hold different representations than their own and to question the hypothesises of others. Collective key moments were recognized as being particularly propitious for more equal and less hierarchical exchanges compared with that which occurs in more traditional systems of interaction.

From the viewpoint of the organization and the analysis of collective actions, some companion modelling experiments enabled a better identification of the difficulties of certain functions within organizations. These experiments also permitted knowledge to be acquired about the modes of functioning and the attitudes of those in charge, as well as their own attitudes regarding other hierarchical levels. This was notably the case in the Ter'aguas study (see the Appendix). This experiment focused on reinforcing the negotiation capacities of community leaders in the Sao Paulo metropolitan region of Brazil to facilitate their participation in collective decisions regarding complex questions of joint management of water and land in the peri-urban zone. In discussion platforms, their involvement remained limited by sharp social inequality, asymmetric information, a lack of training, a hierarchical distribution of power with other stakeholders and competition between leaders. In addition, the authorities' strong tradition of paternalism encouraged a wait-and-see attitude and patronage cultivated by the opportunist, short-term strategies of local politicians.

The process was developed in two stages: the first stage was dedicated to the collective development of different tools[3] that would facilitate discussions on different aspects of the issue and accompany the reconstruction process. The second stage, inspired by the ARDI method (Chap. 3), relied on several activities and tools developed during the preceding stage. Several workshops were held, including an organized session on a computerized role-playing game gathering community representatives and public authorities (principally from the water authority and the municipal government). This role-playing game, named Ter'-aguas, permitted the simulation of collective decision-making processes and the visualization of their impact on the region. This process was tested twice: to support the preparation of a municipal master plan in the north of the Embu-Guaçu region and to help resolve a conflict between three communities, the mayor's office and the water company over a sanitation project in the Paralheilos region (Sao Paulo municipality). Each game session was subject to monitoring and evaluation by game observation and two short questionnaires completed by players before and after the game. The evaluation was carried out nearly eight months after the last workshop. Of the approximately 35 participants from the two workshops, 24 (leaders and government representatives) were interviewed.

After the game session, the evaluation on the first site immediately noted learning about negotiation mechanisms, particularly the notion of mutual benefit, diversity of stakeholders' interests, advantage of a proactive attitude and the need to take into account the set of issues at stake in the discussion of solutions. Other learning also was mentioned (e.g. on the issues, technical aspects and interdependencies), but they tended to diminish in the long-term evaluation. Participants recognized that having taken part in the entire process encouraged them to think about how to interact with other stakeholders. They particularly became aware of the need to organize better and articulate their expectations, and to become more engaged in a process of dialogue rather than complaining to authorities. Participants mentioned a change in the way they interacted with people in their daily work, notably being more open to listening.

---

[3] Computerized and non-computerized role-playing games, a basic drawing to permit the mapping of subdivisions' 'resources' and a dramatization of a conflict over a development issue in a subdivision (without representing biophysical dynamics).

On the second site, interviews after the game session emphasized a new interest in seeking collective solutions, such as a partnership between the water company and municipal government. By freeing participants from habits and conventions, the game allowed them to explore this previously unthinkable type of alternative. Stakeholders declared that participating in a companion modelling process helped them think about the ways they interacted and about how to build collective solutions. They emphasized the light the process threw on to various stakeholders' attitudes in discussions and the different negotiation techniques used. The game also enabled different stakeholders to be brought together, but the stakeholders remained conscious of the specific character of this rapprochement: on the one hand, the cooperative character of the exchanges was very different from the more conflictual and tense traditional forms of interaction; on the other, community leaders rarely have such easy access to public authorities. In the long term, the learning noted shifted to communication and interpersonal aspects even though a better understanding of the issues at stake and the complexity of the situation were still mentioned by some. Community leaders emphasized the acquisition of interpersonal skills, such as taking a position in relation to other stakeholders, involvement and engagement in the analysis of different aspects of an issue, and the need to articulate better and defend their viewpoint. The search for a solution was now seen as a process that involved different steps and stakeholders and required a preliminary search for, and use of, information. Representations of modes of interaction were modified. The most active leaders were also led to rethink their role in relation to their association or community, or the advantage of individual action (of leaders) in relation to community needs. However, while these representations clearly had evolved, few concrete changes in practices were noted. The leaders' stance in relation to inhabitants, inspired strongly by the prevailing thought that the disadvantaged members of the population needed to be educated, did not seem to have changed. Institutional actors mentioned, however, a greater ability to listen to communities and an increased awareness of local stakeholders' viewpoints when working on solutions.

Organizational Learning

The evaluation of the Lingmuteychu case study more specifically highlighted the organizational learning that followed the last workshop where it was decided to institutionalize a committee to manage the natural resources of the watershed. This institutionalization was made possible by the involvement of institutional representatives in the process and by the emergency context that promoted the formalization of the results of the ComMod process. For example, certain results were integrated into the statutes of the watershed management committee that was created 6 months after the final workshop. Funding was sought with the help of the research and development service in order to implement without further delay the first collectively defined action plan (e.g. restoration of canals, community plantations, etc.). This rapid implementation of collectively approved actions thus helped to reinforce the legitimacy of the newly created institution. A calendar of quarterly meetings permitted the monitoring of activities, the adaptation of their planning and the maintenance of a cooperation arena at the scale of the watershed.

The Méjan case study (see the Appendix) also illustrates the organizational learning of stakeholders. Located in the Parc National des Cévennes in southern France, Causse Méjan constitutes one of the last high steppes of western Europe. Current grasslands are the result of efforts over several centuries to develop grazing and agriculture. This agro-pastoral development has resulted in a decrease of natural forests. Beginning in the 1970s, the national forestry policy promoted the afforestation of certain sectors of the Causse with Austrian black pines. The sexual maturity of these afforestations extremely accelerated reforestation through the spontaneous encroachment of the pine trees and worried national park workers. The ComMod process was initiated with all of the concerned scientific department members and field agents. A simulation model was built based on the available literature and field agents' knowledge to understand the overall functioning of the site (Étienne et al. 2003). The simulation tool then was used as such or in the form of a role-playing game to discuss collectively the probable dynamics of the pine trees with farmers, forestry agents and national park workers (Étienne and Le Page 2004).

Simulations of scenarios and role-playing sessions allowed livestock breeders, farmers, forestry agents and national park workers to discuss and acknowledge the future pine tree encroachment process. The ComMod process led to the implementation of a joint local development plan enabling the collective protection of open areas from reforestation. Simulations emphasized the importance of grouping neighbouring farmers to define a united strategy before developing contracts. They also showed the advantage of taking a long-term perspective and to start planning activities to continue after the end of the local plan. This last point resulted in establishing contracts between 28 farmers, the national park and some forest estate owners. However, while the experience allowed concrete organizational learning, structural obstacles persisted impeding the adoption of new practices and a shared strategy to prevent the spread of pine forests. The evolution of agricultural practices on the Causse remains subject to economic realities and the vested interests of agricultural and forestry sectors. As a matter of practice, livestock farmers claimed financial aid to accompany change. A lack of synergy between public stakeholders due to political or institutional reasons did not allow a continuation of the local development plan when funding ended in 2004.

Learning Dynamics During a Companion Modelling Process

In a consultation process, the various types of learning described above combine in a dynamic and progressive manner so that stakeholders with different perceptions and interests are able to hold dialogues, understand each other better, and even reach an agreement on certain points. While there is little information on learning dynamics in the evaluations, it generally seems that this dynamic is linked to the manner that collective key moments of exchanging viewpoints are alternated with periods of more individual reflection, periods during which participants of the ComMod process may be in contact with other stakeholders and other participatory or non-participatory processes.

The Mae Salaep case study (see the Appendix) in the north of Thailand illustrates such a learning dynamic in a consultation process regarding a conflict over access to irrigation water between different types of farmers in a village community. In this village, gravity irrigation through canals capturing brook water was introduced in the early 1990s with the establishment of litchi plantations. The first planters (the wealthiest farmers) then established the rule 'first come, first served': when a farmer set up an intake on a brook, no one else had the right to place his own intake further upstream. Only a few farmers thus had access to water, and the recent increase in the number of those desiring irrigation resulted in rising tensions within the village. This was the context in which a ComMod process was implemented focusing on the water issue.

During the first game session, the players' actions highlighted the water issue, which created an awareness of the need to collectively resolve this problem (learning about the issue). 'The game enabled the players to understand on their own that it is necessary to change the current rules without our telling them so.' This declaration by a village leader illustrated the experimental nature (as defined by Kolb) of the type of learning in evidence here. Furthermore, through its interactive role-play, the game allowed different participants to gain a better understanding of the situations, problems and perceptions of other stakeholders regarding the water issue. This may at first seem surprising as the members of this small community of a hundred families all know each other. However, as one of the participants said about the game: 'In daily life, everyone goes into the fields. We don't have such an opportunity like this to discuss our problems.'

During the debriefing following the first game session, participants discussed the nature of the problem. In a consultation process, the collective reformulation of a problem is a key step because it determines the manner in which the group will seek a solution. It involves a kind of collective learning about an issue that is related to the collective representations of the issue. In the Mae Salaep case, the following question was asked: is the problem one of availability (water shortage) or of appropriation (unequal distribution of water)? The idea finally chosen by the participants was one put forward by a religious leader in the village. Aware that it was impossible to question directly the 'first come, first served' rule (the local elite would never accept this), he expressed the idea that the problem was linked to a lack of water and suggested constructing a hill reservoir for each village brook in order to increase the overall amount of water available. This would provide an opportunity to discuss rules on how to share water between the beneficiaries of each reservoir. 'Without new installations, the rules will not change' this leader said after the workshop.

Participatory simulation sessions using a multi-agent computer model enabled participants to collectively consider what rules could be implemented if such hill reservoirs were constructed. The type of learning used at this stage corresponds to learning about collective organization. During the workshop, a wealthy farmer imposed the idea that the allocation of water should be based on the surface area planted with perennial crops to be irrigated. 3 Weeks later, however, participants who had continued to discuss the issue reached an agreement on a more equitable form of sharing. Small farmers without access to water thus were gradually able to make their position heard in the consultation, not only through a reinforcement of their individual capacities (e.g. self-confidence, better understanding of the stakes), but also by reinforcing their collective position through the creation of a coalition around a charismatic leader. This corresponds to communicational learning linked to stakeholder networking.

Co-Learning Between Scientists and Stakeholders

Thus far, we have focused on presenting learning from the viewpoint of non-scientific participants. Yet the principles of companion modelling emphasize that the research scientist is himself a participant in the process. As such, he may learn from his intervention like everyone else. In reality, the research scientists acquired knowledge about the social and ecological systems studied in all of the examples presented. More specifically, the commodians improved their skills as far as the facilitation of the companion modelling process and the contextual limits of the ComMod process were concerned.

We considered the example of the Njoobaari case study (see the Appendix). This was one of the first companion modelling experiments, conducted between 1995 and 2003. The work was undertaken in two irrigated systems in the Senegal river valley. Since the 1980s, agricultural irrigation policy based on constructing expensive hydro-agricultural installations to overcome this Sahelian zone's climatic constraints had proven to be a failure. A preliminary investigation was carried out to understand the technical aspects of water management in the framework of a water science thesis (Barreteau 1998; Barreteau et al. 2001). From the field analyses, it appeared that the issue at stake was related less to the quantity of water resources than to the coordination between stakeholders in the irrigated system, from managers up to farmers. A simulator was built to understand how the system operated and to test different combinations of parameters that could facilitate assessments of the viability of an irrigated system. A role-playing game was developed to assist farmers to learn elements of the computerized simulator and discuss the parameters. The farmers' and managers' interest in this tool led the designers to use this role-playing game to help farmers think about their management of water and the credit required to cultivate their plots. A second thesis was undertaken between 2000 and 2003 aiming to analyse the relevance of this type of tool in consultations (Daré 2005; Daré and Barreteau 2003). The use of role-playing games enabled farmers to share the diversity of challenges that they faced to cultivate on the irrigated scheme, to realize and discuss the impact of some people's unpaid loans on the entire group's access to credit (joint surety credit), and to promote exchanges outside the highly codified, customary arenas.

From the viewpoint of research scientists, in addition to sociological surveys to obtain a better understanding of the social systems, the role-playing game was validated not only as a pertinent discussion support, but also as a social investigation tool able to acknowledge the complexity of the system through the analysis of stakeholder interactions in the game and in reality. The methodological results of the analysis and the tools produced served as a basis for other case studies in irrigated systems in Asia and Latin America.

Beyond exploring the diversity of the situations studied, this first analysis of learning in ComMod approaches highlights a common core: learning about the issues at stake and interdependencies, rationalization of various types of knowledge, including both scientific and local know-how, the emergence of new forms of knowledge in the form of technical or socio-organizational solutions, and communicational and organizational learning. These types of learning overlap closely, as do individual and collective learning. This work constituted a first step leading to the recognition of the relevance of our hypothesis regarding the learning process induced by participation in a companion modelling approach. However, some questions remain.

Towards Perfecting the Approach to Consolidate Learning

Improving the Survey Framework

Untangling the web of learning is not easy due to a lack of specificity in the questionnaire used for this analysis. The framework used, together with the theoretical analysis of learning, proved to be relevant when taking into account the diversity of the learning observed. However, the challenge was to maintain coherence between the expectations of companion modelling and the types of learning that we wished to monitor and evaluate. In the ComMod Charter, companion modelling is presented more as a mediation approach between individuals, groups and knowledge than a production process of technical knowledge in the strict sense of the word. This does not mean that there is no learning of knowledge or technical skills. For example, in the case of Mae Salaep, the evaluation of learning revealed that farmers had thought about technical aspects, at times beyond what was expected given the activities undertaken. The evaluator revealed a link between the companion approach and the adoption of erosion control techniques among some farmers although this theme had not been addressed directly in any of the three successive cycles. A more detailed analysis highlighted that this learning was the result of interactions between participants following the first workshop, but the learning was attributed directly to the ComMod process by the farmer(s) interviewed. Thus, even when the ComMod process does not emphasize technical aspects, 'seeds of reflection' are sown through the interactions, consciously or unconsciously, and, in certain individuals, will find a favourable field to sprout and grow. This occurs as if the setting of a scene and the participation in a concrete experience, even if partially virtual, had initiated a Kolb learning cycle in other places and at other times on subjects that may have nothing to do with the designer's intentions. We see here the importance of learning about others, which creates the confidence, habit of exchange and sense of 'between-ness' that facilitates future interactions between participants in a ComMod approach. However, from the viewpoint of research, how should one take into account these seedlings that may have been unconsciously sown yet still have an impact on the process?

In our analytical framework, we also tried to distinguish the results of learning (what we learn) from the process itself (how we learn). This differentiation results not only from putting theory into practice, but above all, the need to improve the learning process within the ComMod approach. While learning techniques like role-playing games or collective key moments are heartily endorsed by all participants, we are not prepared actually to take into account the learning dynamics that punctuate the process. To be able to do so, should we follow the example of educationalists and break down the expected learning in such a way that each element is clarified and examined in the light of the effective stages of the process? In processes addressing situations that are by definition complex, based on interactions between actors and the environment, is it possible to undertake such a breakdown without over simplifying the learning and hiding the complex nature of the systems studied?

Learning that is Individual and Collective or Individual Learning of a Collective Group

The analysis of the material collected showed that some types of learning are particular to individuals and other types to groups. However, as long as learning remains in the field of cognition and does not result in action, it is difficult to distinguish between the two levels. Most of the other types of learning, realized in action, are quite often the result of interactions between individuals or between members of a group. With experience, one understands better the paradox of organizational learning raised by Argyris and Schön (2002).[4] Finally, these types of learning that we have described as being individual are part of individual learning that is acquired collectively, thus drawing close to distributed cognition of learning between group members.

Yet in the companion modelling approach, group learning is essential. One difficulty in the learning realized though multi-actor platforms, in contrast to existing organizations, lies in the need to first establish a common objective. Some social learning approaches favour double loop learning, which requires excellent communication skills, while others prefer to consolidate social ties and create a

4 For some, an organization has neither a mental model nor a memory, and thus cannot learn.

sense of engagement by mobilizing participants around a desired future. When participants reach an agreement on an overall objective, experiential learning processes may take place and skills may be reinforced. This result is compatible with the work of Callon and Latour (1981), which rejected the possibility of separating these two dimensions. The actor-network concept introduces an analytical framework in which technical and sociological components are intimately linked in the same network. By observing how a Swedish compacting machine was adapted to brick production from agricultural residues in Nicaragua, Akrich (1989) highlighted the successive, back and forth movement between technical innovation, social evolution and biophysical environmental transformation.

The monitoring of learning in the case studies showed that it is possible to identify a joint group objective during collective key moments in a ComMod approach through the exchange of experiences and the exploration of scenarios. This is easier than relying on double loop learning based on methods such as Socratic dialogue techniques. Once an acceptable common objective or scenario is defined, each participant is able to readjust their own initial objectives. Participants may thus engage in a process of seeking information and technical and organizational options that would enable the achievement of collective and individual objectives.

We must equip ourselves, therefore, with the means, not only to take into account other types of individual learning, but also to promote them when necessary. The question then becomes how shall collective learning be reinforced when this proves to be necessary? Should we try to seek the involvement of local organizations (which touches on the power struggle issue discussed in Chap. 5) more systematically, or should we seek to render the groups constituted through a ComMod process more enduring (which raises the question of their legitimacy or their legitimization in the local context)?

From Learning to Action and Social Change

This last question brings up the issue of the future of the companion modelling process once the commodian has left the study site and thus, of how the learning achieved is perpetuated. How shall one capitalize on the positive dynamics produced during moments that are highly collective but limited in length? Some case studies implemented a specific monitoring process of collective key moments. The learning was re-evaluated *ex post facto* through the ADD-ComMod project several months and even years later. This monitoring over time allowed us to take into account learning dynamics. What emerged is that little of the learning led to effective change in practices or to concrete action able to initiate more profound social change. The achievement of an Argyris and Schön double loop was not easy and undoubtedly is one of the main issues on which commodians must continue to work.

One avenue to explore is the more systematic involvement in the approaches of local partners (e.g. local NGOs, development organizations, village intermediaries), who would be able to assume the process on their own account, to perpetuate results of group discussions and to translate them into concrete actions integrated into development programmes. The training of commodian apprentices would be necessary for such a long-term effort.

Some tools and methods derived from ComMod approaches can be used to consolidate learning. Role-playing games are powerful instruments that enable players to take into account the diversity of interests at play. This can be done even when the game takes the form of a simple social drama, a dramatization of roles that does not involve interaction with an environment (Camargo et al. 2007). For instance, in the SosteniCAP case study (see the Appendix), a role-playing game was associated with a set of activities aiming for legal restructuring and consolidation of community potable water associations. The role-playing game where the economic operations of these associations were represented enabled learning on how to operate and reinforce social control within certain associations. The relevance of these tools is linked to their participatory development. This participatory development enables the integration of different representations into a flexible approach capable of adapting to local specificities. Some tools produced by development approaches may also be used in ComMod approaches to support certain stages or to prepare the integration of results in collective, action-orientated processes.

However, authors such as Jiggins and Röling (2000) questioned the capacity of social learning approaches to move past socio-political differences, power plays and conflicts. Facilitators of these approaches may be led to support an empowerment process for some participants, or more strategic negotiations, and to mobilize themselves to execute the agreements negotiated (Leeuwis 2004). During the 1980s, gender-related literature showed that women needed first to be made aware of the problem under consideration and of their role, and to share their experiences between themselves, before they were able to participate in discussions involving a wider audience. Womens' groups offer the possibility of exchanging experiences, acquiring knowledge about the issues at stake, understanding different interests, developing communication skills and reinforcing self-confidence prior to receiving support enabling them to engage in negotiations with other stakeholders. Preliminary work with hybrid simulations within homogeneous groups (and not platforms of heterogeneous stakeholders), as in the Mae Salaep case study, can thus allow the same type of learning, which is crucial for an equitable consultation process.

For learning to endure, the institutionalization of results, or, in other words, irreversibly anchoring these results by relying on locally legitimate constraint systems, whatever their form, appears indispensable, as was highlighted in the Lingmuteychu case study. However, this institutionalization is not problem-free: it assumes an organization with development partners that is achieved sufficiently early for the latter to feel like stakeholders in the approach rather than instrumentalized by it. Furthermore, the institutions mobilized must be able to intervene with the flexibility needed to integrate propositions that may take unpredictable and relatively heterodox forms. The structure and orientation of development projects does not necessarily allow such flexibility.

It seems that the passage from learning to collective action is in large part dependent on the context. If the context is not mature, if the protagonists do not feel an urgent need to act, if their room to manoeuvre is too limited, then learning risks becoming diluted over time. Does this mean that to result in concrete action, modelling approaches should only be developed in contexts marked by tension? A review of the various case studies (Chap. 4) indicated that such a short cut would be a mistake. The passage to action can be prepared. However, this requires thinking about how to prolong the duration of the ComMod approach after the project ends and how to diffuse results from the very beginning of the project.

How can we proceed towards the production of skills/knowledge/capacities that can be transferred to people outside the process? What method should be put in place to further the diffusion of knowledge? If one works from the perspective of situation learning, Wenger (1998) and Loeber (2003) have shown the depth of difficulty in trying to extrapolate beyond the initial group. The perspectives and engagements resulting from a social learning process reveal themselves to be, in effect, difficult to transfer as much to all members of the social group represented as to the regulators and decision-makers, particularly when they result from intense interactions. Those initiating projects often neglect the importance of pre-existing institutional rules and power dynamics (Chaps. 4 and 5). Henceforth, it is accepted that no matter what the context, facilitators and participants need to position themselves in relation to socio-political discourse and dynamics. Ignoring this socio-political context raises the risk of widening the gap between the solutions that are locally desired and those that actually are implemented. Furthermore, this discussion on the passage from learning to action, and the initialization of social change, should be debated again in terms of the various stances of research scientists (Chap. 5) and the development of a quality approach (Chap. 8).

From a Virtual World to Reality

Collective key moments are special, concrete experiences during which the essentials of learning occur or are initiated. Chapters 3 and 7 showed the power of the intermediate objects used to put stakeholders into situations favouring interaction, allowing them to exchange viewpoints and to construct a shared representation of the issue addressed. However distant the description of the world in role-playing game sessions may be from the real world, certain stakeholders will apply the learning obtained during the game session directly to their daily life. The danger is particularly real in relation to technical learning, particularly where there is an outreach and villagers support mechanism for technical advice. In interviews for the Thai case study, numerous farmers said: ‘I will do as the game showed me because then my income will increase’. This kind of pitfall is dangerous as the simulations only represent a simplified vision of reality. Considering the level of simplification in co-constructed models, commodians must take special care to prevent this lapse. The game is not reality (Daré 2005). It is a moment apart that allows discussion about reality but the learning that occurs in the game is not a toolkit that can be transposed intact to reality. More in keeping with the ComMod stance, others declared that: 'it made me think about this technical solution that I had not considered before'. The translation, deconstruction-reconstruction of learning, and adhering to the principle of reflection must guide the actions of the commodians. It thus might be fruitful to link the ComMod approach with technical services more capable of transforming the technical learning acquired during the games into valid learning for the reality of farm management, to name one example.

Conclusion

Social learning processes traditionally opt for two types of objectives: (i) reinforcing social ties through an engagement around a desirable future; (ii) double loop learning allowing the individuals' mental representations to be reconsidered. This presumes that the capacities of comprehension, questioning, debate and reflection would be mobilized to highlight tacit knowledge and create a real forum for exchange. The ComMod approach aims for both types of objective. On the one hand, it facilitates the identification of a shared, desired objective through the development of scenarios that can be tested and discussed. On the other, it allows the development of interpersonal and communication skills. Collective key moments occupy a crucial place in this learning, whether individual or collective, because they provide an occasion to explore interdependencies during debates over the construction and evaluation of scenarios.

The companion modelling approach, therefore, appears to be an interesting way to promote a certain amount of conscious learning likely to engage a heterogeneous group in actions permitting more sustainable development. It introduces ways of sharing knowledge, learning and decisions that are innovative for most participants. Commodians effectively incite participants to involve themselves and experiment with new forms of communication based on interactions between stakeholders and between stakeholders and systems, which facilitate learning by experience. These exchanges simultaneously allow learning about the challenges and emerging issues in all their complexity and a better understanding of others and their interests while clarifying the reference frameworks of each. This promotes the reconsideration of these various elements, a prerequisite for discussion and experimentation, and the learning of new technical, organizational and communication rules.

However, the learning dynamics in the ComMod processes deserve to be described better. At the close of our analysis, new questions appear that should lead to more precision on the phenomena of learning in ComMod approaches, notably the passage between virtual and real worlds, the latter of which is by definition more complex, regarding the capitalization of this learning and its diffusion within and beyond the group to decision-makers, and the transformation of learning into individual and collective practices.

Chapter 11
The Companion Modelling Approach: Dealing with Multiple Scales and Multiple Levels of Organization

Raphaèle Ducrot, Aurélie Botta, Patrick d'Aquino, Martine Antona, Géraldine Abrami, Stefano Farolfi, Jean-Pierre Müller, Erwann Lagabrielle and Christophe Le Page

The companion modelling approach initially was conceived and tested at a local level because of the focus on resource users. Based on the premise that different stakeholders involved in resource management operate with different areas of reference and time scales in mind, the approach focuses on the representation of this diversity in its tools and in the form of coordination workshops (d'Aquino et al. 2002b; Étienne et al. 2008c; Le Page et al. 2001). Like many participatory

R. Ducrot (✉)
UMR G-EAU, CIRAD, Maputo, Mozambique
e-mail: raphaele.ducrot@cirad.fr

A. Botta · M. Antona · J.-P. Müller · C. Le Page
UR GREEN, CIRAD, Montpellier, France
e-mail: aurelie.botta@cirad.fr

M. Antona
e-mail: martine.antona@cirad.fr

J.-P. Müller
e-mail: jean-pierre.muller@cirad.fr

C. Le Page
e-mail: christophe.le_page@cirad.fr

P. d'Aquino
UR Green, CIRAD, Dakar, Senegal
e-mail: patrick.daquino@cirad.fr

G. Abrami
UMR G-EAU, IRSTEA, Montpellier, France
e-mail: geraldine.abrami@irstea.fr

S. Farolfi
UMR G-EAU, CIRAD, Maputo, Mozambique
e-mail: stefano.farolfi@cirad.fr

E. Lagabrielle
University of Port-Elizabeth, Port Elizabeth, South Africa
e-mail: erwann.lagabrielle@gmail.com

M. Étienne (ed.), *Companion Modelling*, DOI: 10.1007/978-94-017-8557-0_11,

approaches developed at the local level, the approach quickly confronted issues posed by the institutionalization of participatory processes (Pimbert 2004): the institutional integration of collective learning developed at a local level, taking into account stakeholders not present at the local level, and the need to interact directly with regulators and decision-makers at higher levels. A study group was formed within the ComMod network, in parallel to the ADD-ComMod project, to discuss the novel challenges raised by the application of companion modelling to several organizational levels of action and decision and by changing scales in a companion modelling approach.

In a companion modelling approach, the recognition of diverse viewpoints is a central element. Therefore it must pay special attention to preserving and integrating diverse ways of thinking, and must structure action, particularly collective action, in 'levels'. Knowing that the multiple levels of organization in which a ComMod process operates is just one of several cognitive representations of structuring action within a society, and that this representation is not necessarily shared by all stakeholders, the 'multi-levels' issue may be broken down into three questions.

- Which multi-level representations should be used and how should they be integrated into the approach (which tools and participatory methodology)?
- What impacts do these choices of stance and methodology have on the context?
- How can one position oneself in relation to these questions in a way that would be the most rigorous and consonant with the core principle of companion modelling?

This chapter begins by defining several key concepts and by describing the biophysical and social processes underlying multi-scale issues. We then define the challenges that these issues pose for companion modelling approaches and analyse how the ComMod group has addressed changing scales and multi-level accompaniment in different case studies. We then explore more generally the scientific questions raised by such exercises and identify the main fields of research before concluding with the positioning of the ComMod group.

Analytical Framework

Definitions

This section reviews the key concepts encompassed by the terms 'level' and 'scale'. We treat these concepts thanks to three complementary notions, namely dimension, level and resolution. Lévy and Lussault (2003) define dimension as 'a manner that favours the intersection of partial viewpoints over a vast set of phenomena rather than the partition of a field into fragmented and unrelated areas'. Spatial and temporal dimensions are the two most often used to characterize a

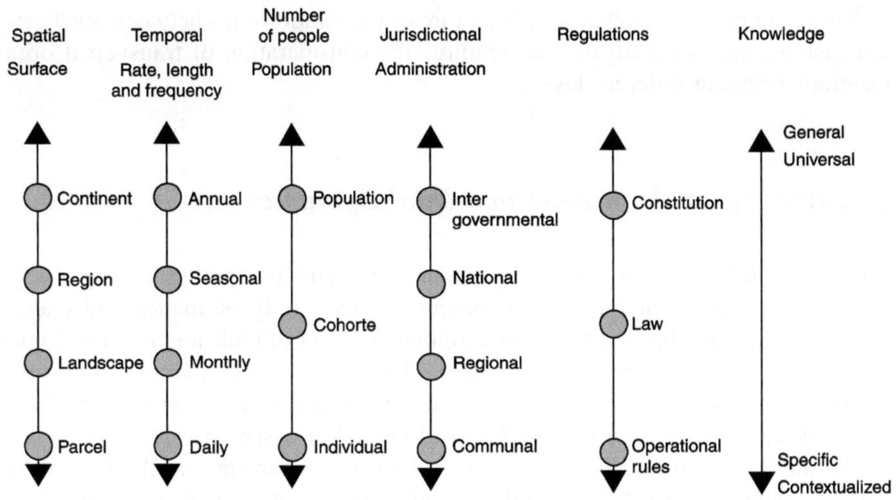

Fig. 11.1 Examples of dimensions and levels potentially relevant to the study of links between society and the environment (adapted from Cash et al. 2006)

process, for example, a biophysical process in ecology often is described in terms of length of time or spatial expanse. However, this definition may designate all other quantitative or analytical dimensions used to study a phenomenon. Following this logic, Cash et al. (2006) cite more specific dimensions to study social dynamics, including, for example, dimensions linked to the institutional structure of societies, and juridical and regulatory dimensions, characteristics particularly of political organizations. Each of these dimensions may be considered on several levels (Fig. 11.1). These levels constitute hierarchies that may be interlocked (e.g. a region is made up of landscapes that are themselves composed of fields) or not (e.g. a constitution sets the framework for possible laws without containing all of them just as a regional administrative authority constrains the operations of communal administrations without being constituted by these communal administrations). Finally, resolution corresponds to the degree of precision of the system description or perception. In spatial representations, it corresponds to the cartographical sense of s scale. A field, like a region, may be studied on the scale of a hectare.

Depending on the set of problems under consideration, these concepts were long considered to be predefined and unique, a particular question falling more specifically under a particular field and treated at a particular level and resolution. Yet due to the diversity of viewpoints of a system, there are in fact a multiplicity of dimensions, levels and resolutions to consider. The representation and sharing of this multiplicity is part of the challenges of collective management. In the remainder of this book, the term 'scale' refers to the ensemble of these three concepts.

These definitions sketch a simple picture of the relationships between levels and dimensions, but our analysis also requires the consideration of transversal organizations bridging different levels.

Questions of Scale in Environmental Dynamics

The acknowledgement of linkages between spatial and temporal scales has always fuelled the analysis of biophysical processes. The study of matter and energy flows, for example, has led to the development of a set of tools and methodologies in ecology and geology, among others. However, taking into account links between levels is much more recent and controversial. It was long thought that a given process should be monitored and discussed at a specific spatial level and time scale. For example, the study of hydrological mechanisms usually focuses on a small-scale watershed level and an hourly time scale when analysing flood processes, yet tends to focus on a large-scale watershed level and a monthly basis when studying allocation mechanisms between different uses subject to seasonal climate variations. This manner of analysing biophysical processes rests on the fact that most of these processes predominate at certain combined levels of space and time. In hydrology, the infiltration of surplus rainfall relates to localised (within 1 km^2) and rapid (about 1 h) phenomena while underground flows concern water tables of several dozen to several hundred square metres over multi-year timeframes (Bloeschl and Sivapalan 1995).

Nonetheless, the major changes affecting our environment increasingly highlight the importance of interactions between levels of the same or different dimensions. For 30 years, numerous disciplines have studied and sought to theorize these phenomena. The theory of hierarchies (Koeslter 1967) applied to ecology (Allen and Starr 1982) describes complex systems as sets of small systems that are merged into a limited number of larger systems. This hierarchy presumes a correlation between spatial and temporal dimensions and levels of the biophysical processes that are described; rapid cycles of small spatial scales are embedded in slower cycles involving larger spaces. This coupling between cycles explains in particular the potential transmission of disturbances along spatial and temporal dimensions. Elmqvist et al. (2003) thus attribute the degradation of the Jamaican coral reef system to the overfishing of herbivores in the region. The disappearance of green turtles followed by herbivore fish weakened the coral reef system to the point that the elimination of the last grazing organism (a type of sea urchin) by a particular pathogen and the passage of a hurricane were enough to allow fatal algae to invade the coral reef. A second essential theory of multi-scale dynamics and complex systems is the theory of emergence. Considering that the articulation between two description levels (micro and macro) of phenomena has different resolutions, the macro-level may only be a simple aggregation of what happens on the micro-level. In this case there is no emergence. However, it is possible that the macro level possesses particular properties that cannot be reduced to, or exist outside of, the elements that make up the system at the micro-level. The phenomenon of emergence is then said to take place.[1]

Questions of Scale in Social Dynamics Linked with Environmental Dynamics

Social processes regulating the relationships between societies and the environment operate in various dimensions (i.e. spatial, temporal, regulatory, demographic), often on several levels, and are linked through these dimensions (Fig. 11.1). The uses of a resource by individuals organized into social groups, the formulation of management rules, or more generally the governance of ecosystems, are thus defined at numerous levels and dimensions.

Yet the analysis of social processes rarely calls on the notion of scale,[2] most often favouring descriptions at a single level. This is due to the difficulty of describing global processes at a reduced level and aggregating local processes at higher levels and indeed, identifying forms of emergence. Numerous questions are raised when considering scales of structure and representation of these social processes.

One question concerns the pertinence of the scales chosen when the structure of the social processes affecting the governance of ecosystems is evolving. The scales conventionally chosen represent a linked stack of spaces and structures of regulation and decision. However, a society's collective modes of action (i.e. transversal, vertical and horizontal) cannot be summed up in this hierarchal structure. On the one hand, contrary to this concept of a set of descending, interlocked regulatory arenas dividing society from top to bottom, society is 'a continuum' (Rosenau 1990, 1992). The shift from a local to a global perspective is a continuous process of interactions that pass through levels of variable and circumstantial aggregation according to the type of interactions being established (Rosenau 1990). On the other hand, networks, lobbies and social or personal ties associating stakeholders motivated by shared action principles (Breiger and Pattison 1978; Vachon 1993; Bakis 1993) contribute as much 'behind the scenes' (Goffman 1979) to multi-level links as do the operational links between hierarchical levels of organization. The combination of these processes and power mechanisms (i.e. confrontation, cooperation and domination) results in an

---

[1] More precisely, Dessalles et al. (2007) stated that the macro-level emerges from the micro-level if the micro-level is constituted by interacting elements whose properties or dynamics are described in a certain theory D, generating a property or global dynamic at the macro-level that can be described by another theory D', such that D' is irreducible to D, in other words, there is no possible way of calculating the elements of D' from the elements of D.

[2] The exception is geography, which uses spatial representations in the form of successive, hierarchical levels of organization, and geographic objects dependent on these organizational levels.

organization of ecosystem governance based on collective groups, institutions and evolving subdivisions (Swyngedouw 2004; Young 2006). Modes of governance also vary in time: some collective groups or institutions (e.g. decision arenas, reference territories) appear or even disappear as certain processes (affecting the rules or the constitution of the collective) are reintegrated into other levels.

A second question concerns the understanding of the mechanisms underlying one level in order to understand the processes that exist at other levels and indeed other dimensions. Numerous authors have examined the mechanisms underlying various levels and have proposed matching analytical frameworks. An example is the IAD (institutional analysis and development) framework, which aims to identify universal mechanisms operating at different levels and dimensions and suggest a taxonomy (Ostrom et al. 1994; Young 1995). The regularity of social behaviour at these different levels creates structures that affect behaviours and their outcomes. This analysis leads to the identification of a core unit of analysis straddling several dimensions, that is, the action arena (Ostrom 2005). In modelling, the agent-group-role concept, inspired by analogies derived from organization theory, aims to describe different levels and entities that hold different positions, either simultaneously or changing over time, in these different levels. Interactions within a system can only take place within a group, interactions between groups being generated by agents who play roles in several groups (Abrami et al. 2005; Ferber and Gutknecht 1998). These different approaches call for a nested scale approach. The question thus looks at how representations of these divisions are constructed.

In the analysis of governance systems, it is now largely accepted that scales are social and political constructs (Adger et al. 2005; Lebel et al. 2006; Meadowcroft 2002). Representations of the dimensions relevant to address society-nature relations thus vary according to stakeholders and are affected by their modes of comprehension and political, economic and scientific simplifications (Bulkeley 2005; Lebel et al. 2006; Meadowcroft 2002; Sneddon 2002). The legitimacy, credibility and means of integrating and understanding knowledge are strongly dependent on the level at which the knowledge was produced and analysed and thus, the type of associated stakeholders. Broadly applicable, generic knowledge produced by formal scientific procedures often is favoured at global levels while lower levels mobilize knowledge based on practice or situation experience[3] that is based on scientific procedures or traditional knowledge (Cash et al. 2006; Weible et al. 2004; Young 2006).

Power relations between institutions and mechanisms of domination, resistance and cooperation contribute to the definition of scales (actual divisions or representations of divisions) and their reconfiguration. 'Politics of scales' illustrate how

[3] More overarching levels sometimes mobilize processes of cognitive hegemony that impose certain models or types of management (Molle 2008). However, this hierarchical authority cannot adequately take into account the complexity of relations between levels as is shown by the difference between how a rule thought out at a higher level is reinterpreted at a local level (Urwin and Jordan 2008).

the choice of scale is thus an inclusionary or exclusionary instrument of power that modifies the means of access to resources and decisions (Lebel et al. 2006). 'Empowerment' processes are thus linked to the acquisition of strategies of scale and capacities to mobilize scales and different levels (Swyngedouw 2004). This capacity depends not only on the capacity of stakeholders to make sense of questions raised, but also on their capacity to develop ties and alliances with stakeholders operating at, or mobilizing, different dimensions and levels (Boelens 2008; Bulkeley 2005).

Companion Modelling and Questions of Scale

Problems of Scale in Modelling

Problems of scale in modelling are well known in the representation of biophysical processes. They derive from the potential inadequacy between the scales taken into account in the different steps of abstraction leading to a model, that is, observation of a process, analysis of observations, then the representation of the process. Numerous methods have been developed to prevent such inadequacies. The non-linear character of certain processes, their different timing, as well as technical measurement constraints, pose problems for data sampling, which become increasingly difficult to resolve as the processes studied become increasingly heterogeneous. Some methods, notably statistical ones, specifically address the representation of this variability within a unit of reference in order to allow the parameterization of one spatial unit to another and to ensure the distribution of parameters or their interpolation under boundary conditions (Bloeschl and Sivapalan 1995). From the perspective of modelling, they thus ensure the transfer of information between the modelling scales and the inquiry scales based on the assumption that a pertinent level exists to address the dynamics studied. The choice of scale, therefore, helps to reduce complexity (Ewert et al. 2006).

The challenge in terms of representing socio-ecological systems is how to better take into account their co-evolutionary character and their capacity to adapt to change (Folke et al. 2007). This assumes that ecosystems, in all their hierarchical complexity, are taken into account. However, the scientific approaches cited above consciously simplify representations through their choice of scales. Having done so, they also largely ignore the strategic value of scales in discussions between stakeholders, as well as the different rationalities implemented in the choice of these scales (Karstens et al. 2007). This new discrepancy is even more important for modelling approaches operating at the interface between scientific and public policies spheres (Sterk et al. 2008).

Problems of Scale in Companion Modelling

Companion modelling is, in addition, faced with the integration of different scales during the confrontation of representations, the development and use of tools and the ensuing discussions of simulations. However, compared with other participatory approaches, companion modelling is challenged in a unique way because the scales to be taken into account are not all determined *ex ante*. The open nature of the modelling process effectively allows participants to introduce new scales during the process. Lastly, the involvement of stakeholders from diverse organizational levels and stakeholders who actually represent several levels simultaneously raise the issue of information transfers between levels: how should the sensitivity of certain information transmitted through the mechanism be handled? All of these questions raise issues of stance and methodology.

Stance

One of the priority challenges of the ComMod approach is to construct an accompaniment that allows the recognition and formalization of different points of view on an issue. When divisions of scale are understood to be a subjective perception and, therefore, able to take multiple forms, the following question is raised: which levels should be taken into account given the fact that the differentiation into levels is subjective and linked to the dominant cognitive model?

All segmentation of society into levels, most often conceived of as a set of nested dolls, is artificial and, therefore, potentially plural. Such structuring can be considered for analytical or operational objectives. However, in participatory approaches, the operational issue is not the only one to be taken into account because these approaches also seek to encourage the participation of different elements of society in the process. In general, one seeks to involve analytical stakeholders from 'internal' levels of society (e.g. spokespersons with different points of view, traditional leaders, etc.) on an operationally conceived level (e.g. management committee, collective, etc.). When a participatory approach addresses several operational levels, it usually reproduces its analytical questioning (which stakeholders should be integrated so that the points of view are representative?) at each level of intervention. However, some agents keep in mind that the 'vertical'[4] organization of a society (e.g. from a territory to a region) may also be considered differently to the operational stack with which they work. They favour in their approach the intervention of networks and other non-stacked structuring modes

[4] In the rest of the text, we only will refer to the 'vertical' dimension to simplify the discussion. However, at each use of the term, we imply that a 'horizontal' dimension to the structure of action exists in addition to the 'vertical', particularly as these two dimensions are themselves very artificial.

without necessitating the use of the stacked representation[5] system of operational levels.

In our operational contexts, the shift to multiple levels means, in particular, the integration of much more differentiated cooperation arenas: in other words, the degree of differentiation of issues and perceptions between the level of a village and the level of national government is by nature much greater than the degree of differentiation between a farmer and a livestock owner or between a village head and the head of a commune encompassing several dozen villages. There is thus a specificity to this question linked to a context that one may refer to as more multi-institutional than multi-level.

Methodology

The Multi-Scale Question Traverses the Set of Processes

The question of which levels and institutions to take into account is raised from the start of the process and is directly linked to the choice of stance discussed above. The formulation in question of a problem and the expectations of the institutions supporting the approach more or less explicitly predefine the entry point, which will restrict the choice of scales.

The linking and integration of points of view continue during the modelling phase due to the iterative and evolutionary character of the ComMod approach. The choice of participants, the formulation of the question to be addressed, the ways the participants' representations are elicited and the conceptual frameworks used all consider the question of scale in a more or less explicit fashion. For example, the ARDI method (Étienne 2006) includes a step dedicated to identifying clearly the levels and territorial and temporal divisions that make sense to different participants.

The representation of multiple levels also affects application choices, particularly the selection of formalisms and tools. Each tool offers different possibilities for representations and partitions between scales. For example, GIS are powerful tools that can be used to represent spatial and territorial dimensions but they are ill suited to capturing temporal dimensions. The latter, which are fundamental to the representation of flows or dynamics, require the use of simulation models. In addition, certain divisions or clusters of levels are more or less possible depending on the tool. Lastly, the implicit or explicit scales included in the tools have an impact on simulation results, their subsequent discussion and the confrontation of perspectives regarding these results. The technical and pedagogical potential of each tool (Chaps. 4 and 7), therefore, must be analysed and discussed according to the chosen scales. The companion modelling approach is characterized by the use of different tools in order to make the most of their complementary aspects and

5 Representatives of level n^{-1} participate in discussions at level n.

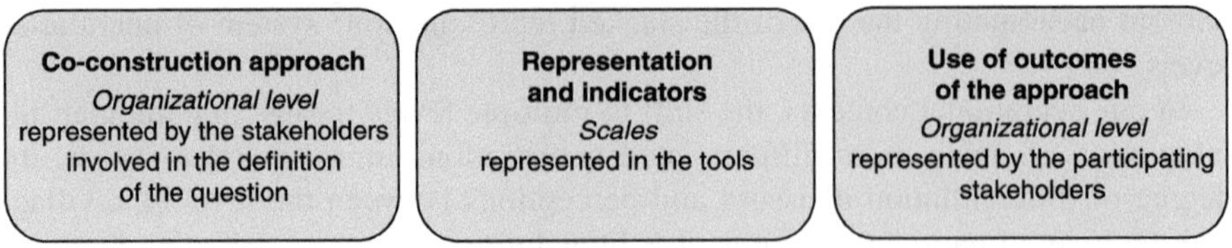

Fig. 11.2 Schema of three ways of taking into account multiple scales in a companion modelling approach

diminish their limitations. Broadly, there are three ways of integrating multiple scales (Fig. 11.2): by integrating stakeholders holding multi-scale representations, integrating multiple scales into the tools used, and lastly, implementing these tools in arenas with multiple mandates.

Different Forms of Shifting Scales

Snapp and Heong (2003) described numerous ways of dealing with shifting scales in participatory approaches. In the ComMod approach, there are two main types. The first is the change of scale that follows when the process, while still addressing the same general issue, is applied to a different level of decision-making. This is referred to as 'up-scaling' when the shift is towards a more encompassing level and 'down-scaling' when descending towards a more local level. The second is the integration of a larger number of stakeholders or situations, in other words, a horizontal dissemination of the approach known as 'out-scaling'. This second type includes the replication of the approach or broadening its reach to encompass other forms of participation that may involve a greater number of people. A particular example would be the local definition of a principle or regulation defined at a higher level. Two routes have been developed by the ComMod group: the diffusion of the approach through training (Chap. 12) and the development of generic, 'de-contextualized' models based on abstract issues and fundamental processes (Chap. 4).

Lastly, the companion modelling approach assumes the existence and succession of learning loops linked to the evolution of the issue and partnerships. The inclusion of multiple levels of organization can be broken down in numerous ways over time, successively using the strategies we have just presented (Fig. 11.3). One of the catalysts for a change in loops may indeed be a change in the decision level, which is translated by a change in the questioning and in the reconstitution of the group of participants.

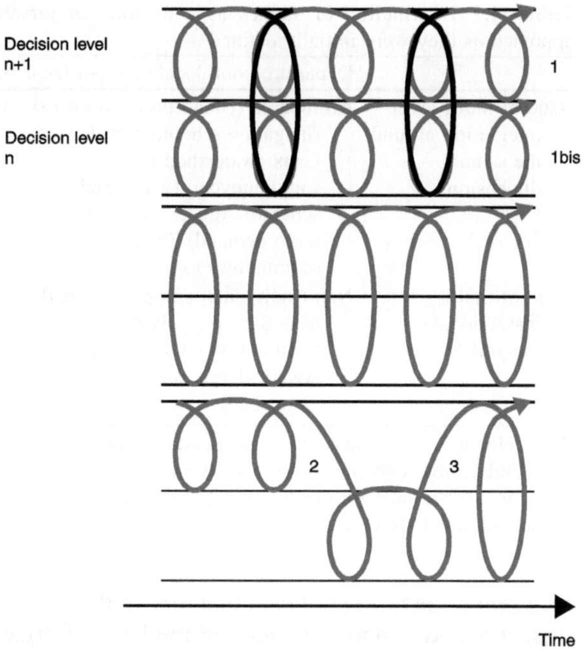

Fig. 11.3 Examples of the sequencing in time of taking multiple organizational levels into account. The numbers 1, 2 and 3 refer respectively to strategies that: take multiple levels into simultaneous account (two types), down-scale, and up-scale

Taking Questions of Scale into Account in the Case Studies

A question Left Out of the Evaluation Methodology

Within the framework of the ADD-ComMod project, the integration of questions of scale was evaluated in 18 study cases, 14 of which were undertaken by commodians. The evaluation methodology did not explicitly take into account questions related to shifting scales and multiple levels. Consequently, the strategies selected, the questioning developed and the problems met were not necessarily laid out in the available material. The issue of multiple levels was only really studied in cases where: the designers of the approach explicitly addressed this subject and were confronted with the methodological difficulties it raises; the concealment of this issue limited the reach of the approach in one way or another; the evaluator was aware of, and interested in, the subject.

A Multi-Scale, Multi-Level or Multi-Institutional Approach?

Approaches that are More or Less Explicitly Multi-Level

Twelve studies explicitly considered questions of scale in the process while only two implicitly took the issue into account. Nonetheless, multi-level integration is a

Table 11.1 Distribution of studies as a function of jurisdictional levels integrated into the approach as they were initially defined

	Departure point local or supra-local	Departure point regional (1)
Issue of multi-level integration around the same dimension	AtollGame (island user) Njoobarri (irrigation scheme plot) Pays de Caux (watershed plot) Lingmuteychu (watershed scheme) AguAloca (watershed large demand) Ter'aguas (community to town)	Domino Réunion (town and region) Nîmes-Métropole (town, inter-town, administrative division) Domino Senegal (regional level)
Issue of multi-institutional integration	Nan (individual, community, park) Mae Salaep (individual, community, collective) Kat Aware (large, multi-sector use watershed)	Ouessant (individual park)

1 A regional level is considered here as an intermediary level between a local and specific (*plot*) level and a more encompassing level (*state, agro-ecological region*). This term does not correspond to a predefined administrative organization (such as UNESCO's definition of a region or a French administrative region)

recurrent issue in a significant majority of the cases studied. It occurs in the choice of spatial and temporal scales and the level of organization of the representations and tools developed, but also directly in the objective of the operation itself. Certain operations aim to help one level of organization to get a better grasp of another level or institutions other than those it is accustomed to considering. In a certain number of cases, and depending on the subject studied, what really is under discussion is a multi-institutional integration rather than a multi-level one (Table 11.1).

Operations principally focus on the 'local' level (the user as an individual), or the 'supra-local' level (groups of users with a recognized representative), and the immediate territorial and jurisdictional levels nearby. Sometimes stakeholders from levels further away are periodically associated. Only four study cases (Nîmes-Métropole, Domino Réunion, Njoobaari and Lingmuteychu) actually integrated more than two territorial and jurisdictional levels. At the most local level, two studies (Mae Salaep and Nan) were interested in the role of community representatives in its interface between users and local authorities.

Frequently, the question raised and its relation to the dynamic of the resource and institutions determines the jurisdictional, spatial and institutional levels considered. With the exception of the La Réunion case study, there appears to have been a lack of specific will to apply the approach to the set of different arenas encompassed by the issue addressed. However, sometimes levels other than those mobilized in the participation arenas were able to be considered in the tools developed. For example, the Nîmes-Métropole study addressed the individual parcel level in a GIS even though the discussion models and arenas did not include this level. The available material did not, however, allow for a detailed analysis of the relationship between the territorial and jurisdictional levels taken into account in the tools and those represented in the discussion arenas of the approach.

Multi-level integration is almost always a simultaneous process. The exceptions occurred when the higher regional and watershed levels were taken into account during the approach (Lingmuteychu) or local authorities were integrated only once the users felt sufficiently comfortable (Mae Salaep and Nan). The levels chosen could lead to a disengagement of stakeholders (and, therefore, their respective levels) whose preoccupations are located at territorial levels other than those considered, or even their withdrawal from the process. This was the case of small farmer representatives in La Réunion.

Working with Public Policy-Makers

Two case studies worked from the start at the public policy organizational level. They chose fairly similar strategies characterized by an upstream mobilization of institutional support for the approach at the level of decision-makers, its insertion into social networks at this level, and the differentiation (to a greater or lesser degree) between arenas in which tools are developed and those in which simulations are discussed. The first relied on technicians operating at the regional level under consideration (if necessary, representatives of users with a mandate at this level), while the second directly mobilized decision-makers (e.g. mayors, etc.). This decoupling permitted more active participation of non-academic stakeholders in the modelling stage, while allowing the effective participation of decision-makers in the approach as long as it was possible to work around their scheduling constraints.

Taking into Account the Temporal Dimension and the Integration of Scales

There are few references regarding dimensions other than jurisdictional or spatial, notably the temporal dimension. Nevertheless, this dimension is integrated, indeed even explicitly taken into account, in numerous simulation tools (i.e. computer models and role games). In AguAloca, for example, a game round representing a climate season was subdivided into two periods (i.e. collective time and individual decision time), while the hydraulic model turned on a monthly basis. One study was able to consider up to four different timeframes for two spatial and jurisdictional levels (i.e. parcel and irrigated scheme) to address the temporality of irrigation coordination. Short timeframes (i.e. day and week) were taken into account in the computerized part of the game while the part played by participants covered longer timeframes (i.e. month and season) (Barreteau and Abrami 2007).

The Modalities of Shifting Scales in the Case Studies

Numerous evaluations refer to questions pertaining to modes of changing scales. Most designers were confronted with these questions during the process and the strategies chosen were rarely supported by theoretical considerations or exchanges of experience. In order to detail these strategies, we describe how these questions were addressed and the problems raised by using the four categories defined in Chap. 4.

Ascending Scales or Up-Scaling

The strategies implemented followed two directions: the institutionalization of an approach developed at a local level in order for it to be replicated more easily (four cases); or the evolution of an approach initiated at a local level towards a higher organizational level with the aim of allowing or facilitating decision-making (four cases).

Institutionalization of an Approach Developed at a Local Level

Institutionalization aims for either: (i) to facilitate horizontal dissemination or the replication of the approach; (ii) the autonomization of participants; (iii) a better appropriation of the benefits of the approach. However, the selected strategy seems to have resulted in real institutionalization in only two cases, Lingmuteychu and Nîmes-Métropole.

In Lingmuteychu, a member of the institution made use of his academic training to develop and test this type of approach in the role of designer. When he was reintegrated into his institution at a decision-making level, he proposed using the companion modelling approach in a systematic fashion to support the implementation of basin committees. The training of staff to implement the approach within the institution was planned.

In Nîmes-Métropole the participation of certain stakeholders as observers in the process enabled its effective appropriation by one of the institutions (the Conseil Général du Gard). However, the cost of this replication to all communes in the department due to the recruitment of a consulting firm now limits its effective implementation (see Box 1).

In the second option, the institutionalization strategy principally rested on the participation of high-level stakeholders either in the approach's design or deployment (as actors in a game where the model is discussed). The weak effectiveness of this strategy was due either to the institutional representative's attitude, which was reserved for personal (Nan) or institutional (Ouessant) reasons, or by the difficulty in mobilizing representatives from a certain level to participate in the approach.

Faced with this difficulty, 'demonstration' games were implemented in higher level arenas with two principal outcomes. In the absence of direct confrontation with representatives from lower levels, the validity of the representation that promotes the issues as they are expressed at a lower level was called into question (Ter'aguas). The tool is seen at best as a training tool for the organization's technicians who operate at a local level. One might ask if the issue must be the institutionalization of the approach or the institutionalization of the topic. In this case, different strategies may be chosen, such as the mobilization and consolidation of a socio-technical network[6] around the question.

Box 11.1: The Nîmes-Métropole Case Study

During the feedback of a ComMod approach dealing with conflicts between foresters, livestock farmers and hunters in a Mediterranean forest (see case study in the Appendix), the Gard district Direction départementale de l'Agriculture et de la Forêt (DDAF) proposed adapting the approach to the issue of forest fires at the urban-wildland interface. Aware of the interest of the Environmental Service of the Nîmes urban community in raising the awareness of the elected representatives to this issue, DDAF persuaded the director and vice-chairman to propose their territory as a case study. In a first phase, the process went through the co-construction of a shared representation of the issue by a pool of technicians from different institutions in the district dealing with activities related to fire prevention. One of the outputs was a virtual map of the area based on three archetypes of village layout that covered the diversity of patterns between natural, agricultural and urban areas. The second phase was devoted to organizing situation action exercises by involving all elected representatives (i.e. mayor and the urbanism or environment deputy) in role-playing game sessions (NîmetPasLeFeu) that encompassed the decisions of farmers at the plot level (abandonment of cropping), elected representatives and urban developers at the village level (new settlements and building permits), and administrators at the urban community level (inter-municipal development project, fuel-break). In a third phase, the Conseil Général du Gard, informed by one of its agents who was invited as an observer in one of the game sessions, volunteered to have the process implemented in all the district municipalities concerned with wildfire risk. After seeking European funding, this institution was able to launch, 1 year later, a new ComMod process for four urban communities located in the north of the district. An accelerated training process was then set up to allow local stakeholders to handle and adapt the method (adjustment of the

[6] For example, in the SosteniCAP case: the collective development of work on the question of mobilizing numerous stakeholders; seminars presenting outcomes to regional decision-makers relying on this work.

conceptual framework and the land map to the local context), with the progressive withdrawal of the researchers (transfer of the facilitation role to consultants) and the progressive empowerment of the Conseil Général to lead the process (appointment of a technician for the monitoring of the operation).

Evolution of an Approach from a Local Level to a Higher Level with the Aim of Consolidating Decision-Taking

In the Mae Salaep study case, an official was invited to participate in a role-playing game session at the request of villagers. However, this participation (in the form of an observer in the game, but an actor in the debriefing discussions) did not have the intended effect and was perceived as a failure due to the poor preparation of the official and an unfavourable institutional contexte (see Box 2). In contrast, in the Lingmuteychu study, the approach was part of a favourable institutional context because the authorities sought to promote the creation of a management body at the watershed level, a jurisdictional level that the approach progressively was considering. Officials (regional level) were invited to participate at a key moment in the process, but they were neither participants nor observers as a particular role (moderation) was assigned to them. This strategy allowed them to orientate the taking of decisions towards the creation of this management body.

In the Larq'asninchej case study, the change of work scales during the process was not recognized as such, but as a simple integration of new stakeholders no longer able to ignore the process due to strong local mobilization and the participants' involvement in the field. Yet even if there had been no real change in scale in the approach or outcomes, the issue had been completely appropriated at this organizational scale due to the engagement of key stakeholders and a campaign to disseminate the theme despite its initial emergence at a very local level.

Box 11.2: The Lingmuteychu Case Study in Bouthan

This Commod process was initiated to deal with a conflict over the sharing of water between two villages during rice transplantation. At the same time, the authorities were looking to implement participative bodies for natural resources management. In a first step, the approach dealt with the way water was shared between the two irrigated areas of the villages. The tools—a role-playing game and multi-agent simulation model—emphasized the relationship between the collective and individual processes of water management in two irrigated schemes sharing the same resource. After a first series of games and simulations with participants from both villages, it was decided to expand the work to the seven villages of the catchment area. Representatives of these villages, chosen for their legitimacy, ability to make decisions on behalf of the village, knowledge of hydraulics and ability to participate in public debate were invited to participate in a game session. During this session, at the request of the authority, representatives of the regional level were also invited to participate in the debriefing at the end of the game session as mediators of the discussion. The process led to the creation of a catchment committee and the implementation of small collective activities (e.g. rehabilitation of the hydraulic network, tree plantation). This work was developed as part of the PhD fieldwork of a technician from the Ministry of Agriculture who proposed to the administration after the completion of his doctorate that this approach be used to develop one of the tools (participative management bodies) of new legislation for natural resources management.

Other strategies also merit note. The replication of discussion sessions at the same territorial level enables a 'quasi-statistical' extrapolation. This type of outcome is particularly valuable at a higher level of organization and may facilitate the appropriation and integration of results at this level. This approach was developed particularly in Vietnam where numerous games were conducted at a local level and the outcomes extrapolated to the regional level using GIS (Castella et al. 2007). However, this type of approach raises questions concerning sampling and the representativeness of experiences given the heterogeneity of representations in the same category of users. In the Domino Senegal case study, this ascent in scale was perceived through a regional tool calibrated to the local level. However, the advantage of this approach cannot yet be objectively analysed due to a lack of detachment.

The following issues emerged from the strategies tested.

- The sporadic participation in the companion modelling approach of an individual (as an observer or direct participant) who is removed from local dynamics does not appear to be an effective strategy and may even be dangerous by promoting the unilateral revealing of local strategies.
- Stakeholders from higher levels are often in positions of dominance in relation to other stakeholders, which should be explicitly considered in the process.
- These stakeholders are often also part of complex and/or hierarchical organizations (administrations), therefore, they must hold a clear and well-defined institutional mandate that is compatible with the objectives of the approach and the feedback mechanisms within the organization facilitating institutional learning. This strong institutional support occasionally requires the approach to work upstream, for example, through parallel activities allowing the involvement of the organization.
- The deployment of a tool in a discussion arena mobilizing levels distant from the territorial ones where the tool was designed is also risky because the change in level accompanies a change in representations, and even underlying hypotheses, of the questions and pertinent dynamics that need to be considered. This risk is even more important when direct dialogue with local representatives

is limited or impossible (no local representatives, or difficulty of local representatives in expressing themselves).

- At times, the institutionalization of the theme or the issue under study is more important than the institutionalization of the approach.

The Horizontal Enlargement of the Approach and Work with a Large User Group

The out-scaling strategy followed three directions:

- the mobilization of a sufficiently representative number of participants to ensure the territorial legitimacy of the approach (six cases)
- the replication of the approach in different sites or communities, meaning the deployment of tools in discussion arenas that are different but on the same level (five cases)
- the dissemination of outcomes from operations or discussions to users who have not participated in the experiment in order to associate them with the decision (five cases).

The Mobilization of a Large Group of Different Users

Underlying this objective is either an attempt to reinforce the legitimacy of the operation, sometimes with the aim of facilitating an ascent in scale, or to facilitate a better appropriation of outcomes through wider participation. It may also involve an effort to ensure the understanding of the heterogeneity of representations.

In four cases, this strategy was used during a single stage (generally the discussion workshops and simulations, notably the role-playing games). It plays out through a preliminary 'sampling' of participants that is meant to ensure a 'good', 'geographic' representation or a good territorial cover. For example, designers sought to mobilize at least one elected official or representative from each town in the territory considered to be in the tool deployment arena (e.g. Nîmes-Métropole), under the assumption that the representations used in this stage were valid for all members of this category of participants.

The effectiveness of this approach depends on the capacity of representatives to participate in an exchange of knowledge about the issue, to institutionalize new learning, and to engage the institution or organization in decisions when decisions are expected. It was to ensure these points that in the case of Lingmuteychu different person in a newly included village were solicited, each bringing one of these 'capabilities'. Some advocate the mobilization of 'champions' of the approach and their social network to mobilize participants. However, this type of strategy raises the question of the effective legitimacy of the approach and its capacity to guarantee that the ensemble of heterogeneous representations is considered and that the effective participants contribute only their personal representations. It is this ambiguity that two case studies in Thailand sought to highlight by confronting the representations of leaders in charge of the interface with those of local authorities and a diversity of users.

Given the capacity of the proposed tools, this strategy of conscious selection is only interesting when the target 'population' is relatively limited. Even when this is the case, the repetition of certain stages, such as games, may prove to be necessary (Nîmes-Métropole). When the population is very dense, the simple replication of the experiment can prove to be difficult or even impossible to carry out. Yet the discussion tools used often do not allow the mobilization of a sufficiently significant number of participants to ensure a true representation of the ensemble of users, a point that is likely to affect the legitimacy of results. In the same way, the activities proposed are rarely sufficient to enlarge the participation base and cause a participatory dynamic in the medium term (Larq'asninchej). Lastly, the multiplication of experiences may lead to the development of quite different discussions and scenarios, posing the problem of their confrontation and sharing, not to mention the integration of new representations that were not initially noticed.

Finally, other avenues are proposed such as: (i) a return to simpler mobilization procedures, for example using a technological medium such as the Internet or the telephone accompanied by investigations into the impact of these technologies on group dynamics and outcomes (ii) the autonomous development of the method by non-expert third parties allows the multiplication of the approach but is accompanied by the risk of normative and prescriptive drifts. This second type of approach assumes training on the approach and its ethic is provided (Chap. 12).

The Reproduction of the Approach to Other Groups or Similar Situations

Given the transaction costs of certain stages in the companion modelling process, notably in developing tools, there is a strong temptation to develop 'generic' tools that can be reused in a relatively complete companion modelling approach. Several of the case studies put or will put into place such a strategy based on seeking generic or de-contextualized tools.

In the Ter'aguas case study, the reproduction of the approach was planned at the beginning of the process given the large number of stakeholders (residents of a suburban zone) affected by the problem. The approach focused on a group of stakeholders to reinforce their capacity to participate although confrontations with other stakeholders were also planned. The generic character of the tools was in the integration into the role-playing game of the heterogeneity of representations concerning the theme studied and on the development of simple tools (e.g. drawings, theatre, cards) that helped to structure rapid discussions around key points of the conceptual model. The latter can be rapidly adapted to the diversity of local situations that have been previously described. The development of different stages in the reproduced approach rested then on the hypothesis that the heterogeneity of representations (within the target stakeholder group) was adequately understood and integrated in the tools so that these a posteriori reconstruction processes did not fundamentally put into question the integrated representation proposed by the game. Nevertheless, the approach required an adaptation of certain tools at the stage when the model was reconstructed to fit local specifications. One might then ask under what conditions such adaptations are possible when the process is carried out by non-expert third parties.

This question was raised in the SosteniCAP case study (a non-commodian participatory simulation) where a partnership with an NGO was developed to reproduce the approach that had previously been simplified to facilitate its application. The evaluation showed that this strategy did not produce the desired results. The process carried out by the NGO was much more directive and normative than the designers had planned. On the one hand, the extreme simplification of the approach partially discredited it among stakeholders, while on the other, it led facilitators to reinforce control over decisions at certain stages. These problems probably are indicative of the risks raised by the systemization and broad reproduction of a companion modelling approach and raise questions regarding the specific training of animators (Chap. 12).

The Diffusion of Outcomes to a Large Group of Users

The dissemination of outcomes may prove to be important in guaranteeing the legitimacy of the operation or in promoting the appropriation of results. The difficulty is that a large portion of these outcomes consist of situation learning based on the direct confrontation of one participant with other members of the group and/or tools.

The evaluation of the Nan case study was the only one[7] to refer to a communication strategy through a game in which virtual and real players assembled together and were given the task of explaining the strategy of the game to a village audience. The standard practice of designers generally is to rely on the dissemination, up- or downstream of the approach, of information, elements of the problem, outcomes and possibly, discussions and products of the process by mobilizing their social networks and diverse communication mechanisms. Others choose to present the results of work made in small groups with the aim of validating them in a legitimate 'ratification' arena (e.g. a general assembly of users). To avoid overly long or detailed presentations, spokesmen generally are encouraged to present only the elements that were the subject of debate within the work group (SosteniCAP). However, the validation of decisions is thus conditioned by the legitimacy of the arena itself and/or by the level of participation at the time of the debriefing.

---

[7] Game outcomes were presented also in different arenas in the Domino Senegal study.

Ascent into Generality or Abstraction

Some approaches rely on existing concepts and theories. This was the case of the simplified game used in the Lingmuteychu case study's third series of workshops in which an emphasis was put on the articulation between collective and individual water resource management in a virtual scenario. The CherIng game initially was designed as a training tool (Chap. 12), but was used with local stakeholders to discuss principles of communal resource management by highlighting certain concepts (i.e. cooperation, competition, discussion and communication).

Other approaches emphasize the design of tools that can be contextualized easily and adapted to specific situations. For example, the Domino approach sought to test the possibility of developing a generic land-use allocation model that could be adapted to two situations (La Réunion and Senegal) (see Box 3). In reality, there was movement back and forth between the successive draft designs of the generic model and the specific models. The evaluation noted that the heterogeneity of the areas involved (a French overseas department and an African country) and the diversity of issues allowed the identification of generic elements of the approach but not the construction of a generic tool.

The third option aims to highlight the representation of generic processes, particularly biophysical ones, in the model to enable transposition to other similar situations. This option was chosen by the Pays de Caux case study with a game that could be transposed to other situations involving watersheds submitted to erosive runoff.

However, other approaches could be envisaged. For example, the 'degradation' or 'simplification' of existing contextualized tools by identifying underlying management principles in the Kat Aware series (South Africa). Another possibility is the adjustment of an approach to enable the collective identification and modelling of management principles and testing the tools derived at different organizational levels.

Box 11.3: The Domino-Réunion Case Study

The French law of 'Solidarité et de renouvellement urbain' (SRU) sets new rules for legal land-use plans: planning procedures must include a participatory approach, and a general framework organizes coherence between the plans of the various levels of territorial organization. In Réunion Island, the land-use plans of the region, inter-municipalities and municipalities were under revision. The main challenge from an information system perspective was to develop something coherent from geographical data produced periodically by a number of different institutions and make effective use of it. The Domino-Réunion project aimed to fuel these multi-level debates thanks to the development and use of tools allowing the elicitation and sharing of multiple viewpoints.

To do so, some members of the extension and support services at the various institutional levels and researchers first developed an integrated simulation model. This development took place in an informal framework that allowed the expression and consideration of viewpoints of minor stakeholders (such as agricultural stakeholders). This model was used to build scenarios of interactions between urban, agricultural and nature conservancy stakes and to simulate the long-term resultant land-use map. In parallel, the region started its own legal participatory approach to revise its regional land-use plan. Both projects collaborated in a second phase. The model was revised and used to illustrate prospective scenarios that the region had pre-defined with its partners. The methodology allowed the fuelling of institutional debates with a dynamic prospective approach and to institutionalize part of the debates that had taken place in the first phase less constrained by political struggles.

In this project, to take account of the multiple levels of management and to develop the data necessary for the land-use allocation processes, we relied on both the models themselves and the involvement of stakeholders from development through to the use of the tools.

The effective advantage of this type of approach, notably for the replication of simulation sessions and the eventual transfer of the approach outside the research world, remains completely open. One also could question the value of having stakeholders who are involved in operations within a given situation participate in the co-construction of a generic model that they may perceive as being distant from their preoccupations.

Definition at a Local Level of One or Several Principles and/or Regulations

Eight studies were aimed explicitly at analysing with users the local impact of a regulation decided at a higher level or at promoting its application at a local level. The initial demand was often prescriptive even if it was at times subtle. Decision-makers considered the operation to be a kind of communication exercise meant to teach users new rules or help them understand previously misunderstood rules, enabling at best some flexibility in the local adaptation of the new regulation, at worst the exploitation of the approach. What often is involved is more along the lines of 'explaining how something should be done to users' than analysing and understanding the appropriation of these rules by users or collectively analysing how to adapt certain rules to local circumstances. However, more specific expectations may be added to this prescriptive demand, such as an analysis of the combination of two regulations (Nîmes-Métropole), or the definition of local criteria of water rights (Kat Aware).

While the approach effectively enables users to better understand the rules, it also allows decision-makers to better understand users and their constraints. Decision-makers mention their own individual learning when directly confronted by users. Communication flows, therefore, are not as one directional as initially imagined. In contrast, as shown in the Ter'aguas case study, the deployment of a tool enabling discussion of a regulation's application in the regulators' arena without the participation of users does not allow this type of learning. Furthermore, the regulators' effective consideration of this knowledge appears limited. A return in the form of institutional learning thus assumes the implementation of a multi-level strategy effectively integrating the regulator level as an organization. Yet this rarely is the case: representatives often are presented as individuals without a real mandate from their institutions. Some only participate as observers while many reveal themselves to be sceptical of participatory approaches.

The analysis of these case studies thus revealed a set of specific issues that represent new scientific challenges allowing a better evaluation of the pertinence of the companion modelling approach in the particular context of changing scales or multiple levels of organization.

Suggestions for Advancing Further

The analysis of the challenges posed by multiple levels and of the difficulties and outcomes of different strategies put forward in the case studies lead to the identification of a certain number of methodological questions. Some questions aim simply to clarify the presuppositions held by the designers of the approach while others are research questions. The latter focus on three ways of taking multiple scales into account in a ComMod process: through the association of stakeholders from different levels in the definition of the issue and the design of tools; through the integration of multiple scale representations into the tools; through the mobilization of these tools and their outcomes with stakeholders from different organizational levels (Fig. 11.2). Some of these questions are specific to the application of the ComMod approach to multiple levels of organization while others are raised systematically but take on a special connotation due to the multi-level context.

Identification and Formulation of Questions

The identification and formulation of questions are closely tied to the scales involved, notably through the different consultation arenas chosen to mobilize and which can call upon different levels and even dimensions, and the choice of participants.

Participation and Formulation of Questions

The management of natural resources raises questions that multiply depending on the scale, level and arenas of governance. The way these questions are defined at different levels, therefore, must be taken into account. It also is useful to verify whether the hypotheses regarding a system's operations and the perceptions of different stakeholders are modified by the scales of work mobilized as priorities and the divisions made. The initial formulation of the project effectively may predefine a privileged discussion arena, the associated scale, and the dominant cognitive domain due to the complexity of the issue. However, the participating parties (whether within the arena or in relation to the question as formulated) may not be affected by the issue at the same territorial or organizational level. The approach must analyse and take into account the manner by which these different participants identify the question, which may lead to a modification of the discussion arena (Chap. 5).

When addressing a question, each participant, including practitioners and research scientists, also holds his own representation of which scales are pertinent and of how they are structured. The dominant cognitive model of the intervention and the choice of participants can help reinforce a conception of the structure of scales (e.g. political-administrative), or indeed certain organizational levels and thus, certain actors. Yet the approach's choice of scales and divisions, which probably will result in compromise and/or competition between participants, may also be a mechanism of exclusion. Thus the process might influence power struggles around scales: it might thus contribute either to the reinforcement or legitimization of a level or arena, or to a better understanding of other levels based on the confrontation of stakeholders' points of view.

Choice of Participants and Transparency

One of the principles of companion modelling is transparency for all participants. In a 'multi-level' approach that is undertaken in a parallel or sequential manner, the objectives of the links and interactions between the levels that the approach seeks to develop to address a given environmental problem must be explicit.

Some stakeholders may not wish to interact with other levels due to power relations (Chap. 6). Other stakeholders who are to be associated when there is a change in scales or levels may choose to simply 'participate as an observer' in the simulation stages. Yet the experiments analysed show that this type of participation rarely produces encouraging results. Consequently, one should consider how to mobilize best the representatives of these multiple levels around their centres of interest and ensure that each one's involvement in the process is real depending on the progress of dialogue. Two aspects must be taken into account; (i) the specific activities enabling the mobilization of institutions and actors involved that are not necessarily linked to participatory modelling or simulation activities; (ii) tools that can address the ongoing changes introduced by the integration of new actors and levels.

Multiple Representations and Their Set of Tools

The question of multiple representations is not specific to changing scales but is amplified by the multiplication of levels and dimensions to be taken into account and their articulation within the same representation or within a set of coordinated representations. This complexity of representations and their formalization in a multi-level context poses strategic questions regarding modelling choices. We distinguish three issues: (i) the development and use of 'generic' representation; (ii) the choice of an integrated tool versus the choice of several tools; (iii) the choice of formalisms in the tools.

Multiplicity of Knowledge and Move Towards Abstraction

Scientific knowledge leans towards the generic or that which is applicable to diverse contexts. A move towards abstraction is expected to provide an opportunity to derive tools that can be used in different contexts and at different levels. This kind of questioning, mentioned in Chap. 4, may be applied in particular to the out-scaling process. The first question raised when generic representations are discussed regards what is meant by generic. Is it 'inclusive' representations, allowing all variants of a process to be r-epresented regardless of the context, or 'exclusive', where only invariants of these processes are represented (whether through intersection or abstraction)? This initial question greatly affects how corresponding tools are built and the added scientific value that may result. It would seem that the first option is richer, allowing a complete representation of a system even if all possible variants are not represented. Yet the use of these generic tools on specific cases raises numerous practical questions, for example, regarding what methodology should be followed, and ethical questions, notably regarding their capacity to allow specific representations to emerge.

Between a Toolkit and an Integrated Tool

Faced with the diversity of partners and the specificity of issues at each decision-making level, two extreme strategies are possible: a single tool integrating all of the representations or one tool per issue and per audience.

The integrated tool option poses a double design challenge: to integrate several levels, dimensions and resolutions, making clear the links between these three aspects of the question, and integrate rationales or heterogeneous cognitive systems. This tool must effectively be capable of representing the key specificities of each participating party's point of view while also being understandable and meaningful to all participants. This strategy has the advantage, however, of gathering all representations into a single coherent frame. This was the option taken in the Domino case study in La Réunion.

The toolkit option was used explicitly at multiple decision levels in two study cases, one in Senegal (Domino), the second in Vietnam (Castella et al. 2007). In the two cases, the reason given was the need for tools adapted to different audiences and different facets of the same problem. The disadvantage lies in the coherence of these different tools, particularly when they do not derive from the same domain model (Chap. 4). It is perfectly possible to combine the two strategies according to the context and evolution of demand.

Multiple Formalisms and Multi-Formalism

Formal tools of expression, often very different, are associated with various scales and particularly dimensions. Among these formalisms are ontologies and more generally formalisms of knowledge representations (e.g. conceptual graphs) produced by artificial intelligence, spatialized and non-spatialized information systems, differential equations and computer models (e.g. UML, cellular automata, MAS). Consequently, the question of integrating knowledge (integrated tool) or of establishing coherence in multiple forms of knowledge (toolkit) technically is defined as a problem of articulation of the different formalisms used. We then find ourselves confronted with either choosing a unique pivot formalism in which knowledge from different scales must be expressed, or to work on multi-formalism platforms while ensuring bridges between discourses. This question, in addition to the one posed in the preceding section, is raised in all ComMod experiments but is even more delicate when the group of participants is very diverse.

Learning in the Approach and Transmission of Outcomes

Finally, a multi-level companion modelling process raises questions that specifically concern the learning permitted by the approach and the transmission of outcomes in arenas and multiple organizational levels.

Social Learning

Knowledge acquired through a companion approach consists less of stable, transferable learning than an ephemeral product of interactions between the various stakeholders involved. Consequently, this knowledge is only mobilized by participants during the process, which by definition represents only a limited number of people in relation to the population concerned. Some of these participants may be charged with representing their institution. The question, therefore, is raised regarding how this knowledge can be integrated into these institutions, or, in other words, its institutionalization. Here we find methodological questions concerning the dissemination of the approach among a large number of participants and the integration into the approach of participants from levels other than the ones initially mobilized.

In addition, one might ask how choices of scale affect learning processes. Certain ComMod tools integrate several spatial, jurisdictional and temporal scales. To what extent can participants take into account the sum of this complexity? How do the choices of scales in the representations mobilized limit social learning processes? Lastly, how do these choices affect power relations?

Transmission of Outcomes Between Levels and at the Conclusion of a Project

Whether the transfer concerned the tools or their outcomes, numerous case studies chose to use, or were led to mobilize with a given audience, tools that were built with other audiences. This was the case particularly when the tool served as a mediator between different groups. Regardless of whether the tools themselves are modified or whether their outcomes alone are used, this procedure raises numerous questions. What information is transferred? How can one enable the new audience to appropriate the tool's underlying hypotheses, regardless of the stage of its development? How can the first set of partners be assured that their choices and representations will not be misunderstood when transmitted? More prosaically, how can they be convinced to accept this transmission? The latter question raises the issue of objectives (e.g. information, validation, appropriation, etc.) and the legitimacy of those involved. It is important to keep in mind that these information transfers do not play a neutral role in power relations between organizational levels and participants.

Lastly, there is the question of the restitution of the approach's outcomes. This question already is delicate in a simple ComMod approach but becomes more complicated in a multi-level approach. How can one take into account the inter-level sharing of representations, debates, decisions and lastly, social learning which may include institutional reorganization?

Conclusion

The results presented in this chapter show diverse methodological approaches that display few invariants, the development of the theme having barely begun. The true multi-level organization may be thought of as an internal process, natural to a society that uses different, stacked official regulatory levels to insert the results of internal, backstage consultations. On the one hand, one also may hypothesize that the way consultations are organized within these official arenas can influence and gradually modify the society's actual internal consultation modes, for example, by favouring a gradual change in the weight of various challenges, values, groups, etc. One returns here to the challenge of a certain form of participation that seeks to insert gradually into the society a greater number of 'regulatory and counterweight operations in response to power struggles' (Viard 1994).

In a companion modelling approach, all *ex ante* structuring into levels must be assumed. If we return to the fact that the difference in our problem between our 'uni-level' and 'multi-level' interventions lies in the larger disparity of frames of reference (e.g. issue, appropriate procedures, values evoked, etc.), this fundamental difference might better be expressed by the term 'multi-institutional'. An institution is defined as a set of actions or organized practices that are stable and recognized by all, in other words, a field of convergent and legitimate social relations defined by rules, codes of conduct and behavioural norms, but also by the way these conventions are applied (Ostrom 1990; Putnam 1993; Weber and Reverêt 1993; Clarke 1995).

Faced with the consideration of several scales, the ComMod stance may be interpreted as a mediator between levels and dimensions. By facilitating an understanding and expression of issues at several levels, the approach promotes exchanges between levels, and thus could be viewed as feeding a mediatory organization. The latter contributes to, and is influenced by, the 'politics of scale' (such as defined by Lebel et al. 2006), and more generally by interactions between levels of organization. This is made possible thanks to two main resources: multi-level participatory strategies during co-construction and the use of representations, and the integration of multiple scales in the tools (Fig. 11.2).

Chapter 12
Transferring the ComMod Approach

Michel Étienne, François Bousquet, Christophe Le Page and Guy Trébuil

As the ComMod group is composed exclusively of research scientists, the facilitation of a ComMod process is for them an ongoing experiment that has lead to a diversity of practices. This penchant for experimentation has enabled them to define more precisely the stance and functions of facilitation during various stages in the implementation of the process (Chap. 3), and to identify and test methods and 'ways of doing'. These are meant to guarantee that stakeholders are able to share knowledge about the management of renewable natural resources in a given area, stimulate the co-design of a development project in this area, and/or facilitate communication between parties often in conflict. In numerous cases, project holders or certain participants expressed the desire to apply a similar approach to another issue or field, or to generalize the use of a tool on a population that was broader than initially covered by the process in which they had participated. In other cases, a project with expatriate research scientists came to an end; to guarantee the continuation of the process, a local corps of young teacher-researchers needed to be prepared to take over. Lastly, there were times when a development project's principal objective was to train such a corps of teacher-researchers, with the possibility of doctoral research over several years.

Most often, people interested in knowing more about companion modelling discovered it through an example of its application. Although they were not part of the application, they nevertheless could find elements that were sufficiently

M. Étienne (✉)
Ecodevelopment Unit, INRA, Avignon, France
e-mail: etienne@avignon.inra.fr

F. Bousquet · C. Le Page · G. Trébuil
UR Green, CIRAD, Montpellier, France
e-mail: francois.bousquet@cirad.fr

C. Le Page
e-mail: christophe.le_page@cirad.fr

G. Trébuil
e-mail: guy.trebuil@cirad.fr

M. Étienne (ed.), *Companion Modelling*, DOI: 10.1007/978-94-017-8557-0_12,

interesting to provoke a desire to discover, or better understand, the main principles of the approach. In the case of the pedagogical heads of teaching modules, there was a wish to adopt an innovative approach to open new perspectives for their students. Beyond this first, conventional stage of transmitting knowledge on a new approach, we also rapidly were confronted with requests to appropriate the approach. These requests generally came from people who had participated in its application, either closely or at a distance, and wished to apply a similar approach to other questions or field but who felt insufficiently autonomous and wished to acquire the necessary 'know-how'.

This type of wish raises the issue of transferring the approach to some of the stakeholders with whom we habitually work. The concerns often expressed by participants in ComMod processes relate to the difficulty of facilitating the collective moments that punctuate the approach and of mastering the design, implementation and use of models that serve to support the process (i.e. conceptual model, computer model, role-playing games). The first concern relates principally to social dialogue, the second is more involved with the technical mastery of the tools. These two stumbling blocks are joined by the main difficulty of making a radically new scientific stance known and appreciated. The basic principles instilled before the start of a companion modelling process are difficult to practise in the field by those raised on conventional scientific ways of thinking and immersed in a cultural context that takes these principles into an unfamiliar framework. This chapter examines three paths that were recently explored to familiarize some of our partners with the application of a ComMod approach: using face-to-face or distance learning teaching modules, organizing professional development training programmes, and accompanying some individuals in the application of a ComMod process.

Teaching the Approach

Teaching modules were developed for degree programmes, the professional training of research scientists or technicians, and on-the-job training for young research scientists and doctoral students. The first set of modules enabled potential practitioners to become aware of the approach and its possible applications in their field. The second allowed our approach and the use of models in mediation contexts to be debated and compared with other ways of viewing and practising participatory research. The last made it possible to render students progressively more autonomous and critical of stances and tools used in the framework of what was often a long operation (upto 5 years in certain cases), in which they were the main designers and facilitators.

There is no consensus within the ComMod group on whether it is necessary to differentiate the types of audience during professional training. Some consider that the principle of companion modelling is itself to promote the sharing of heterogeneous skills, and it thus is anachronistic to tailor professional training for particular audiences. Others consider that research scientists and development professionals do not refer to the same worlds or share the same expectations and, consequently, they cannot be trained in the same way. Research scientists plan to reinforce references to theoretical aspects, and also seek clarification of the scientific approach, information acquisition and validation issues, discussions of the implementation, verification, adjustment and validation of models, in addition to reflections on the research scientist's stance. On the other hand, development professionals prefer to emphasize technical aspects, the way of implementing and facilitating a process, the choice of partners, cost and length issues, or adjusting to regulatory procedures. Finally, in projects involving teachers, they focus on the pedagogical advantages presented by the originality of the ComMod approach and tools used, discussion on the acceptability of levels of simplification imposed by the chosen teaching curriculum, or adjustments to specific learning questions.

The issue remains open to debate and there has not yet been a sufficient opportunity to gain a perspective on mixed training to reach a decision on this point of divergence. Nevertheless, numerous experiments show that the training dynamic with a mixed audience can be rich when the schedule allows significant amounts of time for research scientists and development professionals to work together to apply new knowledge acquired on a topic to a situation they have chosen together and in which everyone is interested.

Degree Programmes

The degree programmes in which we are involved essentially are for students in higher degree programmes (i.e. final year of engineering school, second year of a research or professional Master's programmes). The principal objective of the training is to help students understand that it is possible to model complex systems with local stakeholders by helping them collectively to represent interactions between ecological and socio-economic dynamics. We also aim to make them aware of the ComMod stance and help them discover the principal tools used in the approach. This last task is difficult to achieve but it is important for students to understand that this original approach corresponds to a particular mode of co-construction and use of models rather than their simple production. When time allows, activities take place in the form of interactive workshops. In two to four course hours, activities generally consist of an introduction to the approach followed by a detailed presentation of one or two examples of its application. Interactive workshops are based on the presentation of actual experiences (to explain how a given complex system was modelled), coupled with the manipulation of a model that is easy to grasp (e.g. forest/fire/fireman type) or a rapid role-playing game (harvesting a resource by an actor). The teaching exercise then consists of gradually making the situation more complex, either by rendering the representation of stakeholders or resources increasingly sophisticated, or by introducing new stakeholders or resources.

In longer training, two formats were tested based on a common structure of four modules that are linked together differently, and which are more or less developed depending on the type of training and its length (from several days up to 2 weeks):

- illustrated introductory course on the characteristics of the approach and its application
- description of a conceptual model or the co-construction of this conceptual model
- context immersion through a role-playing game corresponding to this model
- visualization of dynamics on computer simulations.

In the first format, after a brief introduction to the process, the pedagogical approach is focused on a detailed analysis of an actual, completed case study. Students first try to understand the complex system by representing it in the form of a conceptual model that dynamically describes interactions between actors and resources in a given territory. They then visualize these simulated dynamics by means of a multi-agent model and propose management alternatives (e.g. exercise on development scenarios), or play the role of the stakeholders represented in the model and experience interactions with other stakeholders or with the environment (role-playing sessions). In the second format, after the core introductory course, the pedagogical approach chosen for one-to two-week training favours the progressive construction of applications, individually or in small groups during the afternoons, inspired by morning presentations of case studies.

ComMod teaching games were specifically developed with this aim (see the Appendix for a description) and were adapted to the main themes addressed in the courses of study taught by ComMod group members. A tutorial programme was developed for those whom a teacher thought could be more independent (see references at the bottom of the corresponding case studies). These ComMod games can be divided into three categories according to what guides the designers in their development: a game designed specifically to teach, a game designed specifically to organize workshops with stakeholders of a particular system but later used in training sessions (Njoobaari, Don Hoi Lord), or games that from conception claim the right to ambivalence (SylvoPast, ButorStar). The first category groups together games that are either generic or abstract but can be contextualized easily (CherIng), games adapted to students' specializations that either are completely made up (YeunEllez), inspired by a game designed by stakeholders of a system studied (MéjanJeuBiodiv), or contextualized based on a more abstract game (PâtureLesCommuns or ReHab inspired by CherIng).

Within the companion modelling community, there is a synergy to develop both games to teach students and games to facilitate consultation and dialogue between stakeholders in a system studied. At times a game designed to teach the approach (CherIng or RuisselPois) is the foundation for a more sophisticated game destined for stakeholders in the system represented (Radi or Pays de Caux). From a pedagogical point of view, having students play a game that was played by stakeholders of a reference system opens particularly interesting perspectives on the link with reality that can be addressed during the debriefing. Thus in a game like

SylvoPast, which is based on a relatively abstract representation of a Mediterranean forest area, students assuming the roles of woodman, shepherd and hunter implemented strategies in the game that were similar to those used by stakeholders in reality (Étienne 2003).

The one-day module includes an introduction to companion modelling, a short example of its application adapted to the central theme of the course of study, an exercise in the co-construction of a conceptual model (i.e. ARDI method to determine the stakeholders and resources available, their respective dynamics and interactions) representing an issue with a strong tie to the curriculum, and a pedagogical game illustrating the conceptual model. The game may be chosen from the following themes: runoff, fire, biodiversity and agroforestry.

In the 2-day module, it is possible to add a clarification exercise of stakeholders' decision-making rules, the selection of relevant viewpoints to understand the dynamics of a modelled system, the construction and simulation of a scenario, and a search for abnormal behaviour in the model and the causes. Exercises alternate with the use of a model and role-play. The territory and application example chosen may be adapted to a particular theme or environment (e.g. coastal zone, steppe zone, Mediterranean forest or humid zone).

Over 1-week, it is possible to integrate learning on a UML diagrammatic formalism and, while retaining the preceding format, favour group work on development issues (e.g. biodiversity management, risk prevention or multi-functional development). In certain cases, the questions will be identified and chosen according to the issues on which the students are working elsewhere. In other cases, priority will be given to the application of the process at three different levels of resolution (i.e. landscape, watershed and farm) on a question that is presented and which the students must appropriate.

In 2-week modules, the second week is spent on students learning to construct a computer model able to serve as an intermediate object in a companion process. The training covers the construction of a spatial environment based on GIS data, the modelling of a process spatialized by a cellular automaton, modes of formalization, modelling

Box 12.1 **MéjanJeuBiodiv sequence**

2.5-day workshop on biodiversity and companion modelling

The first half-day is dedicated to an introduction to companion modelling and an example of its application. The second half-day is dedicated to an exercise of co-construction of a multi-agent model representing the biodiversity of Causse Méjan. The third evaluates the co-constructed model and the selection of viewpoints and indicators. The fourth presents the structure of the model implemented and simulation of the actual situation according to different points of view. The fifth half-day associates a role-playing game and computer simulations with test scenarios meant to improve biodiversity management.

(i) Modelling interactions between ecological and social dynamics
Alternating lectures on companion modelling with demonstrations of a simple model of fire and firemen. The application example is chosen according to the students' course of study (e.g. fallow land process on Hussant island, harvesting razor clams in Thailand, hunting, grazing and harvesting reeds in the Camargue).

- Principles and contribution of multi-agent modelling (1 h)
- Companion modelling and biodiversity management (1 h)
- Application to a concrete case (1 h)

(ii) Co-construction of a model of biodiversity management on a steppe
Co-construction work in groups of six to eight students on a conceptual model of a MAS describing a steppe area based on a set of photographs in the framework of implementing the ARDI method.

- Identification of stakeholders and resources (1 h)
- Identification and description of natural dynamics (1 h)
- Interactions between ecological and social dynamics (1.5 h)

(iii) Evaluation of modelling
Critique of interaction diagrams between groups, development of a joint conceptual model, followed by group discussion of the most relevant indicators in visualizing model outputs.

- Comparison of conceptual models (1 h)
- Development of a joint conceptual model (1 h)
- Selection of indicators and viewpoints (1 h)

(iv) Use of modelling
Learning how the conceptual model is translated into computer language in the Cormas platform. Then the use of MAS to simulate an actual situation and evaluate its impact according to different viewpoints.

- Model presentation (1 h)
- Simulation of model (1 h)
- Identification of each stakeholder's problems (1 h)

(v) Development and simulation of scenarios
Simulation exercise for participants to identify and discuss what does not work, and to propose a solution for these dysfunctions. Use of a MAS collectively to simulate solutions proposed.

- Role-playing game (1.5 h)
- Simulation of proposed scenario (1 h)
- Analysis of effects of these scenarios (30 min)

a negotiation, the development of logical stages clarifying the decision process, the identification of indicators and the coding of spatialized points of view, the validation of models and tests of sensitivity. Each topic lasts for half a day during which theoretical examples alternate with concrete examples and practical exercises. The entire week is structured around a project to manage the grazing of several herds, a scenario that is made more precise and complex as the learning continues (see PâtureLesCommuns in the Appendix).

Many of these training courses are evaluated both on the pedagogical aspects of the companion modelling tools used and the learning fields observed (Étienne et al. 2008d). The enjoyment found in learning by doing, learning by playing, and learning while simulating is cited repeatedly by students. Becoming aware of the complexity of interactions between human activities and ecological dynamics is often identified as a strong point of the pedagogical approach. Lastly, numerous students announced that they recognized the need for consultations in this type of problem, and the capacity of a computer model to support consultations between different stakeholders.

Professional Training for Research and Graduate Teaching

The involvement of ComMod group members in training sessions for researchers and other research seminars aims both to train research scientists and university staff in the approach and to confront the approach with those of other groups working on participatory methods of research or the use of models in consultation processes. We will address these two components in order.

Training

To make the approach known, the first companion modelling training developed out of a 2-week training session entitled 'Simulation of complex systems: multi-agent systems (MAS) and renewable resources management', implemented by the Renewable Resources and Environment Management (Gestion des ressources renouvelables et environnement—GREEN) team at CIRAD beginning in October 1997. Between that date and March 2006, 12 sessions were organized in Montpellier and nine sessions (in English) abroad. The format of the session evolved over the decade, notably in the way the role-playing games were integrated. Initially presented as tools allowing interaction between stakeholders in order to co-construct and validate computer simulation models, they gradually were introduced as one of the principal tools structuring the companion modelling approach by drawing on the diversity of their use modes within actual projects. Today, it is clear that specific training in companion modelling must be offered independently of training focused on teaching the basics of simulation tools.

A demand for specific training in companion modelling rapidly emerged. A first attempt was made between 2001 and 2004 with scientists from various disciplines in south Asia through an interuniversity project financed by the Asia IT&C programme of the European Union. Ten one-week sessions were organized with the premise that the mastery of companion modelling requires multiple skills and in order to reflect the principle of expressing multiple viewpoints in the teaching method. In each, a professor gave a course on a given field (e.g. multi-agents, simulations in social sciences, watershed management, communication sociology in agriculture, etc.). Approximately 100 people participated in these courses. Some of the most industrious began to develop applications to test whether the approach was suited to an actual system that they needed to manage. About 12 case studies in Asia thus are linked to this cycle of brief training. Following this series, six students undertook doctoral studies on the subject. The lessons learned from the project were described by Bousquet and Trébuil (2005).

In parallel, sessions focused on understanding and mastering the Cormas multi-agent simulation platform developed by CIRAD were organized beginning in April 2002. In the same spirit, a collective training effort focused on the 'role-playing game' tool because it constitutes an original element in our approach and is enthusiastically received in the field but involves some very special skills. Having also been the subject of complaints and criticism, we rapidly undertook rigorous formalization and theoretical referencing work. Relying on the works of numerous scientific groups (particularly those connected with the international journals, *Artificial Societies and Social Simulations* and *Simulation and Gaming*) and the opinions of those who have read numerous scientific articles on the use of our role-playing games (Barreteau et al. 2001; d'Aquino et al. 2003; Étienne 2003; Mathevet et al. 2007), we set up, with the assistance of FormaScience, INRA's training department, a training session for researchers on 'The use of role-playing games in companion modelling. Simulation exercises for stakeholders to share representations and simulate dynamics'. This training has been organized four times in France and twice overseas (Bolivia and Bhutan). Participants were research scientists from CIRAD, INRA, l'institut de recherche pour l'ingénierie de l'agriculture et de l'environnement (CEMAGREF), CNRS and numerous universities (e.g. Paris and Brest), who are working on natural resource management, territorial development and sustainable development issues. The selection of candidates has always favoured a cross selection of disciplines to improve the quality of exchanges during the training and to make the most of exercises planned for days of practical work. The training takes place over four days, during which lectures on theoretical aspects alternate with practical workshops and role-playing. The training proposes to explain the stance, methodology and application conditions of role-playing games. This can be broken down into four objectives: sensitizing and initiating participants in the use of role-playing games in companion modelling; clarifying the operating conditions of role-playing games to support territorial and environmental decision-making processes; illustrating the use of the approach using concrete and diverse applications; providing the methodological basis for the design of this type of interactive tool that facilitates communication between stakeholders, the exchange of viewpoints and collective learning. In the case of the 1-week training in Bhutan, young researchers and development officers worked every afternoon in small groups to construct different versions of possible role-playing games (Box 12.2). The theme of the games was conflict over the use of high altitude pasture lands in Radi, where the tool was to be implemented the week following the training with numerous trainees working in the region.

Confrontation

The confrontation component consisted of bringing our approach into debate in a series of training sessions for researchers and seminars organized by different institutions. The debate focused either on the tools, the approach or the stance of researchers. In this way, role-playing games were compared with other platforms of negotiation between stakeholders in the NSS Dialogue workshop on 'Contribution of modelling to the management of natural resources: dialogue between disciplines', or with other survey

> Box 12.2 **Training course on role-playing games in companion modelling**
>
> The training comprised four modules and three workshops.
> Module 1: Methodological framework.
> Lectures at the beginning and end of the training identify the stance and principles of companion modelling, present the theoretical framework used and its methodological implications, and recall the ethical rules developed by commodians in the ComMod Charter.
>
> Module 2: Typology of role-playing games.
> Briefings and examples of case studies to give a broader view of the various types of games and the particular features of companion modelling role-playing games.
>
> Module 3: Association of role-playing games with models of resource dynamics.
> Succinct lectures present models of resource dynamics, the way these are integrated into multi-agent simulation models applied to the management of renewable resources and the different modes of association between the multi-agent simulation models and role-playing games (see Chap. 4).
>
> Module 4: Design of role-playing games.
> Briefings and simple exercises run through the stages of game design, present the most commonly used formalization tools and, since 2006, several monitoring and evaluation methods.
>
> Workshop 1: The CherIng game.

This pedagogical game was designed as a simple role-playing game allowing the rapid understanding of the tool's principle characteristics and its gradual adaptation to questions that are increasingly complex. It is based on the simulation of a harvest of a virtual resource (the Ing) in a territory where the resource is present at different levels of availability. There is a double objective: (i) to 'break the ice' and allow participants to get to know one another in a relaxed manner; (ii) to discover in a short amount of time (1 h) the principles and essential stages of a role-playing game used in companion modelling: briefing, alternating between individual and collective decision stages following collective discussions, and debriefings on the results obtained from both ecological and social perspectives.

Workshop 2: Game practice.
Participants simulate a situation by practising one of the role-playing games developed by the ComMod group. The game session is then the subject of a debriefing during which the observation methods and analytical tools are discussed. The choice of game is a function of participants' expectations, its effectiveness in conveying the design and animation difficulties of a ComMod game, and the presence of a teacher with good experience (i.e. command of debriefing and feedback on actual experiences).

Workshop 3: Game design.
The workshop either takes place in one day or is divided over several days in afternoon sessions putting into practice the lectures and examples provided in the morning. It takes place in four stages, in three separate groups, and on an imposed question and territory. Participants first analyse diverse documents (e.g. maps, statistics, stakeholder files, atlas, photographs, etc.) to understand the context and the problem set. They then select which information is relevant to the question raised and may ask for complementary information from the animator of the training. Lastly, they put into use the tools and methods proposed during the preceding days to freely design their game. The workshop ends with a presentation of the games developed by each group, followed by a general discussion. The question addressed is usually a question that has already been addressed by the ComMod group and has been the topic of a role-playing game (e.g. grazing, tourism and conservation on Ouessant island, farming, hunting and the protection of the great bittern in the Camargue).

methods during a CNRS session on 'Survey techniques: scientific update on the method and tools'. It also was compared with conventional, multi-criteria analytical methods during an INRA session on 'Approaches and methods for multi-criteria evaluation of the sustainability of livestock and cropping systems'.

In parallel, we were invited to numerous training sessions for researchers and seminars as representatives of an original approach. The approach's capacity to promote dialogue between disciplines was discussed during the session 'Interdisciplinarity between biotechnical and social sciences: agriculture, environment, territories, public policies and sustainable development'. The originality of the approach in relation to other participatory modelling approaches was debated within the multi-agent modelling community during a CNRS session on 'Modelling and complex multi-agent system simulations for human and social sciences: principles and methods of design and use', within the research development community during an INRA session 'Which participatory research for which development dynamic?', and during a research seminar of the ADD-ComMod programme on 'Taking into account stakeholders and their representations in sustainable development'.

Lastly, the ComMod approach was compared with those for modelling dynamic systems, changing land use (the CLUE model by P. Verburg of the University of Wageningen), and the evaluation of vulnerability (school represented by C. Polski of Clark University in the USA) during a 1-week summer school organized at the University of Sapporo by the coordinator of the Global Land Project.

Professional Training for the Development World

The objective of this training is to help development service technicians and local government agents understand the principles of companion modelling and learn how to animate one or several of the group exercises that punctuate the process. The training often is adjusted to a particular institutional framework, such as the review procedure of UNESCO's biosphere reserves network, or the development of charters in regional and national nature reserves. This type of training was developed with two different perspectives: one was to sensitize, while the second was to 'render autonomous'.

In the first case, activities were based on taking apart a case study in which the process was conducted up to its completion, and on the presentation and discussion of the actual experiences of numerous participants in the implementation of participatory or companion modelling approaches. In the second case, the emphasis was placed on the methods used, as much for the development of a joint representation of the question raised as for the animation of a collective session during which models accompanied the exchanges between participants.

In UNESCO's MAB programme biosphere reserves network, the second option was chosen. The approach first was adapted to the context of biosphere reserves (Étienne 2006), and to procedures imposed during the creation or revision of these 'sustainable development models at the regional scale' (Étienne et al. 2007). It then was transmitted through training modules presenting the overall logic of the approach and emphasizing a particular tool. With the Swedish MAB committee and animators of future Swedish biosphere reserves, the focus of the training was the ARDI method of co-constructing models representing the functioning of the biosphere reserve that was the focus of the training. With representatives of local stakeholders, conservationists and national MAB committee heads of six francophone countries of West Africa, the training focused on role-playing games and the resolution of conflicts over use. With participants in the EuroMAB network, the guideline was the application of the approach to a concrete case of a biosphere reserve review. Lastly, training with French biosphere reserve agents emphasized the specific character of animation during highly collective periods by combining workshops on describing and mastering tools with simulation exercises allowing participants to animate sessions of tool design or use. Training always included a positioning of the approach in relation to other methods currently used by managers (e.g. Agenda 21, management assistance guides, development plans, landscape charters and forestry charters).

Accompanying its Application

Another form of learning about companion modelling may take place during the application of a ComMod approach. This involves either professional or doctoral students who generally already have participated in one of the training sessions described in the preceding sections.

The first option was regularly applied in operations in southeast Asia. People who demonstrated an interest in starting a companion modelling operation were invited to observe various parts of a process already underway and participate in several stages of the approach. The objective was to render the process concrete and provide touch points for practical learning based on imitation, and to enable the candidate to acquire confidence, project themselves, and to construct their own application. Over the last 5 years, six French and Asian doctoral students have undertaken theses on subjects related to the conduct of a ComMod process in a given field.

Most of them decided to do so after having participated in a short training course for higher education teachers-researchers of the type described above. This proved to be, with a few rare exceptions, insufficient to enable the doctoral student immediately to adopt an adequate stance and construct a process based on a solid theoretical foundation and methodology. Thus there was a need to plan for close and intensive supervision of the student, particularly when they have followed an academic course of study and come from a culture very different from our own. In this case, the young researcher-in-training needed to learn multiple skills to become autonomous and this may be beyond the capacity of certain candidates: synthesize information, integrate different kinds of knowledge from different disciplines, different modelling techniques, mastery of multi-agent simulation platforms, gift for communication with different kinds of stakeholders (coming close to the art of diplomacy in conflict situations), animate groups of heterogeneous stakeholders over the long term, interest in the transdisciplinary practice of research, writing reports of varied nature, etc. The mastery of the 'art of the ComMod practice' thus is the result of long and demanding careers that can be very fulfilling when they meet with success, with the scientific publication of results relatively easy. However, this path is reserved only for candidates with advanced potential who are extremely well qualified academically. It is still too early to evaluate how well the young teachers-researchers who underwent training will transfer the approach to their students once they have returned to their respective faculties and universities, or once they have returned to professional life.

Disseminate the Approach

Through the Internet

A website (www.commod.org) was set up to provide a range of information for those seeking to become more familiar with companion modelling. The text of the ComMod Charter is online, as are the case studies, which are presented in a standard format that provides a concise description of the application of the approach, specifies the question raised, the territory concerned, the sponsors (those behind the demand), the research scientists involved, and the bibliographical resources available for further information. Standardized descriptive forms provide specifications of the tools developed (i.e. role-playing games, simulation models).

In order to make it easier to consult all of the documented case studies (the site currently offers 34 listed in alphabetical order), three entry keys are proposed. The first is based on thematic cues (i.e. biodiversity, water, forest, agriculture, livestock, peri-urban, other themes). The second distinguishes cases according to the social dynamics addressed (i.e. markets, credit, migration, creating institutions, learning, cooperation, conflict). The third is geographic.

The website also provides lists of various types of publications referring directly to companion modelling, as well as lists of training modules in which companion modelling has been taught for the past 5 years. These pages distinguish between diploma programmes in France and overseas (targeted audience: students), non-diploma programmes (targeted audience: researchers, managers, decision-makers, etc.), periodic interventions that may or may not be specific to companion modelling, and presentations in research seminars.

The private section of the site, which is only accessible to signatories of the charter, constitutes a space for collaborative work and the sharing of information that effectively supports the animation of the network, notably with an archive of messages exchanged in the discussion forum and access to teaching aides produced by members of the network.

Distance Learning

Between 2004 and 2008, following the training project in southeast Asia described earlier in this chapter, a website was developed (http://www.ecole-commod.sc.chula.ac.th)

to respond to requests for training from numerous countries in the region and to develop tools for a network of researchers in training. This website is divided into three parts that we believe complement the teaching of companion modelling.

- A part divided into six modules: introduction, theoretical bases, method of UML graphic design, multi-agent simulations, role-playing games and the Cormas simulation platform); a second part groups together a set of applications undertaken in the region.
- A role-playing game inspired by the CherIng game that can be played over the Internet.
- A part known as e-governance based on the principle that stakeholders who have participated in a ComMod approach should express their views about the process so that potential users in other areas can profit from their impressions and experiences. This was the most ambitious component but we were unable to develop it to the desired level despite having opened an electronic forum to encourage discussion.

The site has only been online for a short time and we have not yet had sufficient time to gain the perspective needed to evaluate its use. The challenge will be to reconcile the use of this distance learning site with the supervision of those who desire to be trained more fully. The Internet tool may also serve to maintain exchanges within a network of people who have experienced a ComMod process. Through the use of the site, a newcomer might thus be integrated into a network and benefit from the experience of its members by following them on some of their field operations.

Through Consulting Firms

In the framework of sustainable development, authorities and technical departments in charge of managing environmental issues must implement participatory field research operations. They often call upon consulting firms for this activity, but such firms rarely are specialized in this type of work. Partners who have participated in a companion modelling process sometimes are also interested in generalizing the approach or applying it to a theme other than the one addressed with commodians. They seek consulting firms specialized in the design, implementation and evaluation of participatory processes.

Several scientists who participated in companion modelling operations through their university study programmes decided to set up a consulting firm known as Lisode, which offers services to public stakeholders that are specially designed to accompany cooperation components of area projects. They can intervene in the application and evaluation of participatory processes associated with the implementation of Schéma directeur d'aménagement et de gestion des eaux (SDAGE), Schéma d'Aménagement et de gestion des eaux (SAGE), Schémas de cohérence territoriale (SCOT), Plan local d'urbanisme (PLU), Plan de prévention des risques

(PPR), Zones d'aménagement concerté (ZAC), Agenda 21, regional nature reserve park charters, etc. They regularly intervene in participatory democracy programmes of regional administrations, such as the organization of citizen conferences or public debates, and propose training modules on mediation and territorial cooperation.

Perspectives

We are working on three new ways of transferring companion modelling, that is, the preparation of a training programme covering the entire process, discussions with agriculture teaching experts on teacher training, and an attempt to develop a distance learning module.

Comprehensive Companion Modelling Training

The need for training on a process that is not tied to mastering a tool led us to develop a new training project addressing the entire companion modelling process. This project aims to combine sessions explaining the principles of companion modelling with sessions on the difficulty of animating the collective exercises that mark the process. Two potential targets were selected: development partners who were likely to appropriate, and who were interested in implementing, the process; and research scientists intrigued by our approach and interested in understanding the entire process.

The FormaSciences team retained and developed the 1-week training session for researchers. The idea was to teach the ComMod approach on the basis of a series of simulation exercises allowing the different stages of the process to be addressed successively. Each teaching module corresponds to a stage described in Chap. 2 and is constructed according to the same sequence:

- an initial simulation exercise
- a debriefing to understand the issues emerging from the simulation
- a formal lecture period (methodological and theoretical references) presenting a number of key points directly linked to the issues selected
- individual work periods in which participants are invited to look through the database of case studies and select documents produced by the ComMod group to research elements related to these key points
- a brief feedback period.

The approach then is applied to a case study chosen from the bibliography put together by the ComMod group according to the profile of participants or the expectations of the sponsor. The training thus includes six modules in which participants will try to put themselves in the position of a commodian (see Box 12.3).

Professional Training for the Teaching World

A special case concerns the experiments underway in French agricultural schools to adapt certain tools that were developed in ComMod processes for more generalized teaching. Three levels were targeted: the final 2 years of the agriculture lycée programme (premiere and terminale), advanced technical diplomas (BTS Management and Protection of Nature, Analysis and Management of Farming Systems, and Forestry Management), and engineering schools. The project is an initiative of SupAgro Florac and the Fondation pour la recherche sur la biodiversité (FRB). It operates from the hypotheses that the ComMod group's methods and aids may be used in different educational situations in agricultural schools to cover three topics:

- modelling: step by step formalization of a complex situation involving the management of a biodiversity issue
- analysing dynamics: observe ecological and socio-economic dynamics through spatialized and non-spatialized computer simulations and try to understand their interactions (i.e. search for information and indicators, interpretation of observations by formulating hypotheses and testing these hypotheses)
- simulating interactions between stakeholders on the same territory: play the role of stakeholder to grasp their means of action, understand their rationality, lead a negotiation, take decisions and measure the impact on biodiversity.

The experiment involved a pilot group of teachers from secondary agricultural schools willing to adopt the aids and methods produced by ComMod, test all or some with their students, and share their experiences. SupAgro Florac, FRB and the ComMod network volunteered to accompany this group by providing support, organizing the group and sharing experiences, monitoring and accompanying field applications, adapting the aids and eventually producing new ones, and promoting and diffusing relevant experiences and products. Education inspectors from the agricultural school system, the Direction generale de l'Enseignement et de la Recherche (DGER) offices and support services, and SupAgro Montpellier were also involved.

Box 12.3 **Training course on the ComMod approach**

Module 1: Organize the process.
The exercise involves formulating a question, defining the specific objectives of the accompaniment, identifying the contextual elements to be taken into account, specifying the source of the demand, organizing the process to be implemented, and selecting different types of participants (e.g. project holder, facilitator or stakeholders) by identifying their level of involvement and what makes them legitimate participants.

Module 2: co-construct a representation of the system.
The exercise consists of making explicit the various points of view on the question and considering the breadth of knowledge on the topic. This

leads to choosing a methodological framework that will enable a structured set of terms and concepts about the stakeholders (decision-making rules, roles) and resources (processes, adjustment procedures) to be produced. It imposes the definition of a spatial–temporal framework (i.e. choice of spatial expanse and simulation timeline). Lastly, to remember the co-construction work, participants must imagine a method to clarify the hypotheses used in the different submodels (biophysical and social processes), keeping track of the choices made among the diverse solutions put forth and their justifications.

Module 3: implement the model.
The workshop makes it possible to acknowledge the variety of tools that can be used (e.g. computer simulator, role-playing games, hybrid models) and the platforms and languages most commonly used. It then provides an overview of the various kinds of data that could be integrated into these models: state variables describing the state of a system at a given moment, and parameters linked either to possible scenarios or to biophysical and social processes. It also encourages reflection on the relevant indicators for understanding the effects of decisions taken on the dynamics of a system according to different points of view. Lastly, it leads to discussion on validation problems of the mathematical, statistical and social processes represented.

Module 4: stage and simulate the situation.
Participants must organize a highly collective period in which the model serves as an intermediate object to facilitate cooperation. The workshop focuses on different aspects of animating these sessions by facilitating discussion on the preparation, animation itself (introduction, facilitation, immediate debriefing), co-construction of scenarios, observation (attitudes, lectures, aids used), and results analysis (consequences for the environment, management of resources or the social system; learning, decisions, and negotiations between participants).

Module 5: handing on the process.
The exercise focuses on both the identification of future bearers of the approach and on the training modes planned (teaching module, self-teaching, accompanying an actual case). It also addresses the advantages of being involved in a network to ensure methodological support (ComMod network, network of participatory approach practitioners) and guarantee the diffusion and appropriation of the approach (creation or activation of local networks). The issue of transferring and adapting tools is also raised.

Module 6: monitoring and evaluating.
Tools for monitoring and evaluating are proposed and tested on a selected case study.

Three pedagogical components were developed with teachers to evaluate their contribution to professional learning on area management in these study programmes, and to constitute the outline for a future collective sustainable development week that is in the process of being incorporated into teaching references:

(i) formalize a joint representation of a complex system: group work on a diagram of social-environmental interactions according to the ARDI method
(ii) place students in simulated situations so that they can experience this complexity: role-playing games around biodiversity management
(iii) use the modelling of this complexity to imagine alternative management: development and computer simulation of scenarios providing insight into the dynamics of a system according to different management modes.

The obvious interest aroused by the proposed methods and process is due to the fact that they consistently rely on creative approaches and naturally induce active teaching positions. Although many teachers think of this as the pedagogy of their dreams, they generally do not find a way to integrate it into their usual routines, particularly when the subject involved demands strong technical and scientific skills. ComMod's accompaniment of this innovative teaching component should result in the institutionalization and spread of the approach in the entire agricultural profession.

Distance Learning

The virtual University on Environment and Sustainable Development (l'Université virtuelle environnement et développement durable—www.uved.fr) is an association that aims to share the contents of training courses from more than 40 French-speaking partners. A web teaching course entitled 'New Participatory Approaches to the Management of Renewable Natural Resources' is in progress. It is formatted through a 2-week Master's 1 course, providing methodological skills arranged in three blocks:

(i) participatory approaches
(ii) modelling complex systems
(iii) the ComMod approach.

These methods can then be applied to five issues related to natural resources management:

(i) integrated management
(ii) land tenure
(iii) land planning
(iv) water management
(v) agro-biodiversity management.

An exercise based on a role-playing game is also proposed. This training course can be used either as a self-training process monitored by a tutor or in a classroom.

Conclusion

The transfer of companion modelling is undertaken through very different frameworks and on extremely varied themes. It includes academic education, targeted training and broad mediums such as the Internet. Nevertheless, a certain number of principles and tools remain common to all of the activities undertaken.

The teaching of companion modelling prioritizes how to pose questions (Fleury and Fabre 2005). It concentrates on seeking the significance of data that lead to debate and on an approach that must help to identify action choices. To do so, it relies on four operations linked to socio-scientific reasoning (Sadler et al. 2004): recognize the complexity inherent in an issue; examine the issue from several angles; admit that the issue should be the focus of further enquiry; demonstrate scepticism of information presented. However, it also adds two fundamental elements of reasoning: identify risks and uncertainties; take into account values and ethical principles in decision-making.

Learning through role-playing also is a constant feature of our pedagogical approach. Within a creative setting, it leads participants to understand ecological functions, complex interactions between nature and society, and the advantages of consultations—and indeed cooperation—between stakeholders involved in the game. Participants can be placed in an original context in which attitudes, behaviours and situation actions are revealed. However, learning only occurs when teaching methods allow learners to reflect and progressively conceptualize while keeping in mind each individual's capacity for abstraction.

Lastly, participatory modelling is at the heart of our pedagogical approach. We hypothesize that the co-conception of a model and the joint use of model simulations facilitate understanding of complex systems and encourage collective reflection on management alternatives (Hare et al. 2003). Teaching through modelling stimulates both a learner's capacity for abstraction through formalization and their imaginative capacities via the possibility to project into the future (Lane 1992).

Chapter 13
Conclusion

Companion Modelling: An Adaptive Approach?

Michel Étienne

This book is the result of a long and thoughtful examination of companion modelling, a stance and research practice shared by a group of 50 research scientists and tested in some 30 applications on issues of renewable resource management. This work sought to understand and discuss how researchers handled the tension between a homogeneous stance, one agreed by all the signatories of the ComMod Charter, and a diversity of practices resulting from a conscious, pragmatic choice to adapt to the context and challenges posed by issues arising from the field. The exercise allowed us formally to identify a set of characteristics common to the application of the approach, highlight certain divergences and enrich the experience of the group. It also allowed us to identify the weaknesses, vague areas and sensitive points on which more thought needs to be given.

The Strengths of the Approach

Various chapters of this book showed that companion modelling is itself a process, one that lies at the intersection between research and collaborative decision-making processes, interacting with both. These interactions are not necessarily long, but they are repeated several times, and in an intense manner. This generates an adaptive approach that allows the people involved, and the way they are involved, to evolve according to changes in the accompanying (the ComMod approach) and accompanied (collective decision-making) processes. However, our degree of involvement in each of these two processes is not the same because the accompaniment process assumes responsibility for interactions between research and decision-making. This confers on it a special responsibility that led to the stance set out in the ComMod Charter. This stance results in coherent, principled

M. Étienne (✉)
Ecodevelopment Unit, INRA, Avignon, France
e-mail: etienne@avignon.inra.fr

M. Étienne (ed.), *Companion Modelling*, DOI: 10.1007/978-94-017-8557-0_13,

approaches that emphasize the importance of learning and modelling as catalysts for interaction between the accompanying and accompanied processes.

Approach and Stance

The approach aims to create a mechanism for the collective consideration of a given complex of problems, and a learning mechanism to reinforce stakeholders' capacity to participate in collaborative projects of resource and land management. It encourages the development of overlapping perspectives on these projects through the expression of multiple viewpoints at various moments of interaction, by using different facilitation techniques. In order to make knowledge accessible and understandable to everyone, the approach promotes the recognition of multiple perspectives and the hypotheses that underpin them. It then seeks to question and reinforce the legitimacy of this knowledge through the regular validation or invalidation of these hypotheses by participants. The iterative character of the approach also encourages important gaps in knowledge to be identified and leads to the creation of new knowledge to fill them.

Companion modelling enables the development of interpersonal and communication capacities and facilitates the identification of a desirable common goal by building scenarios that can be tested and discussed. It thus joins the ranks of patrimonial approaches while taking stakeholders' capacity to invest in these approaches very seriously, that is, transparency of hypotheses and adaptation of tools and representations to the participants' mental models. When questioning social relations, the approach facilitates an awareness of interdependencies and the organization of responsibilities. It facilitates the negotiation of resource management principles by collecting and sharing information and discussing possible options.

All this requires commodians to give themselves the means to be transparent, as much in terms of their objectives and behaviour as in the results of their action, by controlling the risks and opportunities induced by the methods used. No matter how in-depth their initial analysis of the context may be, the commodian must be able to understand the direction in which social networks are evolving, how stakeholders' positions change, and which real world issues are likely to emerge during the interactive process with stake-holders. By adapting to the evolution of issues with each iteration, by involving the available stakeholders, and by mobilizing existing or co-constructed artefacts during the process, the commodian intervenes in a collaborative decision dynamic with the support of the context and the stakeholders involved in the dynamic. Herein lie the challenge and the difficulty of companion modelling: to conserve flexibility and an iterative character while maintaining a specific stance and principles.

Discussions reflecting on this topic during the writing of this book highlighted the flexibility of the positioning of a commodian who, while identifying fairly easily with a dominant stance, considers that it is possible to adapt how this is translated according to the context or objective of the intervention without renouncing the basic principles. This evolving and adaptive positioning results in a movement back and forth between dialogue and critical analysis depending on the chosen objective, the stage of the process, the perception of the context of intervention, and indeed, the evolution of power relationships during the process.

Learning

From the very title of the work, companion modelling presents itself as an approach to support sustainable development. It effectively considers that a society's capacity to adapt depends on its knowledge and ability to recognize intervention objectives, and on its capacity to construct a set of options for the management of its resources. The stated objective of companion modelling is to increase the skills and capacities of local communities to be autonomous and assume responsibility for their future. For this, the approach carries out engaged research aiming to empower populations in terms of renewable resource management and the fight against social inequality and vulnerability. The modification of knowledge, or, in other words, viewpoints about the world, is central to the dynamic induced by the interaction between companion modelling and the collaborative decision-making process that it accompanies: improved knowledge of the system, but also a better understanding of stakeholders' expectations and constraints.

This empowerment takes place according to different kinds of learning. The evaluation of the effects of the approach set out in this book clearly demonstrates an in-depth learning of the issues involved and emerging problems, a better understanding of other people and their interests, as well as learning by clarifying each person's frame of reference. The evaluation also shows that the possibility of putting into question these different elements during group meetings is a prerequisite for experimenting with, and learning new technical, organizational and communication rules. These collective key moments occupy a special place in this learning, whether individual or collective, because it is through them that interdependent links are explored during the co-construction of a shared representation of a system or during debates over building and evaluating scenarios.

The ComMod group's desire to transfer the approach and associated tools to the scientific and professional world led to considerable education and training efforts favouring problem-solving and learning through games. Many of its members have proposed a pedagogical adaptation of the three pillars of companion modelling, that is, formalize a shared representation of a complex system, put students into simulated situations for them to experience this complexity and use the modelling of this complexity to imagine alternative forms of management. This other form of learning stimulates both a student's capacity for abstraction through the work of formalization and their imagination by projecting into the future.

Modelling

From a methodological point of view, modelling is central to our approach because it provides the means to record in a common framework heterogeneous knowledge regarding a complex system. Modelling consists of thoroughly analysing representations of actual systems in operation and aims to construct virtual worlds to discover and explore hypotheses.

By definition, companion modelling uses models, or rather families of models. The modelling process is structured around three stages: from a model of the field it evolves into a conceptual model and then a simulation model. More originally, these three stages aim for a sharing of roles between different stakeholders, either from the accompanying or accompanied process, with skills related to the field, the themes at play, the modelling or the design of the tools used. The continuity of models used in a case study largely depends on the conceptual models that are the backbone of the study. There is no one, unique conceptual model, but an evolution of the conceptual model according to the evolution of issues and points of view regarding the system. Throughout, the models must take into account the relationships between stakeholders and their environment. The diversity in the tools used is a function of the degree of realism and the form. Here are three examples:

- the 'disposable' model, co-constructed by a given group to characterize the state of representations and issues at a given moment
- the generic model that allows an ascent into generality working from numerous specific cases treating similar questions and then to generate new, disposable models more easily
- the role-playing games, which are real human multi-agent systems that either will facilitate the communication and understanding of a conceptual model or will serve to support the social validation of a conceptual model or the learning about the processes that render it dynamic.

In between, an entire range of hybrid models has developed, which allow the possibility of easily testing several hypotheses with the model. The versatility of the models used, therefore, remains a constant in companion modelling.

Throughout the process, one type of model will follow another in an order rarely established in advance. The modelling work that defines the period of translating the real to conceiving the virtual world is systematically associated with an effort to validate, verify and discuss that defines the period of interpreting the consequences of this exploration of a virtual world for the real one. Beyond fulfilling the conventional verification/validation step in modelling, this interpretative feedback provides a sense of security to stakeholders by allowing them to reject conclusions that, even if formally valid, do not suit them on the grounds that the exploration is virtual. This safety-net, which is systematically put into place during a collective debriefing, allows risks to be taken in the choice of hypotheses and scenarios to be explored. This is the core of the iterative aspect of the companion modelling approach: the construction of virtual worlds to share representations associated with the exploration of virtual worlds to draw lessons applicable to the real world. The movement back and forth between translation and interpretation is organized around collective key moments that are specific to the application of the approach.

In all cases, this manner of modelling must allow heterogeneous knowledge to be taken into account, which is not a particularly original feature, and to use non-computerized tools, such as simulation models, which is much more so. The important point is that it does not aim for the convergence of the modelling process. Lastly, during the co-construction of these models, multi-level participatory strategies are applied, which leads to an integration of multiple scales in the tools.

Implementation Difficulties

Several chapters in this book mention the weighty responsibilities and constraints shouldered by facilitators of a ComMod approach. Two potentially problematic features of this special role were identified by our partners and through our analytical work, that is, legitimacy and mastery of tools.

Legitimacy

Having protagonists take a seat around the same table and accord their knowledge equal weight is not a neutral act. To take place, there must be an initial sense of trust between participants and at least one commodian. While the legitimacy of an intervention can be developed over time during the process, the initial legitimacy can determine the way the ComMod process plays out.

When the objective of the process is the production of knowledge, scientific legitimacy is prioritized: does the scientific community perceive the ComMod process and the knowledge it generates as valid? When the objective is to accompany a consultation process or even that of decision-making, one is more interested in 'social' legitimacy: what does the process and the person facilitating it represent in the eyes of the -stakeholders of the society in which the intervention takes place?

In the first situation, the commodian must not only be recognized by peers in their field, but also by the other disciplines convened around the table. This legitimization will be even more difficult to achieve when the proposed participatory approach is very different from the common practices of the invited research scientists, when the commodian pushes them to the frontiers of their disciplines, or when the commodian does not clearly point out their disciplinary standards.

In the second situation, the commodian must construct legitimacy with the stakeholders by examining standards and practices, the procedure put into place, and thus the quality of the process initiated. They must demonstrate skill at encouraging participants to open up, and must be able to respond to certain social expectations, such as the emergence of consultation arenas, by finding a way for the neglected interests of 'social juniors' to be expressed, or by modifying the symbolic field of power. At the service of collective groups, the researcher only participates in the companion process for a limited research period, after which the process must continue on its own. The period of accompaniment consequently is not the same as the time required to implement the decisions produced by the process.

In both cases, legitimacy is difficult to achieve and must rely on an explicit and regular reference to the principles set out in the charter on the transparency of the process, clarifying hypotheses and clearly declaring the uses of the models developed. Legitimacy may be reinforced by assuming a critical stance at several key moments in the process: at the start when setting up partnerships in the flexible choice, adapted to the context, of how to collect data and set up consultation arenas; during collective key moments around the animator or process coordinator's central role; at the end of the process through a marked effort to diffuse the outcomes of the research and their appropriation by stakeholders in the ComMod process.

Mastery of Tools

The design, implementation and use of computer models are central elements in the ComMod process. However, these require the presence of a modeller who also accepts the role of mediator, someone who can lead stakeholders progressively to express themselves, articulate their knowledge and explain their practices, furnishing logical and structured statements that can be used easily in the constitution of a simulation model. If the transparency of the translation process is respected, if levels of uncertainty or lack of confidence are clearly identified and accepted, the model is socially validated. However, this process raises questions regarding the singularity of the models produced, their validity linked to a particular context, and the technical skill required to design and use the tools produced.

Prospects for Further Development

The slow maturation of this work enabled numerous methodological elements to be identified and many questions to be raised. First, we will address the methodological progress achieved on the implementation of the approach, its evaluation and the manner of making it known. We then will debate two possible directions in which we might further develop our work: towards more encompassing decision levels and towards a charter better able to take into account the diversity of stances.

Towards Well Formalized Methods

Following an in-depth analysis of frameworks (Chap. 1), an improved version was proposed in which the implementation of a ComMod loop was refined to 12 phases. Even if they are not all used systematically or follow a different order, they represent a standard succession or a kind of complete model for the companion modelling approach:

- sensitizing those involved in development issues to the ComMod approach and its possible applications in local problems
- definition of the question raised between project holders
- inventory of scientific, lay or expert knowledge, available through surveys, diagnostic studies and analyses of the literature
- eliciting knowledge for the model through surveys and interviews
- co-construction of the conceptual model with stakeholders concerned by the issue
- choice of a tool (computerized or not) and implementation of a model
- calibrating, verifying and validating the model with local stakeholders
- definition of a scenario with local stakeholders
- exploratory simulations with local stakeholders
- diffusion among stakeholders who have not participated in the process
- monitoring and evaluation of the effect of the process on the practices of participants
- training stakeholders interested in using the tools developed.

The conclusions of the analysis examining participants' perception of the technology used (Chap. 7), and expectations in terms of training on the approach (Chap. 11), demonstrated that the principal difficulty lies less in the transfer of facilitation techniques, which can be reinvented to suit the issue addressed, but in the transfer of the stance. A methodological guide (Daré et al. 2009) was designed to encourage reflection on the implementation of companion modelling. It emphasizes the need to make explicit the various questions raised by companion modelling, and illustrates the diversity and potential compatibility of solutions found. The guide is structured around seven questions and draws from over 30 case studies for illustration.

- How can companion modelling help me address a subject that challenges me?
- What contextual elements are useful in starting the project?
- What procedural elements should be mobilized to start the process?

- Which elements of knowledge about the system and which procedural elements should be researched during the application?
- How should the process be evaluated?
- How can one complete one loop and potentially open a new project?

Lastly, the monitoring-evaluation process that we propose (Chap. 6) insists on the advantage of developing coordinated, multi-headed, transdisciplinary facilitation. It emphasizes the impossibility of simultaneously animating and observing a collective key moment. It raises the question of the neutrality of the evaluator who must maintain a certain removal from the project while possessing sufficient knowledge of the issues at stake. It requires one to consider how to reconcile the independence of observers with the coordination of the work of the designer(s) and facilitator(s) of the approach.

Towards Policy Decision

By facilitating the understanding and expression of the issues involved at several levels, the approach encourages exchanges between levels of organization and may, therefore, be seen as sustaining a mediation institution. However, much work remains to be done in terms of capitalizing on learning and on its diffusion beyond the group involved in the ComMod process. This change in scale must be reflected in a sustainable transformation of individual and collective practices, and in the establishment of bridges towards policy decision-makers.

This last change of level, known as 'upscaling', raises numerous questions for our group. How can the actors in these decision-making spheres be mobilized? How can the quality of participatory approaches undertaken at local levels be conserved, and how can one remain in line with the needs of the corresponding populations?

Towards a Quality Approach

Just as experimental science relies on laboratory notebooks and ethnography on field notebooks, commodians felt the need to produce a record to provide the various participants, commodian as well as lay, academic, expert and institutional, in a companion modelling process, the means to interpret what happens without hindering their involvement in the process. Above all, such tools for reflection must address the two spheres in which we wish to be evaluated in the application of our approach: in the scientific sphere, the tools must furnish the means to refute us; in the sphere of collective action, they must acknowledge the plurality of ways of engaging with the world that exists among the stakeholders involved.

Such an institutionalization derives from putting in place an approach that is comparable to the 'quality' processes developed in the business world. However, in our case, what is involved is applying a norm aiming to guarantee that what is done is made explicit in order to keep a record, not a norm stipulating how to conduct the process itself. We propose to continue working with the 'logbook' principle, while reflecting on the best way to record what occurs and is exchanged in the interactions between accompanied and accompanying processes, which involve a more emotional dimension.

Towards a New Charter

The first version of the ComMod Charter was based on a number of founding case studies. The translation of its principles in the implementation of the case studies analysed in this project allowed us to appreciate their adaptability. The charter now must evolve into a new version, taking advantage of the lessons learned from this project and the numerous comments received.

The charter may be considered as a statement of principles by commodians, particularly during collective key moments where the multi-disciplinary and multi-stakeholder character of our approach is fully realized. It indicates a field of possibilities, and, in so doing, a field of impossibilities. It does not award a ComMod label but proposes a common reference, allowing everyone to appropriate it, theoretically, empirically and ethically, according to their professional, disciplinary and social frames of reference.

Depending on which dimension of sustainable development a commodian prioritizes (i.e. social, ecological and economic), different stances are assumed. They give more or less weight to the context, references to theoretical or disciplinary fields to which the commodian belongs and to the values and political, ethical and ideological stances that characterize them as an individual.

Lastly, this work is the result of an extensive examination of the practice of companion modelling. What do the authors of this work have in common and what makes them continue to proclaim themselves as commodians? What leads them to believe that they all belong to the same research community despite their different stances and diverse and varied ways of proceeding? The ComMod group provided itself varied means to respond to these questions: logbooks to note reflections *in itinere*, the Montfavet framework for *ex post* reflection, the Canberra Protocol for reflexivity via the questions of an outside evaluator, and the comparative analysis of case studies. These analytical endeavours are cumbersome, yet the agreed investment in their format merits the discussion of the relevance of their institutionalization. Should the new charter determine how to organize reflexivity?

Appendix
Descriptions of 27 Case Studies

Title	Country	Agricultural issue	Environmental issue	Socio-economic issue
AguAloca	Brazil	Irrigated crops	Global change	Peri-urban, land tenure
Camargue	France	Livestock farming, fishing	Biodiversity	Multiple uses
Domino Réunion	France	Tropical crops	Sustainable development	Regional development
Don Hoi Lord	Thailand	Fishing	Over-harvesting	Tourism
Frêne	France	Livestock farming, forestry	Biodiversity	Urban development
Kat Aware	South Africa	Irrigated crops	Global change	Water legislation
Lam Dome Yai	Thailand	Rice cropping	Soil fertility	Agricultural income
Larq'asninchej	Bolivia	Irrigated crops	Floods	Peri-urban
Larzac	France	Livestock farming, forestry	Biodiversity	Collaborative management
Lingmuteychu	Bhutan	Irrigated crops		Collaborative management
Luberon	France	Livestock farming, forestry	Biodiversity	Timber market, meat food chain
Mae Salaep	Thailand	Fruit trees and tea cropping	Erosion	Credit, ethnic minorities
Méjan	France	Livestock farming, forestry	Biodiversity	Agri-environmental measures
Nan	Thailand	Forestry	Biodiversity	Ethnic minorities
Nîmes-Métropole	France	Vineyards and cereal cropping	Wildfires	Peri-urban, land tenure
Njoobaari Ilnoowo	Senegal	Irrigated crops	Wastage of water	Indebtedness
Ouessant	France	Livestock farming	Biodiversity	Tourism
Pays de Caux	France	Cash cropping	Erosion	Peri-urban
Radi	Bhutan	Livestock farming	Erosion	Collaborative management

(continued)

(continued)

Title	Country	Agricultural issue	Environmental issue	Socio-economic issue
SAGE Drôme	France	Irrigated crops		Equipment
SosteniCAP	Bolivia	Irrigated crops	Wastage of water	Peri-urban
SugarRice	Thailand	Cash cropping	Intensification	Rentability
Tarawa	Kiribati	Coconut cropping	Groundwater	Drinking water
Ter'aguas	Brazil	Irrigated crops	Water pollution	Drinking water
Ubon Rice Seeds	Thailand	Rice cropping	Genetic diversity	Seed market
Ventoux	France	Livestock farming, forestry	Biodiversity	Tourism
Vosges du Nord	France	Livestock farming	Biodiversity	Urban development

AguAloca: Managing Water in a Peri-Urban Catchment Area

Question asked: how can water quality and agriculture be integrated in managing multiple reservoirs in a peri-urban catchment area?

Region: Alto-Tietê Cabeiceras catchment, Sao Paulo, Brazil.

Sponsor: University of Sao Paulo.

Researchers involved: R. Ducrot and L. Clavel (CIRAD UR G-EAU), V. Barban (Instituto Polis), S. Sendacz, C. Rabak and J. Sichman (University of Sao Paulo) and Y. de Carvalho (Agência Paulista de Tecnologia dos Agronegócios);

Objectives and progress: The Alto-Tietê Cabeiceras is a complex catchment area part of the water supply system of the Sao Paulo metropolis including huge hydraulic infrastructures (i.e. five reservoirs, large channels and a pumping station) covering several municipalities. The water quality in the final reservoir is constantly being degraded by eutrophication, which results in increased treatment costs. Studies have shown that this pollution is due mainly to the modalities of managing water transfers between reservoirs, whilst traditionally the managers point out illegal urban development and diffuse agriculture.

In this context, a four-stage ComMod approach was applied as part of the dialogue within a catchment committee about integrated management of water resource while water quality and agriculture aspects.

(i) Thematic studies (e.g. hydrology, land economics, agro-economy, social sciences, geography, etc.) improving the understanding of water management and pollution dynamics at the scale of the catchment area.
(ii) Co-constructing a discussion tool in the form of the AguAloca role-playing game, mobilizing a multi-disciplinary team and a small group of managers.

(iii) Specific work with farmers to increase their ability to face up to the tensions related to water issues in this catchment area and take part in discussions.
(iv) Using the AguAloca game twice: once with engineers from the water company, the water management department and municipality representatives and once with several members of the Alto-Tietê basin committee.

The approach helped participants to gain a better understanding of the meaning of the 'integrated water' management concepts at the scale of a catchment area, by giving them a chance to analyse the interdependences, particularly in terms of water allocation and quality. It also drew attention to agricultural and rural issues in this peri-urban region. The game appeared to be an interesting way of introducing the complexity of catchment area management to newcomers to the discussion bodies and of preparing the stakeholders in the use of simulation tools calibrated to reality.

For further information: a seminar presentation (Clavel et al. 2008).

AguaAloca (Brazil): in the suburbs of Sao Paulo, a role-playing game was designed to discuss about the impact of illegal urban development on drinking water quality in the reservoirs.

Camargue: Multi-Uses and Biodiversity of Reed Marshes

Question asked: how to encourage consultation on biodiversity, water and land management?

Region: reedbeds in the Vendres and Charnier-Scamandre ponds, Camargue, France.

Sponsor: Conservatoire des espaces naturels en Languedoc-Roussillon (CEN-LR), Joint Union of the Lower Aude River Plain (Syndicat mixte de la basse plaine de l'Aude—SMBVA), Joint Union of the Gard Camargue (Syndicat mixte de la Camargue gardoise—SMCG), Fondation de France, and ADD-ComMod project.

Researchers involved: R. Mathevet (CEFE CNRS), B. Poulin (Tour du Valat) and C. Le Page (CIRAD GREEN).

Objectives and progress: This project covers two wetland areas that have suffered tremendously in recent years. It involves two Mediterranean reed marshes used for many purposes and with a great natural and cultural heritage value. The representatives of SMBVA and SMCG took part in the European Life Nature project designed to improve the management of reedbeds (wetland dominated by *Phragmites australis*) to preserve the Eurasian Bittern (*Botaurus stellaris*), a vulnerable heron at the European scale. They tested the ButorStar role-playing game developed under the educational module of this project. Thus made aware of the possibilities, they were interested in testing the tool and the ComMod approach in their respective areas. The ADD-ComMod project and the call for projects by the Fondation de France in spring 2005 formalized this project. Experiments with a companion modelling approach were applied in both these areas: one in a context of advanced consultation and the other in a context of 'paused' consultation. It involved checking that the role-play sessions set up with users could encourage or revitalize the multi-stakeholder consultation by widening the vision of participants through improving their understanding of the wetland, its dynamics and its specific interdependences.

Two role-play sessions were organized in 2006 with 12 users from the Vendres pond (i.e. hunters, livestock farmers, owners, technicians, NGOs). They were asked to complete a questionnaire to assess:

(i) the acquisition of knowledge on the functioning of the ecological system
(ii) the awareness of effects of user practices on other uses and the environment
(iii) the awareness of the interaction mode of participants in terms of negotiation/consultation as well as the advantages and disadvantages of the approach.

A role-play session was organized in 2007 with eight stakeholders from the Charnier-Scamandre ponds. The result of the game session highlighted the need for raising awareness among the local elected representatives and to work closely with decision-makers to improve the negotiation process related to collective decisions in the management of the whole wetlands.

The approach resulted in a collective awareness of the many implications of inappropriate water management and/or the development of certain activities on the reedbed and its avifauna. All users were interested in repeating the experiment. Several of them acknowledged that it had transformed significantly the social links and perception of both hunting and nature conservation.

For further information: an article on the role-playing game (Mathevet et al. 2007) and its application in the Vendres pond (Mathevet et al. 2008).

Camargue (France): reed beds are multiple use wetlands where common rules of water management have to be set up to conciliate production and biodiversity stakes.

Domino Réunion: Mediation and Foresight on Regional Land Use for Réunion Island

Question asked: how to articulate regulatory systems by facilitating the sharing of viewpoints and strategies between stakeholders?

Region: Réunion Island.

Sponsor: CIRAD, Réunion Region.

Researchers involved: A. Botta, W. Daré, S. Aubert, X. Augusseau and J.-P. Müller (CIRAD GREEN), G. Lajoie, D. Payet and R. Courdier (University of Réunion).

Objectives and progress: The various regulatory plans for land use on Réunion Island were under revision. It was decided to prioritize relevant issues at the scale of the regional development scheme (SAR), whilst paying special attention to their articulation with regard to other planning levels (at the scale of the municipality and micro-region). The lack of availability of stakeholders as they were called on for most of these planning processes and the initial low legitimacy of the coordinators of the Domino project led to the approach being divided into two phases according to the associated partners.

Phase 1 was developed between researchers and advisers of agricultural and rural development (i.e. Chamber of Agriculture, Association for Promotion in Rural Environments and the Sugar Cane Steering Committee). This group focused on three tasks simultaneously.

(i) The collection of available datasets and the development of new ones on: the dynamics of natural area conservation, urbanization and agricultural valorization of land, and the land use practices of the main local stakeholders (i.e. farmers, urban developers, foresters and elected representatives).
(ii) The collective development of a prototype of a conceptual model that represented the current functioning of land use and identified levers to account for a range of potential dynamics for the next 15 years.
(iii) The implementation of this prototype as a MAS and the validation of the implementation choices by the entire group.

In parallel, the SAR cell led a legal participative approach to review the SAR. In phase 2, our group signed a contract with the SAR cell to adapt the prototype to the specific questions that had emerged in the legal approach. The final version of the model, therefore, was constructed and validated by our group and officers from the Regional Council, Agency for the Observation of Reunion, Development and Habitat (AGORAH) and the design offices responsible for environmental aspects. It was then used to simulate development scenarios constructed in the legal approach.

Our approach associated and connected the main viewpoints of the agricultural, urban and environmental worlds, thus highlighting the need for arbitration to manage urban development up to 2030, to defend the agricultural space and to maintain a natural area that preserves biodiversity. It strengthened the place of agricultural issues in the regional land debate thanks to the participation of agricultural representatives in the design of the tools. It took into account the impacts of general direction choices in the SAR at the scale of communities of communes and municipalities.

For further information: two articles on the position of sociologists in the approach (Daré et al. 2007) and participatory process (Daré et al. 2008), and two conference papers on the model (David et al. 2007) and on multiple scaling (Botta et al. 2009).

Domino (La Réunion): the review of the land planning documents was based on the connection of the different regulations by sharing viewpoints and strategies among stakeholders.

Don Hoi Lord: Managing the Razor Clam Fishery

Question asked: how to encourage the various stakeholders involved in the fishery discuss possible management methods?

Region: Don Hoi Lord sandbar, a classified Ramsar site in coastal central Thailand.

Local partner: Chulalongkorn University.

Researchers involved: K. Worrapimphong and N. Gajaseni (Chulalongkorn University), F. Bousquet and C. Le Page (CIRAD GREEN).

Objectives and progress: The Don Hoi Lord site 80 km to the west of Bangkok in Thailand is a site of recognized ecological interest (Ramsar site for the conservation and sustainable use of wetlands). This sandbar, enriched by the accumulation of nutrient sediments carried by the Mae Klong river, is also a craft fishing ground for a species of razor clam (*Solen regularis*), a delicacy much appreciated by the Thai people. For several years activity in the fishery (which has operated for over 80 years with no form of regulation) has been intensifying and the resource seemed threatened. In 2003, research work started by implementing a companion modelling approach with the goal of stimulating exchanges between the various stakeholders (i.e. mainly fishermen and local authorities) and exploring collectively various management options for the fishery. The approach was broken down into three phases.

(i) Fieldwork to compile additional data on the biology of the species, of which little is known. At the same time, various stakeholders involved in the fishery (i.e. fishermen, traders and local authorities) were interviewed. This information was incorporated into a multi-agent model developed to integrate existing knowledge and propose a first representation of the Don Hoi Lord razor clam fishery.

(ii) This model was simplified into a role-playing game with the goal of interacting with local stakeholders and to test whether they were interested in manipulating this representation of the fishery operation. A gaming session was organized with 11 fishermen from the same village. Having played out a first scenario and discussed the validity of the model, the discussions turned to defining alternative scenarios based on introducing management rules (in particular a rotation of reserved areas). These scenarios were also played out. During the debriefing, the participants expressed their interest in the approach by requesting that a new role-playing session be organized involving more stakeholders (especially fishermen from other villages).

(iii) A second gaming session was organized a little later with more participants: 10 fishermen from two villages, one trader and representatives from local authorities and the fisheries department. Additional scenarios were identified and played out. The local authorities subsequently invited the researchers to present the approach to local political decision-makers. The provincial governor stated that he was very interested in the approach.

The approach facilitated sharing of knowledge and viewpoints between scientists and fishery stakeholders. It is currently continuing with a new campaign of data collection in the field (i.e. razor clam biology, fishing practices, market availability and seasonal price fluctuations). This will lead to a new version of the model incorporating this information that will serve to support a new series of participative simulation workshops.

For further information: a Master's dissertation thesis (Worrapimphong 2005), a PhD dissertation thesis (Worrapimphong 2010) and an article on the combined use of role-playing games and computerized simulation (Worrapimphong et al. 2010).

Don Hoi Lord (Thailand): in this Ramsar site, the intensification of the exploitation of razor clams is threatening its sustainability and urges to discuss about different management options with all the concerned stakeholders.

Frêne: Spontaneous Reforestation of the Bigorre Mountains

Question asked: how can the spontaneous reforestation of mountain landscapes be controlled?

Region: Parc National des Pyrénées, Villelongue municipality, France.

Sponsors: INRA and the French National Research Agency (ANR).

Partners: Centre for Pastoral Resources and Land Management (Centre de ressources pastorales et de la gestion de l'espace—CRPGE), DDAF and agricultural advisory services (Chambre d'Agriculture des Hautes-Pyrénées—CDA), and the Villelongue municipality.

Researchers involved: A. Gibon, C. Monteil, C. Simon, S. Ladet, D. Sheeren, G. Balent and A. Gavaland (UR Dynafor) and M. Étienne (INRA Ecodevelopment Unit).

Objectives and progress: In the early 2000s, the scientific department of the National Park of the Pyrenees initiated research into the processes and consequences of spontaneous reforestation of valleys in its area with an INRA research group and the agricultural and rural development agencies of a Pyrenean district (Département des Hautes Pyrénées). In relation to agricultural decline, ash (*Fraxinus excelsior*), a species associated with traditional agropastoral systems, has been colonizing agricultural land for several decades. The aim was to assess the processes of reforestation and their consequences for rural development, to provide methods for controlling its impacts on biodiversity and landscape amenities, and designing development options for agricultural and forestry activities. Ash colonization processes and their consequences on biodiversity were investigated between 2003 and 2005; spatio-temporal changes in agricultural land use and management strategies at the farm and landscape levels were studied concomitantly.

To assist local landscape stakeholders and rural development policy decision-makers in their search for sustainable development pathways, a companion modelling approach was started in July 2006 to develop an integrated simulation tool for the relationships between ecological and social systems. The participatory building of a MAS simulation model was based on further elaboration of knowledge gained previously. It currently includes:

The co-construction of the conceptual model of the main stakeholders, resources, dynamics and interactions involved in land management and landscape reforestation (ARDI method).

An investigation of the functions and representations of traditional barns far from the villages; a change in their use (from agriculture to second homes) having appeared during the co-construction process as a major source of change in agricultural land-use practice.

A progressive development of the simulation tool, in a process where design and implementation periods amongst researchers alternate with periods of communication with the stakeholders for assessing and further building progress in the construction of the conceptual model.

For further information: a book chapter presenting the approach (Monteil et al. 2008) and a paper on the model (Gibon et al. 2010).

Frêne (France): in the district of Villelongue, reasons, processes and consequences of ash trees overspreading on ancients croplands were studied and modeled.

Kat Aware: Multi-Use Allocations of Water in South Africa

Question asked: how can local water management be improved by taking into account the concerns of various users?

Region: Kat River valley, South Africa.

Sponsor: Kat River Water Users Association (WUA).

Researchers involved: S. Farolfi (CIRAD G-EAU) and J.P. Muller (CIRAD GREEN).

Objectives and progress: In accordance with the new South African Water Act, the Kat River WUA will ultimately be responsible for managing the water resource in the catchment area. This will mean that water licences will be allocated to the various users with a price fixed for the resource. The decision-making process will move from the Ministry of Water Affairs and Forestry (current decision-maker) to the WUA when the latter will have its catchment management plan (CMP) validated. The members of the WUA, therefore, find themselves with a major responsibility to be addressed without the actual means of doing so. There is a tremendous need for tools and approaches facilitating the implementation of a consultation and negotiation process between representatives of very different water users and aiming at the preparation of the CMP.

A companion modelling approach was set up with the local WUA around a basic question: how to improve local management of water resources in the Kat river basin whilst listening to the various user groups? The approach was broken down into four phases.

(i) Integrating available scientific knowledge on the state and uses of the resource within a multi-agent model (KatAWARE).
(ii) Sharing this representation of the Kat system with the local stakeholders (WUA) and incorporating local know-how and criticisms in the model.
(iii) Running simulation exercises for agents taking part in the approach with the support of a role-playing game; two role-playing game sessions have taken place.
(iv) Developing possible scenarios for allocating water resources via a new multi-agent model and let the stakeholders discussing them.

The approach culminated in improved knowledge of the hydrological and socio-economic consequences of adopting alternative water allocation strategies in the Kat basin. The stakeholders involved also interacted considerably, especially during the role-playing sessions, which made dialogue and discussion on water management scenarios between participants much easier. The WUA is currently drawing up the catchment area management plan and the water allocation scenarios constructed with the KatAWARE model are used as a basis for negotiating and defining the strategies to be proposed in the plan. Given the good results from the project, the Water Research Commission, which financed the project, proposed to extend the approach to other South African catchment areas to improve the negotiation process and collective decision-making within new

WUAs. Currently a WRC project inspired by the Kat experience is running in the Inkomati Water Management Area, and a new negotiation platform called Wat-A-Game and originated from the KatAWARE RPG was developed in the Sabie catchment.

For further information: three articles on the approach (Farolfi and Rowntree 2007; Farolfi et al. 2008; Farolfi et al. 2010) and an article on the relations between role-playing games and game theory (Dinar et al. 2008).

Kat Aware (South Africa): in the Kat River valley, participatory workshops were organized in order to improve local water management and allow different types of users to give their opinion.

Lam Dome Yai: Rice Production and Labour Migration in Northeastern Thailand

Question asked: what are the effects of young worker migration on rice-growing practices under a high risk of drought?

Region: Lam Dome Yai watershed, Ubon Ratchathani province, Thailand.

Local partner: University of Ubon Rajathanee (UBU).

Researchers involved: F. Bousquet, C. Le Page and G. Trébuil (CIRAD GREEN), W. Naivinit and M. Thongnoi (UBU).

Objectives and progress: Rainfed lowland rice growing is practised in northeastern Thailand a region characterized by a six-month dry season and coarse textured soils. The drought risk is high at the beginning of the wet season. The seasonal labour migrations have for decades been an adaptive mechanism widely used by smallholders to face up to climatic risks and low rice productivity in the poorest region of the country. With certain migrations becoming definitive and authorities wishing to improve rice growers' access to irrigation, an improved understanding of the interaction between labour management on and off the holding and the use of land and water is crucial to designing well-adapted hydraulic schemes and foreseeing to whom they would benefit. Warong Naivinit carried out his doctoral research on this subject by applying the ComMod approach to understand this interaction in the central part of the Lam Dome Yai watershed in southern Ubon Ratchathani province.

The approach was implemented through several iterative loops between the field and the laboratory between 2004 and 2008.

(i) Review of the literature on the local agrarian system, especially rural worker migrations. Construction of the first UML diagrams to conceptualize the interactions between the agricultural system components.
(ii) Design, test and use of a first role-playing game in the Ban Mak Mai village. This game represented the researcher perception of the interaction under study during key phases of the annual rice-growing cycle (labour management at transplanting, harvesting and in the dry season) and interactions between types of rice-growing holdings. It was replayed subsequently with the migrants and the gaming sessions led to modifications of the initial UML conceptual diagrams.
(iii) Construction of a second game, which was more explicit regarding the cropping calendar and a computerized application playing the game under Cormas. The observation of farmers' adaptation to very dry periods and irrigation (water was drawn from a canal, individual or community ponds).
(iv) Gradual construction of a multi-agent model representing the interaction under study for four different rice farms by alternating programming sessions and the repeated submission of successive versions of the prototype for review by rice growers.
(v) Validation of the improvements of the multi-agent model requested and simulation of scenarios of interest to its users.

The collaborative modelling approach adopted could be spaced out over a long period and was ideal for multiple exchanges of viewpoints on the simulated interaction between the researcher and 11 rice-farming households, some offering a work force, others hiring it out during peak of labour demand periods in the rice-growing cycle. Work on the various effects of the approach on the rice growers who took part heralds the end of this research.

For further information: an article on the process and its simulation tool (Naivinit et al. 2010).

Lam Dome Yaï (Thailand): in this rainfed rice growing lowland, understanding the interactions between labor and land and water use is crucial to design well-adapted hydraulic schemes.

Larq'asninchej: Urbanization and Irrigation Channels in Bolivia

Question asked: how to facilitate discussion between irrigators and urban dwellers to deal with irrigation channels in a peri-urban area?

Region: municipality of Tiquipaya, metropolitan region of Cochabamba, Bolivia.

Sponsor: University Mayor of San Simon.

Researchers involved: N, Faysse (CIRAD UR G-EAU), D, Vega and R, Peñarietta (University Mayor of San Simon, Centro Agua).

Objectives and progress: The Tiquipaya valley in the urban area of Cochabamba in Bolivia has for 15 years been facing rapid and uncontrolled urban expansion, which affects the traditional collective irrigation systems for the region's agriculture, which criss-cross the valley. Degrading infrastructures, obstructing canals and access roads and evacuating wastewater into canals affect the functioning of the irrigation systems and deter farmers from carrying out regular maintenance. The proposed approach, based on recognition of the dual irrigation and drainage function of canals in this area, aimed to facilitate negotiations between farmers and urban residents on the issue of maintaining irrigation canals.

The approach was organized in five stages.

(i) Detailed socio-technical diagnostics so that the community could select two or three priority work themes. A non-computerized role-playing game was developed using a stakeholders' analysis.

(ii) Game sessions that united farmers and urban residents to facilitate the exchange of perceptions on problems encountered in a ludic, tension-free framework and to help identify possible solutions.

(iii) A participative inspection of canals that identified the problems in the field and initiated a discussion on potential solutions. Developing a community map identifying problems identified to select collectively solutions that were validated during restitution to the entire community.
(iv) Technical draft solutions were developed with the support of the project consultants and submitted to the authorities for funding.
(v) Agreements on the future maintenance of the infrastructures were put together within communities and between communities and the local authorities.

Although the infrastructure projects could not find funding and were not pursued, the approach as a whole drew attention to the dual role of canals in the area and modified the perceptions of local stakeholders and the authorities. Thus the Tiquipaya municipal authority has allocated a small budget item to maintaining irrigation canals and the federation of irrigator associations uses the agreement as a basis for discussion with the municipality towards a larger scale agreement on the future management of irrigation infrastructures in the municipality.

For further information: an article (Vega et al. 2006) on the website http://www.negowat.org.

Larq'asninchej (Bolivia): in the suburbs of Cochabamba, it is necessary to facilitate the consultation between irrigators and urban dwellers to avoid conflicts over irrigation channels

Larzac: Forest Dynamics in the Causse du Larzac

Question asked: how best to take advantage of pine encroachment?

Region: north–east Larzac, Lozère, France.

Sponsor: Société Civile des Terres du Larzac (SCTL).

Researchers involved: M. Étienne and C. Simon (INRA Ecodevelopment Unit), L. Dobremez and H. Rapey (CEMAGREF), G. Guérin (Institut de l'Elevage) and O. Picard (Institut pour le Développement Forestier).

Objectives and progress: Interested in the companion modelling process that was applied on the Causse Méjan (Méjan case study), the SCTL requested INRA researchers in Avignon to use a similar approach in collective reflective thinking on how best to exploit the woods that have developed throughout its land over the last 50 years. The forest development plan had three objectives: control the current dynamics of pines; provide an income from pine timbering and oak coppicing; improve the pastoral production of wooded rangelands.

The approach was broken down into four phases.

(i) Mapping the management entities of all 30 farms in the SCTL, describing the grazing and work timetable and explaining the activities taking place in the forest.

(ii) Incorporating these data into an agent-based model developed by the researcher group then validating this model with the farmers and managers of the SCTL.

(iii) Simulating the probable dynamics of the area mapped if current practices are maintained over the next 20 years.

(iv) Inventory of forest plots for each farmer and the development of scenarios for forest resources exploitation according to various options: taking farmer priorities only into account; promoting the most productive wood; and global reflective thinking about sustainable management of the entire forest mass.

The approach culminated in a collective awareness of the many implications of pine encroachment on SCTL land and developing long-term management scenarios for this spontaneous afforestation. The search for a compromise between the overall SCTL objectives deliberated at the forest scale and the individual objectives of farmers at the farm scale produced a wealth of debate and a meeting of different viewpoints.

Following the support for this project, one farmer was trained in identifying productive stands and timber marking, and a team of loggers was put together to carry out the forestry projects with volunteer farmers and several management operations are in progress.

For further information: a dissertation by a forestry engineer (Simon 2004) and an article on the approach and scenarios (Simon and Étienne 2009).

Larzac (France): the Société Civile des Terres du Larzac set up a collaborative reflective thinking on how best to exploit their woods, while reinforcing the sustainability of the farms of its adherents.

Lingmuteychu: From a Water-Sharing Conflict to the Creation of a Watershed Management Institution in Bhutan

Question asked: how to solve an old conflict between two villages over sharing irrigation water and how to upscale a ComMod process to improve community-based management of natural resources among seven villages?

Region: Lingmuteychu catchment in west-central Bhutan.

Local partner: Renewable Natural Resources Research Center (RNR-RC), Bajo.

Researchers involved: F. Bousquet, J. Queste and G. Trébuil (CIRAD GREEN), T.R. Gurung and A.K. Bhujel (RNR-RC) and Gyenbo Dorji.

Objectives and progress: Rice growers in Lingmuteychu for many years have to share a limited amount of water to irrigate their highland terraces at 1200–2200 m asl. At this altitude, the period propitious to rice transplanting is very short and stirs up repeated disputes about timely access to the precious resource between the seven stepped village communities. Customary rules cannot solve these disputes whilst cropping practices change with the adoption of commercial crops. An initiative encouraging community-based management of renewable natural resources (CB-NRM) was launched by the Ministry of Agriculture and Lingmuteychu was chosen as one of the pilot sites. Having been trained in companion modelling, the local team leader, Tayan Raj Gurung decided to use the approach in an attempt to mediate the dispute between the two highest villages. The process was then extended to all seven villages in the catchment area, then to two more sites in the eastern part of the country. Two phases can be distinguished in the implementation of the approach in two distinct cycles.

First cycle (2003–2004)

(i) Review of the literature on the local agrarian system and, in particular, the existing knowledge about the conflict on agricultural water use to build a conceptual model.
(ii) Design of a first role-playing game on water sharing between two villages for a better understanding of the decisions of various types of farmers in each village, at the village level and between the two communities.
(iii) Following a test with RNR-RC researchers, the first workshop was held based on gaming sessions under three different modes of communication demonstrating the effects of farmers' decisions on the satisfaction of water requirements and cropping results.
(iv) Six months later, the second workshop was held with an improved version of the game (including the possibility of exchanging water against labour requested by the players) and more diverse participants and observers. A modification to the collective rules for sharing water was adopted, but not formalized and not applied the following year.
(v) Construction of a multi-agent model similar to the game and simulation of numerous scenarios of exchange protocols among farmers analysed in T.R. Gurung's Master's thesis.
(vi) Preparation of a second cycle at the whole subwatershed scale.

Second cycle (2005–2006)

(i) Design and testing (with RNR-RC researchers) of a very abstract role-playing game on water sharing among the seven stepped villages.
(ii) Workshop based on this role-playing game was held with village representatives according to varied modes of communication among villages (first each village played autonomously, then inter-village communication was allowed, finally the roles were switched between upstream and downstream villages).
(iii) Facilitation of plenary discussions aiming at the creation of a subcatchment resources management committee and adoption of a common action plan.
(iv) Support in the drawing up of the by-laws of the watershed committee and search for funding to implement the committee's first action plan.

The collaborative modelling could be spaced out over a long period at this site due to the intensive involvement of Bhutanese colleagues from RNR-RC Bajo and the agricultural extension worker, a key facilitator in the process, based at the site. Funding from the United Nations Development Programme (UNDP) was used to rehabilitate irrigation channels, protect springs and plant community forests in 2006–2009.

For further information: an award winning article on the approach is available (Gurung et al. 2006).

Lingmuteychu (Bhutan): in this mountain catchment, authorities wished to solve an old conflict between two villages over sharing irrigation water.

Luberon: Livestock Farming and Landscape Dynamics

Question asked: to what extent is livestock farming likely to be present where the society will expect it to be in the region in the future?

Region: 22 municipalities in the Luberon Regional Nature Park, France.

Sponsor: Luberon Park, Centre d'Études et de réalisations pastorales Alpes-Méditerranée (CERPAM) and Institut de l'Élevage.

Researchers involved: M. Étienne and J. Lasseur (INRA Ecodevelopment Unit).

Objectives and progress: Twenty years of cooperation between INRA in Avignon, CERPAM, the Institut de l'Élevage and the Luberon Regional Nature

Park have produced numerous studies on the technical and organizational aspects of converting livestock systems and adapting them for use in natural areas (i.e. grasslands, forest rangelands and Mediterranean forests). Faced with the scale of forthcoming changes in public policy, social mutations or climate change, the group wished to develop an approach and a tool to put accumulated data covering the entire regional park into perspective.

The INRA Ecodevelopment Unit suggested implementing a companion modelling approach covering the long-term analysis of the ability of livestock farming to take part in projects for managing the region in terms of environmental issues and local development on a time scale of the next 10–15 years. The project aimed at co-constructing with partners in the field various scenarios depicting the changes in livestock activities and their interactions with landscape dynamics. The intention was to put together spatial representations of issues raised by each stakeholder and formalize various scenarios in its spatial dimensions developed by the working group.

The approach is in progress and is broken down into four phases.

(i) Integrating the available scientific and technical knowledge on the dynamics of vegetation, forestry and livestock practices and the structure of agricultural farms by co-constructing a conceptual model representing the complexities of the question asked.
(ii) Implementing the model with the Cormas platform in three contrasting land units covering 22 municipalities in the regional park.
(iii) Situated simulation exercises for the livestock farmers taking part in the area modelled, as part of a role-playing game.
(iv) Improving the model using behaviours observed during the role-playing game sessions and simulating a few likely scenarios for transforming agriculture over the next 15 years.

The first simulations showed the need to incorporate forestry activities as another decisive factor in the landscape dynamics and thus invite partners from public (ONF) and private (Centre régional de la propriété forestière) forests.

For further information: a seminar presentation (Napoléone et al. 2008) and an article on the approach (Lasseur et al. 2010).

Luberon (France): on a large area of the Natural Regional Park, agents of forest and livestock technical services set up a collaborative reflection on the capacity of sheep farming to maintain open lands in a sustainable way.

Mae Salaep: Management of Land Degradation Risk in a Highland Catchment of Upper Northern Thailand

Questions asked: cycle 1: what crops and cultivation practices could reduce the risk of soil erosion? cycle 2: what rules for the allocation of rural credit could lead to a more equitable access to perennial commercial crops? cycle 3: how should water be managed for a more equitable access to irrigation?

Region: Mae Salaep village, Mae Fah Luang district, Chiang Rai province, northern Thailand.

Local partner: Chiang Mai University (CMU).

Researchers involved: G. Trébuil, F. Bousquet, C. Barnaud (CIRAD GREEN), B. Ekasingh, P. and T. Promburom (CMU).

Objectives and progress: In the sloping highlands of upper northern Thailand conflicts over the management of renewable resources are becoming more frequent. The ethnic minorities populating these highlands are blamed by the dominating Thai lowlanders for increasing the risk of soil erosion by concentrated runoff due to their farming practices. The highlanders are facing the threat of their farmland being seized by the authorities for reforestation, and so forcing them to migrate. The recent decentralization policy allowing an increase in the participation of citizens in the management of local resources represents an important opportunity for these communities. However, the problems are complex and stakeholders involved in the use of renewable resources are diverse. In this context, in 2002, a ComMod process was initiated with the Akha community of Mae Salaep village.

In this experiment, the learning process focusing on a given problem in a ComMod cycle raised new problems and questions that were examined in the following cycles. For this purpose, each time the role-playing game and associated computerized multi-agent model were adapted. The first cycle aimed at understanding the interactions between crop diversification and the risk of soil erosion. At the end of this loop, the villagers identified perennial crops as a promising way to mitigate the problem from both ecological (lower risk of land degradation) and economic (higher cash incomes) perspectives. They also underlined that poor villagers could not invest in such crops because of their lack of access to credit and requested that this issue be addressed in a second ComMod cycle. Moreover, since these perennial crops are irrigated, their expansion in the village catchment created social tensions over access to agricultural water. On the villagers' request, this question was addressed in a third cycle.

The main phases of a ComMod cycle were as follows.

(i) Analysis of secondary data and field interviews to fill knowledge gaps on the key problem and the related key question to be examined.
(ii) Construction or modification of the role-playing game and computerized multi-agent model based on the same conceptual model.
(iii) Participatory field workshop combining gaming sessions and plenary discussions to validate the model and facilitate exchanges about the problem among participants, followed by individual interviews to obtain a better understanding of the participants behaviours and opinions, then computer simulations for the collective exploration of scenarios selected by the players.
(iv) Monitoring of the effects of the ComMod process on the participants in terms of learning, communication, change in behaviour, decision-making and practices.

This ComMod process increased the villagers' awareness of the necessity to solve collectively the different problems examined during this process. A better understanding of the other participants' perspectives on these problems was also achieved as well as the collective identification of socially acceptable solutions.

For further information: cycle 1: a conference paper on the process (Trébuil et al. 2002b) and a book chapter on the first multi-agent model (Trébuil et al. 2005); cycle 2: a journal article on the adaptive nature of the process (Barnaud et al. 2007) and another on the computer multi-agent simulations (Barnaud et al. 2008c); cycle 3: a PhD thesis (Barnaud 2008), a journal article on the effects of the process (Barnaud et al. 2008b) and another one on the power relations issue (Barnaud et al. 2010).

Mae Salaep (Thailand): in this mountain catchment, a collaborative and iterative learning process was run on soil erosion risk, rural credit allocation and equitable access to irrigation.

Méjan: Invasive Pines on the Causse Méjan

Question asked: how to prevent pine encroachment on high biodiversity habitats?

Region: Causse Méjan, Lozère, France.

Sponsor: Cévennes National Park.

Researchers involved: M. Étienne and M. Cohen (INRA Ecodevelopment Unit) and C. Le Page (CIRAD-GREEN).

Objectives and progress: In 2000, the Scientific Service of the Cévennes National Park expressed its concerns to its scientific committee about the development of pine encroachment on native grassland habitats and the threats affecting many heritage issues identified in the Causse Méjan plateau. A prospective study was planned under the research programme 'Recreating Nature' entrusted to M. Étienne, a member of the scientific committee. He suggested setting up a companion modelling approach with all local stakeholders focusing on three questions. How to forecast changes in a natural space of anthropic origin under contrasting strategies of regional development? How to account simultaneously for heritage and productive issues? How to facilitate consultation between stakeholders subject to the same ecological dynamics?

The approach was broken down into four phases.

(i) Integrating available scientific knowledge on the state of natural resources, the dynamics of the priority spaces to be protected and the practices of the main local stakeholders (i.e. livestock farmers and foresters) in the form of an agent-based model.
(ii) Sharing this representation of the Méjan system with local stakeholders and incorporating local know-how in the agent-based model.
(iii) Situated simulation exercises for stakeholders involved in the Causse Méjan dynamics in a fictitious area yet close to their reality to make them react to the probable encroachment of pine trees, as part of a role-playing game involving farmers, foresters, park officers and a public policy sponsor (Chamber of Agriculture).
(iv) Developing possible management scenarios for the region and a collective evaluation of their consequences based on the different viewpoints of each category of stakeholder via a new agent-based model.

The approach culminated in a collective awareness of the many implications of pine encroachment in the Causse Méjan and the setting up of a collective action encompassing the community of municipalities, farmers, Chamber of Agriculture, Cévennes National Park and forest agencies in a local concerted management plan; this led to the introduction of contracts between 28 farmers, the park and certain forestry owners.

Following this project, the Cévennes National Park decided to extend the same approach to the Causse du Sauveterre and the Larzac livestock farmers asked to adapt the approach to a valuation of their pine and oak stands (see Larzac case study).

For further information: an engineer's dissertation on the model (Cohen 2001), an article on the model and scenarios (Étienne et al. 2003), and an article on the approach and role-playing game (Étienne and Le Page 2004).

Méjan (France): the collective awareness of the multiple consequences of pine encroachment on native grasslands lead to setting up a collaborative management plan.

Nan: Dispute Between Communities and a National Park

Question asked: how to facilitate a concerted process on managing forest resources between communities and a national park?

Region: Nantaburi National Park, Nan, Thailand.

Sponsor: Chulalongkorn University and Royal Forests Department.

Researchers involved: C. Barnaud, C. Le Page and G. Trébuil (CIRAD GREEN), P. Dumrongrojwatthana and N. Gajaseni (Chulalongkorn University).

Objectives and progress: The northern Thailand highlands are home to ethnic minorities whose agricultural and food practices depend greatly on forest resources and who are accused jointly by the Thai government and the majority of degrading the forests. These highlands are governed by powerful environmental policies, particularly in the form of national parks from which theoretically all human activity is banned, which generates major disputes with the communities. At the same time, the Thai government is introducing decentralization policies and is highlighting the need for citizens to participate in resource management. In this context, a ComMod approach was initiated in the province of Nan during the setting up of a national park to facilitate a concerted process between park officials and the villagers of two Mien communities located on the park boundary.

The main phases in the approach were as follows.

(i) An initial in-depth agrarian and institutional analysis to identify, among other things, the main restrictions in implementing a fair concerted process.
(ii) Constructing a role-playing game embracing all the dynamics and problems associated with setting up the national park.
(iii) A participatory workshop in each village based on the use of the role-playing game so that the 12 participants could criticize the proposed game and discuss amongst themselves the question of the national park.
(iv) Restitution of these gaming sessions to park officials using a multi-agent computer model replaying the game to make them discuss it amongst themselves.
(v) A meeting in each village based on using a hybrid multi-agent system (MAS)associating real players and artificial agents, so that all villagers could participate in discussions on the park and prepare for a potential meeting with the park officials.
(vi) A participatory workshop uniting park officials and the villagers from two villages combining gaming sessions and multi-agent computer simulations for collective exploration and discussion of possible scenarios.

The effects of this ComMod process were monitored and assessed throughout the process. This process was an opportunity for the villagers to understand the diversity of interests existing in the villages over the park issue and to appreciate the need for collective reflective thinking on a negotiation strategy with park officials. It subsequently spurred park officials and the villages to mutual understanding and awareness that dialogue was not only possible but also potentially beneficial.

For further information: a PhD thesis (Barnaud 2008), an article on the methodological importance of an initial analysis (Barnaud et al. 2008a) and an article on common values (Ruankaew et al. 2010).

Nan (Thailand): in order to avoid conflicts between communities and the agents of a new National Park, a participatory process was set up to better understand the diversity of point of views.

Nîmes-Métropole: Fire Prevention, Abandonment and Urbanization

Question asked: how to make elected representatives more aware of forest-fire prevention?

Region: Municipalities in the northern Nîmes area, Gard, France.

Sponsor: Nîmes-Métropole Urban Community.

Researchers involved: M. Étienne and M. Bourgeois (INRA Ecodevelopment Unit).

Objectives and progress: In December 2005, at a presentation of the SylvoPast case study outputs, the Gard DDAF suggested that it be adapted to the forest-fire prevention problem at the interface of urban and natural zones. The Nîmes-Métropole Urban Community Environmental Service, keen to raise the awareness of its elected representatives to this issue, offered its territory as a test zone.

The approach was broken down into four phases.

(i) Compiling the available mapping data on the forest, the dynamics of urbanization and the practices of the main local stakeholders (i.e. farmers, urban developers and foresters).

(ii) Developing a virtual map representing three typical adjacent municipalities in the northern Nîmes area and validation of this map by a group of technicians covering the main activities found in this area.

(iii) Co-constructing, with the same group, a conceptual model representing the current functioning of the area and the likely dynamics over the next 15 years, then implementation of this model by INRA researchers as a MAS.

(iv) A situated simulation exercise for elected representatives from the 14 municipalities involved to discuss forest-fire prevention issues in conjunction with urbanization during five sessions of role-playing games (NîmetPasLeFeu) involving an urban developer, three mayors, a DDAF officer and a representative of Nîmes-Métropole.

The initiative culminated in a collective awareness of the implications, in terms of fire hazard, of expanding urbanization at the expense of agricultural fallow lands and natural areas. The importance of reflecting collectively on the setting up of prevention systems was identified clearly, but the elected representatives debated at length on the failure to integrate these systems with town planning projects and the lack of financial resources to maintain them.

At the initiative of the Gard General Council, the process is being standardized in all municipalities concerned by the problem of forest fires. It has also been taken up by the Languedoc-Roussillon Regional Council with elected representatives and technicians from Mediterranean basin regions under the Incendi InterReg project.

For further information: an engineer's dissertation on the co-construction approach (Bourgeois 2006) and an article on the involvement of local stakeholders (Étienne et al. 2008b).

Nîmes-métropole (France): in the districts surroundings the city of Nîmes it is urgent to make elected representatives more aware of forest fire prevention issues at the wildland-urban interface.

Njoobaari Ilnoowo: Viability of Irrigated Systems in Senegal

Question asked: what is the influence of organizational patterns within irrigated systems on their viability?

Region: Senegal river valley.

Sponsor: CIRAD.

Researchers involved: O. Barreteau (CEMAGREF) and W. Daré (CIRAD GREEN).

Objectives and progress: In the early 1990s, everyone agreed on the relative failure of irrigated systems in the mid Senegal River valley; a reclamation rate and yields far less than forecast. Attempts to explain this through disciplinary viewpoints were doomed to failure. CIRAD took up this question by raising the hypothesis of the influence of coordination patterns within irrigated systems. Through two PhD theses, we were involved firstly from a research perspective by exploring various scenarios for coordination patterns between farmers. This involvement produced an agent-based model and a role-playing game, and contributed to the conception of the companion modelling stance. We then sought to understand what happened in the role-playing games in order to test their relevance as a discussion medium and a way of investigating the social reality.

The approach was broken down into three phases.

(i) Developing three successive versions of agent-based models representing an archetypal irrigated system in Senegal: one based on water flows within the irrigated scheme, a second considered the questions of loans and exchange of services between irrigating farmers and a third took rules for change into account.
(ii) Translating the third agent-based model in a role-playing game to restore phase one to the farmers interviewed.
(iii) Exploring the game thus constructed and its ability to fuel discussion on actual farmer issues.

The approach set the first milestones in the companion modelling approach and accounted for the use of game settings by farmers to explore collective sensitive questions, even taboos in their own irrigated systems.

For further information: two theses (Barreteau 1998; Daré 2005), an article on the role-playing games (Barreteau et al. 2001), and an article on the link between game and reality (Daré and Barreteau 2003).

Njoobaari (Senegal): in the mid Senegal River valley, it is necessary to encourage a better coordination among farmers and to discuss about the sustainability of irrigated systems.

Ouessant: Shrub Encroachment and Biodiversity

Question asked: what are the links between access modes to resources and biodiversity?

Region: Ouessant island, France.

Sponsor: French Institute for Biodiversity (IFB).

Researchers involved: M. Étienne (INRA Ecodevelopment Unit), F. Bioret, F. Gourmelon and M. Rouan (University of Brest), C. Kerbiriou (Study Centre for the Ouessant Environment—CEMO, Ouessant) and H. Levrel (Muséum national d'Histoire naturelle—MNHN), Paris).

Objectives and progress: In September 2002, during the annual meeting of French biosphere reserves, it was decided to mount a research programme around shrub encroachment in partnership with four biosphere reserve managers. In the Irish Sea biosphere reserve, an interdisciplinary research team was made up of researchers from the University of Western Brittany (UBO) (a plant ecologist, geographer, ethnologist and computer scientist) joined by an economist from the MNHN and an ornithologist from CEMO. Following funding from the IFB, M. Étienne proposed gathering each individual's approach by co-constructing a MAS representing the interactions between nature and societies in the land part of the island.

The approach was broken down into five phases.

(i) Co-constructing the conceptual model using the ARDI method by identifying the main stakeholders, resources and dynamics in the shrub encroachment process on the island.
(ii) Implementing the multi-agent model and identifying the interactions required to strengthen available knowledge.
(iii) Developing complementary interdisciplinary research on the interaction between grazing and shrub encroachment, turf harvesting (turf is used as a fuel for the preparation of a traditional stew) and stakeholders perception of the shrub encroachment.
(iv) Development of scenarios for the likely changes on the island depending on whether or not livestock rearing is maintained or whether or not biodiversity issues are taken into account.
(v) A situated simulation exercise on a virtual territory but close to their reality based on a role-playing game involving several types of residents on the island (e.g. elected representative, biodiversity reserve official, cattle farmer, native of Ouessant, holiday home owner, etc.) to make them react collectively to the shrub dynamics.

The approach resulted in a close and successful collaboration between researchers from different disciplines and the implementation of interdisciplinary work culminating in shared perspectives on the question of the biodiversity and access to resources. It has been extended into work on opening to the general public (project with the Océanopolis Museum) and towards teaching (use of the BrouteLaMotte role-playing game as a teaching tool).

For further information: an article on the ARDI method (Étienne 2006), an article on the modelling work (Rouan et al. 2009) and an article on the approach and the scenarios (Gourmelon et al. 2008).

Ushant (France): the decreasing of Ushant isle population and the consequent reduction of sheep farming leads to a progressive shrub encroachment of the coastal area and threatens strong touristic and biodiversity stakes.

Pays de Caux: Erosive Runoff in an Agricultural Watershed

Question asked: how to commit to concerted management of erosive runoff in an agricultural watershed based on reflective thinking on the implementation of both better agricultural practices and landscape structures?

Region: Pays de Caux, France.

Sponsor: INRA researcher.

Researchers involved: V. Souchère, J. Echeverria and L. Millair (INRA Sadapt), F. Bousquet and C. Le Page (CIRAD GREEN) and M. Étienne (INRA Ecodevelopment Unit).

Objectives and progress: In the cultivated silty areas of the Pays de Caux, erosive runoff is a widespread phenomenon notwithstanding low rainfall intensity and a gentle topography. It frequently generates modest damage (e.g. deposits on seedbeds or ephemeral gullies) and more rarely, severe muddy floods. After the construction of storm basins to solve the problem, it appeared necessary to combine a remedial approach with a preventive process to reduce runoff from agricultural land. Various studies showed that these problems were partly linked to the lack of coherent flow management at the watershed scale and to changes in land uses and agricultural practices. As runoff flows down a slope and does not respect the limits of fields and farms, the actions that needed to be undertaken required co-operation between stakeholders. However, designing collective management of agricultural land is all the more difficult because the economic context means that farmers focus primarily on productive and individual strategies. To design collective management of a watershed is a real challenge especially since environmental management is controlled by natural processes, meaning that actors are not free to choose how they wish to cooperate in a given watershed. A companion modelling approach was initiated in 2006 with local stakeholders to test the possibilities of introducing concerted actions to limit runoff at various scales of investigation (i.e. farm and watershed level).

The main phases in the approach were as follows.

(i) Co-constructing the conceptual model using the ARDI method by identifying the main stakeholders, resources, dynamics and interactions linked to the runoff problem.
(ii) Implementing the model via the Cormas platform.
(iii) Simulation exercises for local stakeholders within the framework of a role-playing game involving farmers, agricultural watershed advisers and mayors, to raise their awareness of the problem and help them envisage together possible scenarios for collective management of the erosive runoff within a virtual watershed.

Two sessions of the CauxOpération role-playing game were organized in 2007 in the area of two different watershed management committees. Given the very positive interest from local stakeholders, new sessions of the role-playing game were organized in 2009 and 2010. They served to support a thesis that started at the end of 2008 with the goal of exploring the consequences of choosing this particular tool, a role-playing game, in a decision-making process.

For further information: two Master's dissertations (Echeverria 2006; Millair 2007) and an article on the approach and role-playing game (Souchère et al. 2009).

Pays de Caux (France): in order to reduce erosive runoff due to agricultural practices on silty soils, collaborative actions have to be launched at the farm and catchment level.

Radi: Mediation of a Grazing Land-Use Conflict Between Sedentary and Nomad Herders in Eastern Bhutan

Question asked: how to resolve an open conflict between sedentary Radhips and nomad Merak herders competing to graze on degraded highland pastures?

Region: Sheytimi natural pastures in Trashigang district, eastern Bhutan.

Local partner: Renewable Natural Resources Research Center (RNR-RC), Wengkhar.

Researchers involved: C. Le Page, J. Queste and G. Trébuil (CIRAD-GREEN), T.R. Gurung, P. Chettri and L. Dorji (RNR-RC, Wengkhar).

Objectives and progress: For many years, in winter, Tibetan herders from Merak have descended with their herds of yaks and zomos on to the Sheytimi natural pastures, where access rights have become increasingly complex over recent decades. For their part, the Radhips sedentary herders from Radi village also wish to drive their cattle up to part of these pastures during the summer rice-growing cycle in the rainy season. This dual use has resulted in major degradation to the grazing potential of the Sheytimi, repeated and occasionally violent skirmishes between the two communities, and spectacular land erosion gullies threatening the lower rice terraces of this eastern district rice bowl. The kingdom's highest authorities requested that a solution to this destructive land-use dispute be found. Following several unsuccessful attempts and based on the Lingmuteychu experiment (see above), a participatory modelling workshop was organized at this site with the main stakeholders in the conflict, after a short training on the ComMod approach.

The process took place over two weeks.

(i) Implementation of a one-week training course on the ComMod approach focusing on the role-play tool. The land-use conflict over the Sheytimi pastures was used extensively during this training, culminating at the end of the week in a proposed role-playing game to be used the following week in the field with the Radhips and Merak stakeholders.

(ii) The first gaming sessions were organized with two Rhadip subgroups to illustrate the degradation of pastures. Then, following the arrival of Merak herders, more sessions with two mixed groups each with six herders were held, and a plenary discussion on the degradation dynamics of the resource took place before a new gaming session with all the herders playing together with unrestricted communication was held. The advantage of coordinating herders' decisions and actions to preserve the resource became visually explicit. The day ended with a mutual design of a modified game for use the following day and the farmers suggested a grid corresponding best to their representation of the heterogeneity of the space to be managed.

(iii) The second day of the gaming workshop; the Radhips and the Meraks played separately, and exhausted the simulated forage resource rapidly, whilst the contrasting strategies of the two groups were expressed. The Meraks behaved like true herders, while the Rhadips aimed at occupying land. Deferred grazing for part of the pastures was agreed at the end of a plenary discussion on new pasture management options. A final gaming session was organized with free communication among herders and deferred grazing of selected plots each year to ensure regeneration of the pasture. This time, a more sustainable way to manage the grazing land was simulated.

(iv) The next day, an evaluation of the advantage of the game for the herders, its relation to reality and potential improvements were discussed during individual interviews with players. Lastly, after a demonstration of the computer simulation of the session played the previous day, a plenary discussion attempted to produce a common action plan for the rehabilitation of the Sheytimi pastures and halt land erosion.

Compared with the Lingmuteychu case, which could be developed over a long period of time, only the description of the first key moment of the ComMod process is provided here, organized in a very tense social context but nevertheless triggering creative dialogue between the conflicting parties.

For further information: no publication on this case study is available yet.

Radi (Bhutan): to solve an open conflict between sedentary and nomad herders competing to graze the same highland pastures, a participatory workshop permitted to initiate a creative dialogue among two ethnic groups.

SAGE Drôme: Sharing Water in the Drôme River Basin

Question asked: what are the methods for sharing the water resource for irrigation during low water between farmers in the lower Drôme valley?

Region: Drôme river basin area downstream of Crest, France.

Sponsor: Communauté de Communes du Val de Drôme (association of villages and small towns of the Drôme River basin).

Researchers involved: O. Barreteau and G. Abrami (CEMAGREF).

Objectives and progress: In 1999, the Communauté de Communes du Val de Drôme was in charge of the secretariat of the local water commission of the Drôme river SAGE water development and management schemes. It requested support from CEMAGREF to define 'water use instructions statements' for sharing the water resource in the Drôme basin to comply with the flow rate objectives downstream of the Drôme river as defined in the Drôme river SAGE. CEMAGREF then added a research project to this request under the Environment Decision Consultation programme of the French Department of Environment to test the use of role-playing games and agent-based simulations as support in jointly designing these methods of sharing the water resource. The main use covered was irrigation.

The approach was broken down into four phases.

(i) Developing a model on a spreadsheet, based on volume assessments, only considering one type of crop (corn) and studying the effects of various water-sharing scenarios defined by the local water commission for all the irrigated surface area and the irrigating population.
(ii) On this basis, following public restitutions of work to the relevant stakeholders, a draft water-sharing charter was designed collectively based on an initial proposal by CEMAGREF, but which had changed significantly in the interaction with these stakeholders.
(iii) Testing the development of an agent-based model used to specify the model from the viewpoint of types of action undertaken by agents and to observe the consequences of measures at a more detailed scale in time and space.
(iv) Testing a hybrid role-playing game, Pieplue, aimed at initiating discussion and putting into practice rules on water sharing in a virtual space similar to the irrigated area of the Drôme.

The approach culminated in an official agreement and the disclosure of the possibility of win–win scenarios between the farming sector and the Communauté de Communes du Val de Drôme. However, the role-playing game was found to compete with the activity of the agricultural extension services and had no follow up.

For further information: a paper dealing with the design of the agent-based model as a dialogue support tool (Barreteau et al. 2003), a paper dealing with the design and implementation of a hybrid tool based on a computerized agent-based model and a role-playing game (Barreteau and Abrami 2007), a book chapter dealing with the problems faced in stakeholders involvement in this case study (Barreteau et al. 2006).

Sage Drôme (France): to cope with the increasing demand for irrigation water, the local water commission set up a collaborative process to define how to allocate water resources during the low water period.

SosteniCAP: Municipalities and Drinking Water in Bolivia

Question asked: how to improve the operation of community domestic water supply associations?

Region: peri-urban region of Cochabamba, Bolivia.

Sponsor: University Mayor of San Simon.

Researchers involved: N. Faysse (CIRAD UR G-EAU), R. Ampuero and F. Quiroz (University Mayor of San Simon, Centro Agua).

Objectives and progress: The drinking water in the peri-urban area of Cochabamba is almost exclusively produced, distributed and managed by the district-based autonomous community committees. These committees have endured over time but are frequently unable to improve their management, which remains weak.

The proposed five-stage approach aimed to develop and test a management support procedure for these community committees.

(i) Socio-economic diagnostics of the committee aimed at identifying the operational problems and the perceptions of users and the administration team regarding these problems.

(ii) Collective validation of results and selection of three or four priority themes to be processed by working commissions formed on a voluntary basis and facilitated by the researchers.
(iii) The administrative commission selected an institutional model after analysing the advantages and disadvantages of various options and developed the content of the internal regulations in detail using a model proposed by the work team. The economic commission set water prices by analysing all the costs of the association and income from the tariffs. Scenarios simulated in a spreadsheet were used to test several pricing options to select a suitable tariff. The technical commission checked the reliability of meters and assessed the infrastructures with mounting a call for tenders, etc.
(iv) Each commission's proposals were presented to a general meeting for potential amendments and validation.
(v) At the same time, role-playing sessions (SosteniCAP) were organized, uniting users and the administration team to raise the awareness of users to the overall operation of the association, pricing issues and the role of the different stakeholders within the association.

The approach was applied in four drinking water committees. Apart from acquiring internal regulations, even a legal status, it built up the management capacities of the administration team. The role-playing sessions encouraged improved knowledge on the operation of committees by the users, thereby boosting the social control. Although so far pricing proposals have not been validated collectively, the entire process has led to an improved financial situation (fewer late payments) and a strengthened internal and external institutional legitimacy (with respect to other stakeholders) of drinking water committees in the peri-urban area.

For further information: an article (Ampuero et al. 2006) on the website http://www.negowat.org.

SosteniCap (Bolivia): in the suburbs of Cochabamba, participatory workshops were organized in order to support the autonomous community committees in producing, distributing and managing drinking water.

SugarRice: Land-Use Change in Upper Northeast Thailand

Question asked: what are the effects of extending sugarcane into upper paddies on food safety and income?

Region: Nam Phong district, Khon Kaen province, upper northeast Thailand.

Local partner: Faculty of Agriculture, University of Khon Kaen (KKU).

Researchers involved: G. Trébuil and F. Bousquet (CIRAD GREEN), I. Patamadit, N. Suphanchaimart and C. Wongsamun (KKU).

Objectives and progress: In 2002–2003, Thai university partners, who have been working for many years with local villagers, were concerned by the expansion of commercial sugarcane cultivation into the upper paddies of the traditional, rainfed, lowland rice ecosystem. At the same time, a research team was seeking opportunities to test the use of the ComMod approach to facilitate the understanding of rapid transformations in rice ecosystems. They proposed a collaboration on modelling to their KKU partners to analyse the determining factors and consequences of land-use change in this region. This was the first ComMod case study implemented in Thailand and consisted of five phases.

(i) Co-constructing a conceptual model, mainly formalized in UML diagrams, representing the decision-making processes related to the allocation of a given crop to a plot in a small catchment and the sale of sugarcane by the producers. This model was based on existing knowledge, supplemented by a few interviews with local stakeholders in the sugarcane sector.

(ii) Constructing a role-playing game based on the initial understanding of land-use change by the research team. This game was first tested and calibrated with KKU students.

(iii) A first field workshop was held based on this game played with various types of farmers (some of them quota leaders for the local sugar mill) to share and improve this dynamic representation of their agricultural system. They then requested modifications and the chance to replay the game with other stakeholders in the sugarcane sector.

(iv) A second workshop was held with the same villagers and representatives of the sugar mill, the professional association of sugarcane growers, and a member of the sugarcane board based in the capital city fixing the annual farm gate price for the sugarcane. This workshop ended with the presentation of a computerized multi-agent simulator replaying quickly the gaming sessions practised by the stakeholders and checking the villagers' ability to follow such simulations. Scenarios of interest based on the diversification of agricultural activities were also identified.

(v) Running a workshop similar to the previous one (in March 2003) in a neighbouring village where a diagnostic analysis of the agrarian system had identified an advanced level of agricultural diversification. This workshop was combined with a training course on the ComMod approach and its tools held at KKU.

This process demonstrated the possibility of running a ComMod process in a Thai rural environment and it was followed by further case studies starting up in different regions in the country on other common problems of renewable resource management. The type of role-playing game developed was of great interest to the main Thai economist partner and has been used in several training programmes and other projects. However, the combination of such a game with a computerized multi-agent model, as validated in this case study, was not adopted by the Thai team at KKU despite the availability of local computer resources for this purpose.

For further information: a book chapter on the case study (Suphanchaimart et al. 2005).

SugarRice (Thailand): extending sugarcane into high rice paddies impacts food safety and income of local villagers.

Tarawa: Equitable Groundwater Management on Tarawa Atoll

Question asked: how to manage equitably the limited groundwater resources in the overcrowded low-lying atolls of the Pacific?

Region: Tarawa Atoll, Republic of Kiribati.

Sponsors: Australian Centre for International Agricultural Research and Agence Française de Développement.

Researchers: A. Dray and I. White (Australian National University), P. Perez, C. Le Page and P. d'Aquino (CIRAD).

Objectives and progress: Low coral islands are heavily dependent on limited groundwater resources for freshwater supplies. This study was carried out on the low-lying Tarawa Atoll in the central Pacific Ocean. Tarawa is the capital and main population centre of the Republic of Kiribati. The pollution generated by its 45,000 inhabitants has already contaminated several freshwater lenses. Hence, drinking water is pumped from groundwater protection zones with restricted access called 'water reserves', located on a few remaining scarcely populated islands. Settlements on water reserves are restricted. Water abstraction and supply are controlled by the Public Utility Board (PUB) and various pieces of legislation controlling water management are overseen by the Ministry of Works and Energy, the Ministry of Health and the Ministry of Environment.

The declaration by the Government of Kiribati of water reserves on Tarawa Atoll, over privately owned land, has led to conflicts, illegal settlements and vandalism. Also, water consumption has tended to increase towards Western levels, and human pollution has already contaminated most freshwater lenses. This project aimed to provide relevant information to local stakeholders in order to facilitate dialogue with government agencies and to collectively devise sustainable water management practices. A computer-assisted role-playing game was implemented to fulfil this aim.

The following three-stage methodology was applied.

(i) Collection of local and expert knowledge.
(ii) The blending of the different viewpoints into a computer-assisted role-playing game.
(iii) Playing the game with representatives of the traditional landowners and government agencies to explore better governance scenarios.

Although the game sessions delivered successful outcomes, the final stage of the project was characterized by the upheaval of contradictory government stands that undermined the whole process.

It is argued that heterogeneous viewpoints may be handled in a satisfactory manner during the gaming sessions but that long-term hidden agendas may override the outcomes. Beyond the inherent question of legitimacy attached to such approaches, some players clearly must deal with constraints that are often genuinely considered external to the ongoing negotiation process.

For further information: two articles on the role-playing game (Dray et al. 2006, 2007).

Tarawa (Micronesia): the overgrowing of population urges to discuss about drinking water availability and find a collaborative way to better manage the freshwater lenses.

Ter'aguas: Integrated Water and Land Management in Peri-Urban Areas

Question asked: how to boost the negotiating capacity of local leaders on water and land in the peri-urban area of a large metropolis?

Region: Guarapiranga catchment area, Sao Paulo, Brazil.

Sponsor: University of Sao Paulo (USP).

Researchers involved: R. Ducrot, (CIRAD UR G-EAU), V Barban (Instituto Polis), P Jacobi, W Gunther and M Arteiro (USP), T. Franca and Y de Carvalho (Agency for Agribusiness Technology (Agência Paulista de Tecnologia dos Agronegócios—APTA), B Reydon and K Bueno (University of Campinas).

Objectives and progress: Illegal urban development in catchment areas producing drinking water for the metropolis of Sao Paulo leads to degradation in water quality in the storage reservoirs due to a lack of sanitation infrastructures. The involvement of community leaders in discussions concerning water and land management is restricted by social inequalities, asymmetry of information and powers, competition between leaders and the traditionally paternalistic attitude of authorities. A ComMod approach was initiated to facilitate (boost and reinforce) the negotiating capacity of local stakeholders and facilitate their participation in discussions on the complex issues of water and land management in peri-urban areas.

The main phases in the approach were as follows.

(i) Thematic studies to improve understanding of water and land management dynamics in the area.
(ii) Participative developing and testing of various types of discussion tools (e.g. computerized role-play, tool to identify resources mobilized at district level, etc.) with representatives from different types of settlements.
(iii) Organizing a logical activity sequence called Ter'aguas using several of the tools developed: identifying resources (i.e. land, water, urban infrastructure) of the settlements and their dynamics; identifying the stakeholders involved and their action in the resources; rapid dramatization (stating a fictitious dispute) used to analyse the modalities for stakeholder interactions; Ter'aguas role-play gathering local leaders, representatives of the water company and the municipalities, which simulated the collective decision-making processes and illustrated their impact on the region; collective development of a negotiation strategy by the communities. This process was tested twice: helping to prepare a master plan (Embu Guaçu municipality) and with three communities in the Paralheilos region in dispute with the local authorities and the water company over a sanitation project.

The approach modified perceptions concerning the stakeholders and the functioning of the system. It improved the understanding of participants on the local water and land dynamics and their management, helped to bring local and institutional stakeholders closer together and strengthened the ability of local stakeholders to engage in constructive interaction processes.

For further information: two seminar presentations (Ducrot and Barban 2008).

Ubon Rice Seeds: Constructing a Common Representation of the Rice Seed System in Lower Northeast Thailand

Question asked: how does the structure and operation of the provincial seed system affect the dynamics of the local rice agrobiodiversity?

Region: Ubon Ratchathani province, Thailand.

Local partners: University of Ubon Ratchathani (UBU) and the International Rice Research Institute.

Researchers involved: G. Trébuil, F. Bousquet and G. Abrami (CIRAD GREEN), C. Vejpas and W. Naivinit (UBU).

Objectives and progress: In 2003, following extensive survey work by Thai researchers on the varietal choice and seed renewal practices of rainfed lowland rice growers in the 25 districts of Ubon Ratchathani province, a reduction in rice biodiversity was noted in this cradle of Thai aromatic rice. These UBU researchers were interested in participatory approaches in agriculture and took part in training courses on multi-agent modelling for integrated natural resource management before deciding to attempt, with specialists in modelling complex systems, the co-construction of a multi-agent model representing the structure and functioning of the provincial rice seed system with its main stakeholders. This case study focused on the modelling of communication and seed exchanges among agents in a heterogeneous environment at the provincial scale.

The approach was implemented in six phases over 3 years.

(i) Co-construction of a conceptual model formalized using UML diagrams representing the decision-making processes linked to the choice of varieties and seed suppliers used by the rice growers, and to the production and exchange of seed between the various stakeholders in the provincial seed system. Initial diagrams constructed by researchers based on existing knowledge were then submitted to the seed system stakeholders for review and gradually improved during a series of joint meetings.

(ii) Based on the improved UML diagrams, a first role-playing game on the choice of varieties and seed suppliers by farmers was constructed to share the researchers' representation of the seed system at the village level with various types of rice growers farming at varying distances from seed production and multiplication centres in the province. In this game six farmers from two quite separate villages played in parallel.

(iii) A second game complementary to the first was put together to represent seed supply and demand, this time at the provincial scale, and the meeting of seed requirements by various producers and suppliers through seed orders and exchange. The second game was used with provincial stakeholders in the sector, and then extended to Bangkok-based representatives of the Ministry of agriculture and cooperatives. The post-gaming session discussions led to modifications of the initial conceptual diagrams.

(iv) The construction of a prototype computer agent-based model was based on these improved UML diagrams. The prototype merged the two role-playing games and represented both the creation of the demand for seed at the farm level and communication about seeds and their exchange between the stakeholders in the supply chain. It also distinguished between an area close to rice seed producer and distributor public bodies and an area far away with access to a more restricted number of suppliers only.

(v) The prototype was presented to the main stakeholders in the sector in Ubon province then in the relevant Ministry of agriculture research and extension agencies located in Bangkok. These interactions led to improvements in the computer model regarding the representation of the links among stakeholders, farmers' priorities in accessing seeds, and the dynamics of the demand for seed. Lastly, a few scenarios of interest for the stakeholders were identified to simulate the withdrawal of a key seed-producing public institution, an increase in demand for seed and its production by community seed centres.

(vi) Improvement of the computer agent-based model and simulation of scenarios of interest to users.

The approach allowed the stakeholders in the seed sector to meet, exchange viewpoints on the state of the system and its future, and to become aware of its high degree of complexity.

For further information: a book chapter on this case study (Vejpas et al. 2005) and a conference paper on simulations (Abrami et al. 2010).

Ubon Rice Seeds (Thailand): the reduction in the rice biodiversity measured in this cradle of Thai aromatic rice leads to study the relationships between the structure and operation of the provincial seed system.

Ventoux: Fir Encroachment and Biodiversity in the Ventoux Biosphere Reserve

Question asked: what are the links between access modes to resources and biodiversity?

Region: Lure mountain, Alpes de Haute-Provence, France.

Sponsor: French Institute for Biodiversity (IFB).

Researchers involved: M. Étienne, E. Faugère and E. Lécrivain (INRA Ecodevelopment Unit), B. Anselme, P. Pech and L. Simon (University of Paris I), B. Fady (INRA, URFM) and A. Lyet (CNRS).

Objectives and progress: In September 2002, during the annual meeting of French biosphere reserves, it was decided to mount a research programme around shrub encroachment in partnership with the four biosphere reserve managers. An interdisciplinary research team was set up on the Lure mountain including researchers from University of Paris I and INRA in Avignon (a plant ecologist, geographer, ethnologist, ethologist, forester and modeller). Following funding from the IFB, M. Étienne proposed integrating each individual's approach by co-constructing a MAS representing the interactions between nature and societies on the Lure mountain.

The approach was broken down into four phases.

(i) Co-constructing a conceptual model using the ARDI method by identifying the main stakeholders, resources and dynamics involved in the juniper encroachment and dispersal of fir trees.
(ii) Implementing the multi-agent model and identifying the interactions required to strengthen available knowledge.
(iii) Developing complementary interdisciplinary research on vegetation dynamics, organization of grazing, tree management, fir genetic diversity and the Orsini's viper (*Vipera ursinii*) population dynamics.
(iv) Developing scenarios for the probable evolution depending on whether or not livestock rearing is maintained, confirming global warming effects or implementing a dynamic forestry for storing carbon or for preserving the genetic biodiversity of the fir tree.

The approach stimulated close and successful collaboration between researchers from different disciplines and forest managers and the implementation of interdisciplinary work culminating in shared perspectives on the question of biodiversity and access to resources. The attempts to use the model with local stakeholders to discuss the conflicts of use around Mont Serein or the objectives document for the Natura 2000 site on the Lure mountain were unsuccessful.

For further information: an article on the ARDI method (Étienne 2006) and an article on the model (Anselme et al. 2010).

Ventoux (France): on the Lure Mountain, managers have to find their way between two antagonistic biodiversity stakes: maintaining sheep farming to secure the conservation of Orsini viper, or enhancing fir stands overspreading.

Vosges: Shrub Encroachment in the Northern Vosges Valley Bottoms

Question asked: **what are the** links between access modes to resources and biodiversity?

Region: Valley of the Northern Zinsel, Northern Vosges National Park, France.

Sponsor: French Institute for Biodiversity (IFB).

Researchers involved: M. Étienne (INRA Ecodevelopment Unit), C. Rolland-May and M. Wintz (University of Strasbourg) and A. Schnitzler (University of Metz).

Objectives and progress: In September 2002, during the annual meeting of French biosphere reserves, it was decided to mount a research programme around shrub encroachment in partnership with the four biosphere reserve managers. In the northern Vosges biosphere reserve, several officers from the regional park were involved in the project in conjunction with a small interdisciplinary group of researchers (an ecologist, geographer and sociologist). Following funding from the IFB, M. Étienne proposed amalgamating each individual's approach by co-constructing an agent-based model representing the interactions between nature and societies in a valley bottom representative of the northern Vosges.

The approach was broken down into five phases.

(i) Co-constructing a conceptual model using the ARDI method by identifying the main stakeholders, resources and dynamics involved in the problem of shrub encroachment of a valley bottom.
(ii) Implementing an agent-based model and identifying interactions required to strengthen available knowledge.
(iii) Developing complementary interdisciplinary research on the interaction between grazing and shrub encroachment, on the dynamics of the alder and on stakeholder perception of the fallow lands.
(iv) Development of scenarios for the likely changes in the region depending on whether or not livestock rearing is maintained or whether or not biodiversity issues are taken into account.
(v) A situated simulation exercise in a fictitious region but close to reality to make the main stakeholders (e.g. elected representatives, biodiversity reserve officers, farmers, landowners, residents, etc.) react collectively to the vegetation dynamics through the role-playing game.

The approach culminated in a close and successful collaboration between researchers from different disciplines and managers of the biosphere reserve. The various perceptions of valley bottoms and the term 'fallow land' were understood better. It encouraged the park officers to raise further questions about shrub encroachment and to produce alternative solutions to maintaining the current herd of highland cattle.

For further information: an article on the ARDI method (Étienne 2006), an article on the confrontation between scientific and lay knowledge (Étienne 2009).

Vosges du Nord (France): in the Zinsel du Nord valley, the ecological management of wet grasslands with highland cattle is questioned through an interdisciplinary debate on the concept of fallow land.

Description of Seven Games or Teaching Models Used in the Companion Modelling Process

CherIng is a teaching game for the individual or collective harvesting of a virtual resource with limited growth levels in order to discover the principles and key steps of a ComMod role-playing game (continuing training).

MéjanJeuBiodiv is a teaching game on biodiversity management and interactions between crop farmers, livestock farmers, foresters and conservationists for the conservation of target species with contrasting habitat requirements (Biodiversity Master's degree).

Njoobari is a teaching game to illustrate the modes of coordination between farmers belonging to the same irrigated system in the middle valley of the Senegal

River and generate discussion about the viability of this system (continuing training).

PâtureLesCommuns is a teaching model on the sharing of grazing resources between several livestock farmers to learn progressively about the complexity of the interactions between ecological and social dynamics (Agronomy Master's degree).

RuisselPois is a teaching game on erosive runoff problems and interactions between farmers, mayors and cooperatives in cash crop farming areas (Agronomy Master's degree).

SylvoPast is a teaching game on fire prevention, livestock grazing and hunting during the planning of silvopastoral management in Mediterranean forests (Forestry and Veterinary Master's degree).

YeunEllez is a teaching model focused on the impact of agricultural and urban development on the conservation of natural habitats in order to support students in co-constructing a model of the interactions between ecological and social dynamics (Geo-architecture and Environment Master's degree).

CherIng

Teaching game created by C. Le Page (CIRAD Montpellier), M. Étienne (INRA Avignon) and O. Barreteau (CEMAGREF Montpellier).

A generic game initially focused on the individual or collective harvesting of a resource with limited growth.

Motivation: This role-playing game was developed specifically for the training session 'Use of role-playing games in companion modelling'. It has two objectives: (i) to 'break the ice' and allow participants to get to know each other in a relaxed fashion; (ii) in a short period of time, to discover 'from the inside' the principles and essential stages in playing a companion modelling game: briefing, alternating stages of individual and collective decision-making, and debriefing.

Description and specific features: The style of the game is taken to the extreme in the sense that no realistic context is evoked. A space divided into a minimum of 20 basic portions (numbered cells) holds a resource (known as 'Ing') with an abundance that is non-existent (0), low (1), medium (2) or maximum (3). A multi-agent model developed on the Cormas platform allows the visualization of this abstract territory, which is continuously projected in the game room. Once the territory is adjusted to their number, players randomly seated in the game room harvest the Ing. Their goal is to harvest the maximum amount in a sustainable manner, proceeding as if the Ings represented a vital renewable resource for their community. At each round, the model uses hidden rules to update the abundance of the Ing in each of the cells, then all the players must simultaneously note the number of the cell on which they plan to harvest the resource. When several players choose the same box, the model attributes the harvest to one or two of them randomly selected.

A session comprises five consecutive rounds. The exercise is organized into two sessions. The first 'individualistic' session is played without allowing any time for discussion between players. It allows players to try to understand the dynamic of the system as a function of the harvest rate and adjust their harvest according to that of the others. The second 'collective' session starts with the formation of three groups of players that are kept apart from each other and who have 10 minutes to discuss and elaborate a collective strategy. All the players then go back to their seat to play another five rounds. This session enables more collaborative behaviour to be put in place and to evaluate the capacity of participants to imagine adaptive strategies. Finally, a debriefing is organized as a continuation of the process to compare the two sessions based on indicators calculated and registered on the computer.

This game has numerous variations integrating new functions: for example, adding the role of research scientist to consider a participatory research approach on the management of a limited resource; adding the role of conservationist to consider the effect of public policy on the regulation of resource use; adding a more productive resource, etc.

Application: CherIng was played in training sessions organized in France, Germany, Thailand, Bolivia, the Philippines and New Caledonia.

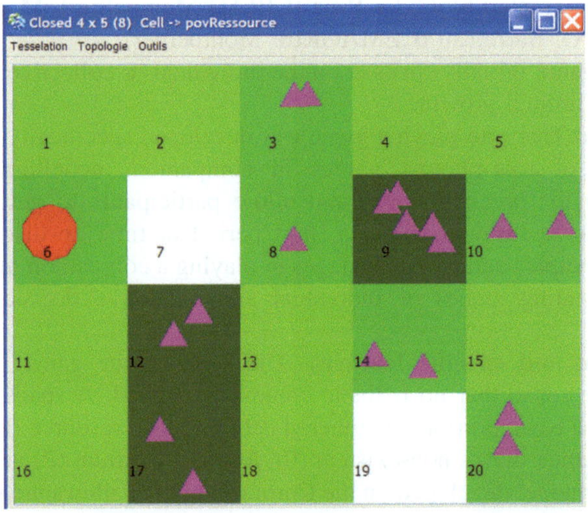

CherIng: the game board permits to simulate the individual or collective harvest of a virtual resource according to different levels of availability and the impact of exploitation on regeneration.

MéjanJeuBiodiv

Teaching game created by M. Étienne (INRA Avignon) and F. Bousquet (CIRAD Montpellier).

The game focuses on the management of biodiversity at different scales and is based on research undertaken on the Causse Méjan (Étienne et al. 2003; Kinzig et al. 2006).

Motivation: The game was created for two Master's degrees on biodiversity management in Toulouse and Paris. It aims to assist graduate students become aware of interactions between human activities and biodiversity dynamics in zones subject to strong socio-economic change. Biodiversity dynamics is addressed at different levels (i.e. gene, species and landscape).

Description and specific features: The game was inspired by the MéjanJeu role-playing game (Étienne and Le Page 2004) with a similar territory to be managed (a portion of the Causse Méjan), a strong process of pine encroachment and endangered plant species and animals sensitive to pine cover (e.g. pheasants eye *Adonis vernalis* and stone curlew *Burhinus oedicnemus*). It adds strong interaction with farming activities (e.g. land clearing and removing of stones from fields) and animal biological cycles (e.g. apollo butterfly *Parnassius apollo* and little owl *Athene noctua*). Students must put themselves in the place of both humans (e.g. crop and livestock farmer, forester, nature conservationist) and animals (e.g. owl, butterfly and curlew). Students who have chosen the role of farmers define the cultivated surface area, the ratio between forage and cereals, and adjust the rangeland stocking rate to the paddocks' productivity and size. Those playing the role of nature conservationist define the plots presenting priority natural heritage objectives and, with farmers and foresters, negotiate a strategy to control pine encroachment and conserve key habitats. Those playing the role of forester manage the forest areas and propose a plan to conserve the genetic quality of pine trees. Working from a short text or a set of pictures, those playing an animal species must understand its biology, define the initial size of its population and locate the nesting places, and then convince the nature conservationist to preserve its habitat.

In the beginning, the game serves to raise awareness of ecological dynamics at play and interactions with farming and forestry activities. During the second stage, it allows students to imagine consultation modes and alternative territorial management scenarios allowing better biodiversity conservation. These scenarios are tested through a computerized simulation during the third stage. Special effort is made to acknowledge the management units of each (i.e. farm, field, paddock, forest stand, home range) and to identify each stakeholder's scale of perception (e.g. the farmer on his farm, the nature conservationist on the entire territory, the forester on pine stands).

A complete tutorial provides a step by step guide to teachers wishing to use the game. It specifies the instructions to give to students, provides guidance to help workshop facilitators, and gives the solutions to exercises.

Application: The game is played in groups of six to eight students; a teacher can animate up to three groups playing simultaneously. It has been used since 2006 with students in a biodiversity management Master's programme at the Universities of Toulouse and Paris VI, a geography and development Master's programme at the University of Paris X, in advanced programmes on agriculture and the environment at École d'ingénieurs en agriculture (ESITPA) in Rouen, and training in modelling complex systems at the University of Wageningen.

MéjanJeuBiodiv (causse Méjan, France): the students get progressively aware of the interactions between human activities and biodiversity, first by co-constructing a conceptual model, then by simulating management options.

Njoobari Ilnoowo

A role-playing game created by O. Barreteau (CEMAGREF Montpellier) and W. Daré and F. Bousquet (CIRAD Montpellier).

The game focuses on the management of an irrigated scheme based on research undertaken in the middle valley of the Senegal River (Barreteau 1998; Barreteau et al. 2001; Daré and Barreteau 2003).

Motivation: This role-playing game was developed to interpret the SHADOC MAS, a computer model built to understand the links between farmer coordination modes on an irrigated system in the middle valley of the Senegal river, and to assess the viability of this system. The designers of the model chose to translate it into a role-playing game in order to limit the risk of technological barriers when presenting the results of the model to the farmers involved in the irrigated system. This restitution also aimed to test the validity of the model from the point of view of stakeholders represented in the model and investigate its capacity to generate discussion between stakeholders actually interacting on the same irrigated system.

Description and specific features: The dynamic of the entire game is based on the sharing of water and credit. The game is played with 10–15 players. It is made up of trilingual cards (French, Pulaar and Wolof) describing the possible behaviour of each player and a set of 'chance' cards reproducing unpredictable aspects of the model that represent farmers' multiple activities. The game requires a space to be divided into two isolated parts, one representing the village and the other, the irrigation scheme (e.g. two classrooms in a school). It also requires a drawing representing an irrigation scheme typical of the middle valley on which the fields attributed to each player are located.

Players are situated in the village space and move on to their fields according to their production objective and the drawing of a chance card. Those who go to their field decide cropping operations, including the irrigation of their plot as a function of their needs and collective arrangements. The consequences in terms of water levels and rice production are presented by simple graphs. There are phases of reimbursement and seeking credit between two campaigns that are key moments of interaction between farmers on irrigated schemes. The roles assumed by players are disassociated from their roles in actual daily life. For the role of farmer, they are randomly drawn, while the roles of chiefs are collectively chosen. Players thus reflect on the overall game rather than on their personal situations.

Application: The game was played first with farmers in five irrigated systems, which allowed the representation offered by the game to be validated and to observe the discussions generated on actual systems. It was repeated with around 12 groups of players in the Senegal valley to work on the interactions between actual dynamics and those of the game. It also was used in training in various cultural contexts with students and research scientists.

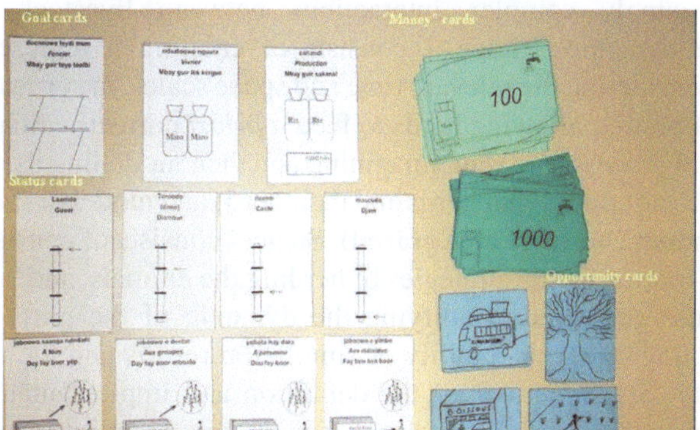

Njoobari (Senegal): to share water and credit, the players withdraw behavior and chance cards that account for the randomness of the model or the pluri-activity of the farmers.

PâtureLesCommuns

A teaching model created by N. Becu (University of Paris I) and M. Étienne (INRA Avignon).

It is a modifiable model focused on the sharing of resources between several stakeholders.

Motivation: The model was created in the framework of an in-depth training module on companion modelling at ESITPA in Rouen. Its intention is to open the 'black box' of agent-based models and help students who have participated in a preliminary, one-week training course on companion modelling to understand the model's implementation modalities.

Description and specific features: PâtureLesCommuns is a teaching model that can be constructed to represent the dynamic of a grassland shared by several livestock farmers during a grazing season. In the beginning of the grazing season, the grass is at a medium level of abundance. Slow but continuous plant growth takes place throughout the season. The grassland is exploited by five livestock farmers who each own 40 cows. The animals move about every day and graze where they have been moved. In the middle of the summer, a fire lays bare part of the grassland. Soil laid bare can regenerate if it is surrounded by a sufficient number of grassed cells. Each livestock farmer can pursue their own pasture management strategy. As the grassland is shared, the livestock farmers will suffer or benefit from the strategies of others.

The exercise takes place in five stages during which students learn to take into account increasingly complex interactions between livestock activities and environmental dynamics. Stage 1 consists of creating a representation of the grassland by considering the relevant time and space scales, the growth modalities of grass and a calibration allowing herds to feed *a priori* correctly. Stage 2 consists of adding a natural process (fire) and imagining and then modelling its impact on grass growth. Stage 3 introduces shepherds and their herds and models a simplistic grazing behaviour (random choice of area grazed). Stage 4 consists of considering and then implementing more realistic strategies of herding the animals, and, according to the grazing pressure generated, modifying the dynamic of the grass resource. This includes reflecting on the variability of simulation results and the management of this uncertainty. Stage 5 involves the definition and implementation of different management scenarios of the rangeland. It is broken down into three parts during which students will be able to play successively on the resource (introducing a more productive grass), the division of space (creating five private farms), and communication between livestock farmers (negotiating herding behaviour).

Application: The approach was tested for the first time in December 2008. It combines the gradual co-construction of a model that acknowledges interactions between stakeholders and the environment with reference to object-oriented modelling theory and computer coding examples. A teacher can supervise 12 students divided into pairs at computer stations equipped with the Cormas platform.

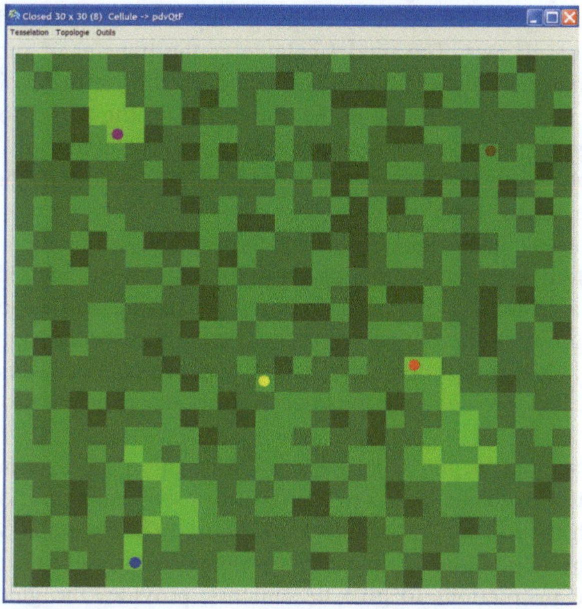

PatureLesCommuns (virtuel): a model that simulates the dynamics of a grazed grassland shared by several livestock farmers is progressively implemented by the students.

RuisselPois

A teaching game created by M. Étienne (INRA Avignon) and V. Souchère (INRA Grignon).

The game focuses on large-scale farming erosive runoff based on research in Pays de Caux (Souchère et al. 2005).

Motivation: This model was created as a teaching tool in the framework of the Master's programme 'Living Science and Technologies', in the course of study at AgroParisTech entitled 'Agronomy, Environment, and Landscape: Ecosystems Sustainability'. It causes graduate students to become aware of the effects of interactions between different types of stakeholders around an erosive runoff problem in cash crop farming areas while sensitizing them to companion modelling.

Description and specific features: The territory is represented by a watershed occupied by three or four farms. The watershed's outlet opens on a village area that potentially could become urbanized. The spatial grid is based on a pixel of 400 m^2 and enables the set of crop fields on the watershed to be visualized. The three or four farmers are played by students who must schedule their crop rotations using three types of crops (wheat, potatoes and peas) and, for peas, select a technical itinerary engendering more or less runoff (rolled or not). The cooperative fixes the purchase price of the three crops, giving a bonus to rolled pea, which is considered to be cleaner. The model calculates the transfer of runoff between a field and the adjacent field(s) situated below it according to the choice of crop and then generates a volume of runoff at the outlet. According to the volume recorded and the damage caused to the potential urban zone, the mayor of the village will ask for an explanation from the watershed technician who will then identify the probable causes of the problem.

The role-playing game takes place in two stages: acquisition of knowledge about the system and scheduling of crop rotation in an individual way over 3 years, then adjustments of decisions after consulting and negotiating among players to propose an alternative scenario. During the second stage, players can implement technical (e.g. grassy borders or water retention basins) or economic solutions (e.g. tax on negative practices, financial compensation for positive practices).

The exercise can begin with the co-construction of a conceptual model representing how students perceive the issue raised using the knowledge they earned through the Master's programme. A tutorial (Étienne and Souchère 2010) provides step by step guidance to teachers wishing to carry out the entire exercise. It specifies the instructions to give to students, provides the guidelines to help workshop facilitators, and gives the solutions to exercises.

In 2007, this game was adapted to raise awareness among actual local stakeholders (i.e. elected officials, farmers and animators of watershed syndicates) regarding the problem of erosive runoff and to help them to devise together collective management alternatives to this problem on the watershed (CauxOperation game).

Application: The game is played in groups of six to seven students, with a teacher able to animate up to three groups playing simultaneously. It has been used since 2006 with students in the AgroParisTech Master's programme and those in advanced courses on agriculture and environment at ESITPA in Rouen.

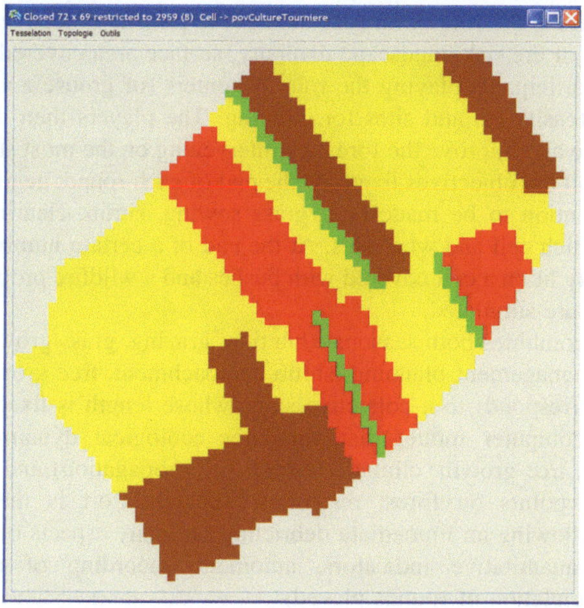

RuisselPois (Pays de Caux): on a simplistic catchment, the students discover the process of erosive runoff and the associated interactions between farmers, catchment agents, cooperative stores and mayors.

SylvoPast

A teaching game created by M. Étienne (INRA Avignon).

The game focuses on the multi-use of a Mediterranean forest and potential conflicts between forest rangers, livestock farmers and hunters, and is based on research on silvopastoral management in Mediterranean forests (Étienne 2003 2006).

Motivation: This game initially was created for silvopastoral training programmes destined for ONF agents. It aimed to help the agents understand the diversity of users' viewpoints on the forest and negotiate with users during the implementation of silvopastoral management plans linked to fire prevention in Mediterranean forests. It was then adapted for a broad student audience.

Description and specific features: The territory consists of a virtual forest that is represented by a spatial grid of 100 cells. Its structure is described according to the combination of three layers of dominant vegetation (i.e. tree, shrub and grass). The teacher may choose between three types of forest stands, that is, cork oak forest over heathland, Aleppo pine forest over kermes oak shrubland, dawny oak coppice over broom shrubland. The participant playing the role of livestock farmer defines the flock's grazing circuit and schedule that will best satisfy the flock requirements. They can increase the flock size and have to pay a grazing duty. The participant playing the role of forester defines the priority objectives for the forest (i.e. reduction of fire risk, landscape diversity, surface area covered by productive forest). The participants playing the role of hunters (of grouse and/or wild boar) choose the intensity of, and sites for, hunting. The players then decide together how to manage and improve the forest by intervening on the most strategic cells in order to best fill the objectives fixed. At the end of each round, negotiation focuses on the intervention to be made (e.g. grass sowing, shrub, clearing, afforesting, felling), on which cell and who pays. At the end of a certain number of rounds, a fire is randomly lit on a cell covered with bushes and a wildfire propagation and its consequences are simulated.

The game combines both seasonal (hunting, grazing, grass growth) and annual time scales (management planning, shrub encroachment, tree growth). The latter necessarily corresponds to a collective stage whose length is fixed. The game is based on a computer model that simulates ecological dynamics (i.e. shrub encroachment, tree growth, climatic hazard, fire propagation) and offers a broad range of viewpoints on forest resources. Special effort is made to include mechanisms allowing an immediate debriefing on many aspects of the game (i.e. dynamic of quantitative indicators, automatic recording of technical acts, accelerated simulation of rounds played).

This game was adapted to forest-fire prevention at the interface between urban and wild zones in a context where farming activities are decreasing (NîmetPasLeFeu game).

Application: The livestock farmer-forester version of the game has been used in 32 sessions with actual stakeholders since January 2001, complemented by sessions undertaken with agronomy students, forestry agents, animal husbandry students, zootechnicians, ecologists and geographers. The livestock farmer-forester-hunter version has been operational since September 2003 and has been used over 80 times with students at École Nationale du Génie Rural, des Eaux et des Forêts (ENGREF), Institute of Tropical Animal Husbandry and Veterinary Medicine (Elevage et Médecine vétérinaire tropicale—EMVT), University of La Réunion, as well as with ONF practitioners and research scientists.

SylvoPast (Mediterranean forest): the students play the role of foresters, hunters or sheep farmers in order to set up a silvopastoral management plan in a virtual forest.

YeunEllez

A teaching model created by M. Étienne (INRA Avignon) and F. Bioret (University of Brest).

It is a modifiable model focused on the impact of agricultural and urban development on the conservation of natural habitats in the Monts d'Arrée region of Brittany (France).

Motivation: This model was created as a teaching tool for a course of study on the relationship between urbanization and landscape dynamics in a Master's programme at the Faculty of Geo-architecture in Brest. It works to confront knowledge acquired by students on the interactions between urban and ecological dynamics through a companion modelling exercise. The approach picks up the principles applied in the Master's programme on the management of coastal areas at the University of Brest (littoral pedagogical model), but substitutes the virtual territory (which is represented by a modifiable but a priori undefined scale map) with a real territory previously surveyed by the students.

Description and specific features: The space corresponds to a territory emblematic of Monts d'Arrée whose land use was surveyed, digitized and then rasterized by students during a preliminary session in the field and supervised laboratory work. The spatial grid is based on a pixel of one hectare and allows all the urban, agricultural and natural spaces (e.g. lake, peatbog, grassland, woodland) described using 12 land-use categories, to be visualized.

The exercise takes place with three groups of up to six students working simultaneously. It starts with the development of a conceptual model taking into account interactions between human activities and ecological dynamics on the Monts d'Arrée area mapped using elements collected by students during surveys in the preliminary field round. Each group must develop, by applying ARDI methodology, a conceptual model representing their idea of the functioning of the socio-ecological system, and must propose decision-making rules clarifying the key processes at play, such as waterflows, vegetation dynamics, curlew population dynamics and urbanization.

After validating and correcting a multi-agent model based on their conceptual model, students put forward a scenario for the development of the territory according to their sensitivity and the principal issue affecting the territory. The three scenarios developed are then simulated and collectively discussed based on viewpoints and indicators co-constructed with the students and covering conservation, economic and social aspects.

Application: This approach requires considerable preparatory work by students and their supervising professors in the field diagnosis, scientific bibliography and mastery of the GIS. It is particularly relevant when the Master's programme is made up of students of different backgrounds who will be able to confront and then share their knowledge and different points of view. This comparative approach is even more effective when the groups are constituted by bringing together students from the same discipline (e.g. biologists, developers and geographers).

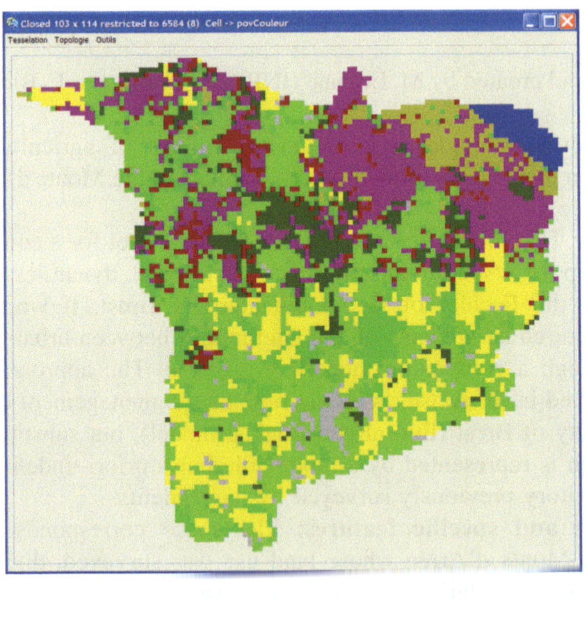

Yeun Ellez (Brittay): the students have to find out and model the interactions between human activities and nature conservation on a place of Arrée mountains where they previously realised a map of land use categories.

References

Aarts MNC (1998) Een Kwestie van Natuur; een Studie naar de aard en het Verloop van Communicatie over Natuur en Natuurbeleid. Wageningen University, Wageningen

Aarts N, Woerkum CV (2002) Dealing with uncertainty in solving complex problems. In: Leeuwis CRP (ed) Wheelbarrows full of frogs. Social Learning in Rural Resource Management. Koninklijk van Gorcum, Assen, pp 421–435

Abel N, Ross H, Herbert A, Manning M, Walker P, Wheeler H (1998) Mental models and communication in agriculture. RIRDC Publication No 98/140. Rural Industries Research and Development Corporation, Canberra, 90 pp

Abbot J, Chambers R, Dunn C, Harris T, de Merode E, Porter G, Townsend J, Weineret D (1998) Participatory GIS: opportunity or oxymoron? Particip Learn Action 33:27–34

Abrami G, Lardon S, Barreteau O, Cernesson F (2005) ORIGAMI, une méthode organisation-centrée de modélisation multi-agent de systèmes complexes. Revue Internationale de Géomatique 16(2):211–232

Abrami G, Vejpas C, Bousquet F, Trébuil G (2010) Participatory modelling and simulation of the rice seed system in Northeast Thailand. In: Coudel E, Devautour H, Toussaint Soulard C, Hubert B (eds) Innovation and sustainable development in agriculture and food (ISDA), 28 June–1 July 2010, Montpellier, France

Adger WN, Brown K, Tompkins EL (2005) The political economy of cross-scale networks in resource co-management. Ecol Soc 10(2):9

AFNOR (2001) Démarche qualité en recherche, Principes généraux et recommendations. Association française de normalisation (AFNOR), Paris, France, 8 pp

Akrich M (1989) La construction d'un système socio-technique. Esquisse pour une anthropologie des techniques. Anthropologie et Sociétés 13:31–51

Allen TFH, Starr TB (1982) Hierarchy: perspectives for ecological complexity. University of Chicago Press, Chicago 310 pp

Allen W, Kilvington M, Horn C (2002) Using participatory and learning-based approaches for environmental management to help achieve constructive behavior change. Publication No LC0102/057. Landcare Research, Lincoln, 51 pp

Amblard F, Rouchier J, Bommel P, Phan D, Varenne F (2006) Évaluation et validation de modèles multi-agents. In: Amblard F, Phan D (eds) Modélisation et simulation multi-agents, applications pour les sciences de l'homme et de la société. Hermes Science Publications, Lavoisier, Paris, pp 103–140

Ampuero R, Quiroz F, Faysse N (2006) Experiencia de apoyo a la gestión de comités de agua potable en la zona peri-urbana de Cochabamba. In: Quiroz F, Faysse N, Ampuero R (eds) Apoyo a la Gestión de Comités de Agua Potable: Experiencias de Fortalecimiento a Comités de Agua Potable Comunitarios en Bolivia y Colombia. ETREUS Impresores, Cochabamba, pp 259–282

Anselme B, Bousquet F, Lyet A, Étienne M, Fady B (2010) Modelling of spatial dynamics and biodiversity conservation on Lure mountain (France). Environ Model Softw 25(11):1385–1398

Anzieu D, Martin J-Y (1968) La dynamique des groupes restreints. Les Presses Universitaires de France (PUF), Paris, France 397 pp
Arborio A-M, Fournier P (1999) L'enquête et ses méthodes: l'observation directe. Nathan, Paris 125 pp
Argyris C, Schön DA (1978) Organizational learning: a theory of action perspective. Addison-Wesley, Boston 344 pp
Argyris C, Schön D (1996) Organizational learning II: Theory, method and practice. Addison-Wesley, Boston 305 pp
Argyris C, Schön D (2002) Apprentissage organisationnel: Théorie, méthode; pratique. De Boeck Université, Brussels 380 pp
Arnstein S (1969) A ladder of citizen participation. J Am Plan Assoc 35:216–224
Aubert S (2006) Production normative et modalités d'application des normes de gestion intégrée de la biodiversité dans un contexte de recherches interdisciplinaires. Université I Panthéon Sorbonne, Paris
Aubert S, Perez P (2007) Le guide de l'évaluateur, Projet ADD-ComMod. CIRAD, Montpellier
Aubertin C, Vivien FD (2006) Le développement durable, enjeux politiques, économiques et sociaux. No 5226. La Documentation française, Paris, 144 pp
Augustin J-P, Gillet J-C (2000) L'animation professionnelle: Histoire, acteurs, enjeux. Débats jeunesses. L'Harmattan, Paris 188 pp
Auray N (2006) Une autre façon de penser le lien entre technique et politique: les technologies de l'Internet et le réagencement de l'activité autour de l'exploration. In: Thévenot L (ed) Politiques du proche. La Découverte, Paris
Avenier MJ, Schmitt C (2007) La construction de savoirs pour l'action. L'Harmattan, Paris 245 pp
Babin D, Bertrand A, Weber J, Antona M (1999) Patrimonial mediation and management subsidiarity: managing pluralism for sustainable forestry and rural development. In: Pluralism and sustainable forestry and rural development, 9–12 Dec 1997, Food and Agriculture Organization/Global Network for Forest Science Cooperation, CIRAD-Forêt, Rome, pp 277–303
Bailey F (1969) Stratagems and spoil: a social anthropology of politics. Blackwell, London 240 p
Bakis H (1993) Les réseaux et leurs enjeux sociaux. No 2801. Que sais-je? Les Presses Universitaires de France (PUF), Paris
Balci O (1988) The implementation of four conceptual frameworks for simulation modeling in high-level languages. In: 20th Conference on winter simulation, San Diego, USA, pp 287–295
Bales RF (1950) Interaction process analysis. Addison-Wesley, Boston 203 pp
Bandura A (1977) Social learning theory. Prentice-Hall, New York 247 pp
Barbault R (1997) Biodiversité: introduction à la biologie de la conservation. Collection Les Fondamentaux. Hachette, Paris 160 p
Barnaud C (2008) Équité, jeux de pouvoir et légitimité: les dilemmes d'une gestion concertée des ressources renouvelables. Mise à l'épreuve d'une posture d'accompagnement critique dans deux systèmes agraires des hautes terres du Nord de la Thaïlande. Thèse de Doctorat en géographie humaine, économique et régionale, Géographie, Université Paris X Nanterre, Paris, France, 407 pp
Barnaud C, Bousquet F, Trébuil G (2008) Multi-agent simulations to explore rules for rural credit management in a highland farming community of northern Thailand. Ecol Econ 66(4):615–627
Barnaud C, Promburom P, Bousquet F, Trébuil G (2006) Companion modelling to facilitate collective land management by Akha villagers in upper northern Thailand. J World Assoc Soil Water Conserv 1:38–54
Barnaud C, Promburom P, Gurung T, Le Page C, Trébuil G (2006b) Companion modelling for collective learning and action in water management: lessons learnt from three case studies in northern Thailand and Bhutan. In: International symposium 'Towards Sustainable Livelihoods and Ecosystems in Mountainous Regions', 7–9 March 2006, Chiang Mai, Thailand

Barnaud C, Promburom T, Trébuil G, Bousquet F (2007) An evolving simulation and gaming process to facilitate adaptive watershed management in mountainous northern Thailand. Simul Gaming 38:398–420

Barnaud C, Trébuil G, Dumrongrojwatthana P, Marie J (2008) Area study prior to companion modelling to integrate multiple interests in upper watershed management of northern Thailand. J Southeast Asian Stud 45(4):559–585

Barnaud C, Trébuil G, Promburom P, Bousquet F (2008) La modélisation d'accompagnement pour une gestion concertée des ressources renouvelables en Thaïlande. Économie Rurale 303–305:39–59

Barnaud C, van Paassen A, Trébuil G (2006c) Power relations and participatory water management: lessons from a companion modeling experiment in a highland community of Northern Thailand. In: International forum on water and food, challenge program on water and food, Vientiane, Laos

Barnaud C, van Paassen A, Trébuil G, Promburom T, Bousquet F (2010) Dealing with power games in a companion modelling process: lessons from community water management in Thailand highlands. J Agric Educ Ext 16(1):55–74

Barreteau O (1998) Un système multi-agent pour explorer la viabilité des systèmes irrigués: dynamique des interactions et modes d'organisation. PhD thesis, École Nationale du Génie Rural, des Eaux et des Forêts, Montpellier, France, 260 pp

Barreteau O (2003) The joint use of role-playing games and models regarding negotiation processes: characterization of associations. J Artif Soc Soc Simul 6(2). http://jasss.soc.surrey.ac.uk/6/2/3.html

Barreteau O (2007) Modèles et processus de décision collective: entre compréhension et facilitation de la gestion concertée de la ressource en eau. Université Paris Dauphine/CEMAGREF, Paris 85 pp

Barreteau O, Abrami G (2007) Variable time scales, agent-based models, and role-playing games: the PIEPLUE river basin management game. Simul Gaming 38(3):364–381

Barreteau O, Abrami G, Chennit S, Garin P (2006) Support to stakeholder involvement in water management: circumventing some participation pitfalls. In: Perret S, Farolfi S, Hassan R (eds) Water governance for sustainable development. Earthscan, London, pp 275–289

Barreteau O, Bousquet F (1999) Jeux de rôles et validation de systèmes multi-agents. In: JFIADSMA'99, Ingénierie des systèmes multi-agents. Hermes Science Publications, Paris, pp 67–80

Barreteau O, Bousquet F (2000) SHADOC: a multi-agent model to tackle viability of irrigated systems. Annal Oper Res 94(1–4):139–162

Barreteau O, Bousquet F, Attonaty J-M (2001) Role-playing games for opening the black box of multi-agent systems: method and lessons of its application to Senegal River Valley irrigated systems. J Artif Soc Soc Simul 4(2). http://jasss.soc.surrey.ac.uk/4/2/5.html

Barreteau O, Bousquet F, Weber J (1997) Modes de gestion et viabilité des périmètres irrigués: questions de représentation. In: Blasco F (ed) Tendances nouvelles en modélisation pour l'environnement. Elsevier, Paris, pp 153–159

Barreteau O, Garin P, Dumontier A, Abrami G, Cernesson F (2003) Agent-based facilitation of water allocation: case study in the Drome river valley. Group Decis Negot 12(5):441–461

Becu N, Barreteau O, Perez P, Saising J, Sungted S (2005) A methodology for identifying and formalizing farmers' representations of watershed management: a case study from northern Thailand. In: Bousquet F, Trébuil G, Hardy B (eds) Companion modeling and multi-agent systems for integrated natural resource management in Asia. International Rice Research Institute, Los Baños, pp 41–62

Becu N, Bousquet F, Barreteau O, Perez P, Walker A (2003) A methodology for eliciting and modelling stakeholders' representations with agent based modelling. Lect Notes Artif Intell 2927:131–148

Becu N, Neef A, Schreinemachers P, Sangkapitux C (2008) Participatory computer simulation to support collective decision-making: potential and limits of stakeholder involvement. Land Use Policy 25(4):498–509

Berkes F, Colding J, Folke C (2002) Navigating social-ecological systems: building resilience for complexity and change. Cambridge University Press, Cambridge 393 pp

Berkes F, Folke C (1998) Linking social and ecological systems, management practices and social mechanisms for building resilience. Cambridge University Press, Cambridge 476 pp

Berkes F, George P, Preston R (1991) Co-management: the evolution of the theory and practice of joint administration of living resources. Alternatives 18(2):12–18

Bertrand G (1978) Le paysage entre la Nature et la société. Revue Géographique des Pyrénées et du Sud-Ouest 49(2):239–258

Beuret J-E (2006) La conduite de la concertation. Pour la gestion de l'environnement et le partage des ressources. L'Harmattan, Paris 342 pp

Bloeschl G, Sivapalan M (1995) Scale issues in hydrological modelling: a review. Hydrol Process 9:251–290

Boelens R (2008) Water rights arenas in the Andes: upscaling the defence networks to localize water control. Water Altern 1(1):48–65

Boiral P, Lantéri J-F, Olivier de Sardan J-P (1985) Paysans, experts et chercheurs en Afrique noire. Sciences sociales et développement rural. Karthala Éditions, Paris, 224 pp

Boissau S, Castella J-C (2003) Constructing a common representation of local institutions and land-use systems through simulation gaming and multi-agent modelling in rural areas of northern Vietnam: the Samba-week methodology. Simul Gaming 34(3):342–357

Boltanski L, Thevenot L (1991) De la Justification. Les économies de la Grandeur. Gallimard, Paris 483 pp

Bommel P, Müller J-P (2007) An introduction to UML for modelling in the human and social sciences. In: Phan D, Amblard F (eds) Agent-based modelling and simulation in the social and human sciences. The Bardwell Press, Oxford, pp 273–294

Bornstein G, Yaniv I (1998) Individual and group behavior in the ultimatum game: are groups more 'rational' players? Exp Econ 1:101–108

Bossen C (2002) The parameters of common information Spaces: the heterogeneity of cooperative work at a hospital ward. In: ACM conference on computer-supported cooperative work. New Orleans, LA, USA, pp 176–185

Botta A, Daré W, Antona M, Leclerc G (2009) Integration of multi-scale stakes in governance by applying companion modelling to land use foresight. In: Anderssen RS, Braddock RD, Newham LTH (eds) 18th International congress on modelling and simulation, July. Modelling and Simulation Society of Australia and New Zealand/International Association for Mathematics and Computers in Simulation, pp 2377–2383

Boudon R, Bourricaud F (2002) Dictionnaire critique de la sociologie. Les Presses Universitaires de France (PUF), Paris 714 pp

Boulding KE (1989) Three faces of power. SAGE Publications, Thousand Oaks

Bourdieu P (1979) La distinction: critique sociale du jugement. Le sens commun. Éditions de Minuit, Paris 670 pp

Bourdieu P (1994) Raisons pratiques. Sur la théorie de l'action. Le Seuil, Paris 254 pp

Bourgeois M (2006) NîmetPasLeFeu, un outil d'aide à la décision pour la prévention des incendies de forêt dans le milieu périurbain nîmois. Mémoire INA-PG. AgroParisTech, Paris 57 pp

Bousquet F, Bakam I, Proton H, Le Page C (1998) Cormas: common-pool resources and multi-agent systems. Lect Notes Artif Intell 1416:826–837

Bousquet F, Barreteau O, Le Page C, Mullon C, Weber J (1999) An environmental modelling approach. The use of multi agents simulations. In: Blasco F, Weill A (eds) Advances in environmental and ecological modelling. Elsevier, Paris, pp 113–122

Bousquet F, Barreteau O, Mullon C, Weber J (1996) Modélisation d'accompagnement: systemes multi-agents et gestion des ressources renouvelables. In: Quel environnement au XXIe siècle? Environnement, maîtrise du long terme et démocratie, 8–11 September, Abbaye de Fontevraud, France

Bousquet F, Le Page C (2004) Multi-agent simulations and ecosystem management: a review. Ecol Model 176(3–4):313–332

Bousquet F, Trébuil G (2005) Training on multi-agent systems, social sciences, and integrated natural resource management: lessons from an inter-university project in Thailand. In: Bousquet F, Trébuil G, Hardy B (eds) Companion modeling and multi-agent systems for integrated natural resource management in Asia. International Rice Research Institute, Los Baños, pp 309–326

Boy D (2003) L'expert citoyen, le citoyen expert. Cahiers Français 316:20–24

Brandon KE, Wells M (1992) Planning for people and parks: design dilemmas. World Dev 20(4):557–570

Breiger R, Pattison M (1978) The joint role structure of two communities elites. Sociol Methods Res 7:213–226

Brown JS, Collins A, Duguid P (1989) Situated cognition and the culture of learning. Edu Res 18(1):32–42

Brundtland GH (1988) Notre avenir à tous. Rapport de la Commission mondiale sur l'environnement et le développement. Les Éditions du Fleuve, Quebec, 454 pp

Bruner J (1960) The process of education. Harvard University Press, Cambridge 97 pp

Bulkeley H (2005) Reconfiguring environmental governance: towards a politics of scales and networks. Political Geogr 24:875–902

Burnouf J (2008) Archéologie médiévale en France. Le second Moyen Âge (XIIe-XVIe siècle). La Découverte, Paris, p 170

Caillois R (1967) Les jeux et les hommes. Gallimard, Paris

Callicott JB, Crowder LB, Mumford K (1999) Current normative concepts in conservation. Conserv Biol 13(1):22–35

Callon M (1986) Eléments pour une sociologie de la traduction. La domestication des coquilles Saint-Jacques et des marins-pêcheurs dans la baie de Saint-Brieuc. L'Année Sociologique 36:169–208

Callon M, Lascoumes P, Barthe Y (2001) Agir dans un monde incertain. Essai sur la démocratie technique. Collection La couleur des idées. Le Seuil, Paris 358 pp

Callon M, Latour B (1981) Unscrewing the big leviathan; or how actors macrostructure reality and how sociologists help them do? In: Knorr KD, Cicourel A (eds) Advances in social theory and methodology. Toward an integration of micro and macro sociologies. Routledge & Kegan Paul, London, pp 277–303

Camargo ME, Ducrot R, Jacobi P (2007) Using role-playing game for capacity building on water and land management: comparing some Brazilian experiences. Simul Gaming 38(4):472–493

Carnevale JP (2006) Creativity in the outcomes of conflict. In: Deutsch M, Coleman PT, Marcus EC (eds) Handbook of conflict resolution: theory and practice, 2nd edn. Jossey-Bass, San Francisco, pp 414–435

Cash DW, Adger W, Berkes F, Garden P, Lebel L, Olsson P, Pritchard L, Young O (2006) Scale and cross-scale dynamics: governance and information in a multilevel world. Ecol Soc 11(2). http://www.ecologyandsociety.org/vol11/iss2/art8/

Castel R (2002) La sociologie et la réponse à la demande sociale. In: A quoi sert la sociologie? In: Lahire B (ed) La Découverte Poche, Sciences humaines et sociales 184, Paris, France, pp 67–76

Castella JC, Kam SP, Quang DD, Verburg PH, Hoanh CT (2007) Combining top-down and bottom-up modelling approaches of land use/cover change to support public policies: application to sustainable management of natural resources in northern Vietnam. Land Use Policy 24(3):531–545

Cerdan O, Souchère V, Lecomte V, Couturier A, Le Bissonnais Y (2002) Incorporating soil surface crusting processes in an expert-based runoff model: STREAM (Sealing and Transfer by Runoff and Erosion related to Agricultural Management). CATENA 46:189–205

Chambers R (1983) Rural development: putting the last first. Longman, London 246 pp

Chambers R (1994) The origins and practice of participatory rural appraisal. World Dev 22(7):953–969

Chambers R (1994) Participatory rural appraisal (PRA): analysis of experience. World Dev 22(9):1253–1268

Chambers R (2006) Participatory mapping and geographic information systems: whose map? Who is empowered and who disempowered? Who gains and who loses? Electron J Inf Syst Dev Ctries 25(2):1–11

Chambers R, Pacey A, Thrupp LA (1989) Farmer first: farmer innovation and agricultural research. Intermediate Technology Publications, London 218 pp

Checkland P (1981) Systems thinking, systems practice. Wiley, Chichester 330 pp

Checkland P (1995) Model validation in soft systems practice. Syst Res 12(1):47–54

Checkland P, Scholes J (1990) Soft system methodology in action. Wiley, Chichester 329 pp

Chess C, Purcell K (1999) Public participation and the environment: do we know what works? Environ Sci Technol 33(16):2685–2692

Cibien C, Étienne M (2008) La démarche MAB: la voie de l'optimisme. Réserves de Biosphère, Notes Techniques 3. UNESCO-MAB, Paris, pp 12–15

Cicin-Sain B, Knecht RW (1998) Integrated coastal and ocean management. Concepts and practices. Island Press, Washington DC 517 pp

Claeys-Mekdade C (2001) Qu'est ce qu'une population concernée? L'exemple camarguais. Géocarrefour 76(3):217–223

Clarke SE (1995) Institutional logic's and local economic development: a comparative analysis of eight American cities. Int J Urban Reg Plan 4:513–533

Clavel L, Ducrot R, Sendacz S (2008) Gaming with eutrophication: contribution to integrating water quantity and quality management at catchment level. In: IWRA. 13e congrès mondial de l'eau. International Water Resources Association, Montpellier, 13 pp

Cohen M (2001) Gestion sylvopastorale des pins sur le Causse Méjan: une approche multi-agents du maintien des milieux ouverts. Mémoire INA-PG. AgroParisTech, Paris 52 pp

Colin L, Petit V (2008) La vidéo participative: outil d'accompagnement du développement local? Étude de trois processus de concertation en Bolivie, en Équateur et au Mali. PhD thesis, CIRAD GREEN, AgroParisTech, Paris, France, 658 pp

Collectif ComMod (2005) La modélisation comme outil d'accompagnement. Natures Sciences Sociétés 13:165–168

Collectif ComMod (2009) La posture d'accompagnement des processus de prise de décision: les références et les questions transdisciplinaires. In: Hervé D, Laloë F (eds) Modélisation de l'environnement: entre natures et sociétés. Quae Éditions, Versailles, pp 71–89

Collinson M (2000) Understanding farmers and their farming. A history of farming systems research. In: Collinson M (ed) A history of farming systems research. Food and Agriculture Organization/CABI, Wallingford, pp 1–4

ComMod Group (2003) Our companion modelling. J Artif Soc Simul 6(2). http://jasss.soc.surrey.ac.uk/6/2/1.html

Conein B, Jacopin E (1994) Action située et cognition. Sociologie du Travail 94(4):475–500

Cooke B, Kothari U (eds) (2001) Participation: the new Tyranny? Zed Books, London, 224 pp

Corcuff P, Lafaye C (1996) Légitimité et théorie critique: un autre usage du modèle de la justification publique. MANA Revue de sociologie et d'anthropologie, Volume 2, Pouvoir et légitimité, pp 217–233

Cornwall A, Jewkes R (1995) What is participatory research? Soc Sci Med 41(12):1667–1676

Costanza R, Ruth M (1998) Using dynamic modeling to scope environmental problems and build consensus. Environ Manag 22(2):183–195

Creighton JL (2005) The public participation handbook: making better decisions through citizen involvement. Jossey-Bass, San Francisco 261 pp

Curnan S, LaCava L, Sharpstee D, Lelle M, Reece M (1998) Evaluation handbook. Philosophy and expectations. W.K Kellogg Foundation, Battle Creek. http://www.wkkf.org/pubs/Tools/Evaluation/Pub770.pdf

Dagron AG (2001) Making waves: stories of participatory communication for social change. The Rockefeller Foundation, New York

Daniels SE, Walter GB (1996) Collaborative learning: improving public deliberation in ecosystem-based management. Environ Impact Assess Rev 16(71):103

d'Aquino P (2002) Accompagner une maîtrise ascendante des territoires: prémices d'une géographie de l'action territoriale. HDR, Géographie, Université de Provence, Aix en Provence, France, 342 pp

d'Aquino P (2007) Empowerment and participation: how could the wide range of social effects of participatory approaches be better elicited and compared?. CIRAD, Montpellier

d'Aquino P (2009) Les enjeux de la participation aujourd'hui: une approche explicite du contexte social. Exemple de la gestion autonome progressive. Cahiers Agriculture 18(5):433–440

d'Aquino P, Barreteau O, Étienne M, Boissau S, Aubert S, Bousquet F, Le Page C, Daré W (2002a) The role playing games in an ABM participatory modeling process: outcomes from five different experiments carried out in the last five years. In: Rizzoli AE, Jakeman AJ (eds) International Environmental Modelling and Software Society, Integrated Assessment and decision Support, 24–27 June, Lugano, Schweitzerland, pp 275–280

d'Aquino P, Le Page C, Bousquet F, Bah A (2002) A novel mediating participatory modeling: the 'self-design' process to accompany a collective decision-making. Int J Agric Resour Gov Ecol 2(1):59–74

d'Aquino P, Le Page C, Bousquet F, Bah A (2003) Using self-designed role-playing games and a multi-agent system to empower a local decision-making process for land use management: the SelfCormas experiment in Senegal. J Artif Soc Soc Simul 6(3). http://jasss.soc.surrey.ac.uk/6/3/5.html

d'Aquino P, Seck SM, Camara S (2002) Un SIG conçu par les acteurs: l'opération POAS au Sénégal. L'Espace Géographique 1:23–37

Daré W (2005) Comportements des acteurs dans le jeu et dans la réalité, indépendance ou correspondance?: Analyse sociologique de l'utilisation de jeux de rôles en aide à la concertation. PhD, Ecole nationale du Génie rural, des eaux et forêts, Paris, France, 402 pp

Daré W, Aubert S, Bah A, Botta A, Diop Gaye I, Fourage C, Lajoie G, Leclerc G (2008) Difficultés de la participation en recherche-action: retour d'expériences de modélisation d'accompagnement en appui à l'aménagement du territoire. Vertigo 8(1):22

Daré W, Barreteau O (2003) A role-playing game in irrigated system negotiation: between play and reality. J Artif Soc Soc Simul 6(3). http://jasss.soc.surrey.ac.uk/6/3/6.html

Daré W, Boutet A, Barreteau O, Ferrand N (2004) Le point de vue des sociologues sur l'usage de méthodes de modélisation d'accompagnement pour la gestion collective de l'eau. In: XVIIe congres international des sociologues de langue française. 'L'individu social: autres réalités, autre sociologie? Tours, France, pp 1–24

Daré W, Ducrot R, Botta A, Étienne M (2009) Repères méthodologiques pour la mise en œuvre d'une démarche de modélisation d'accompagnement. Cardère Éditions, Laudun 127 pp

Daré W, Fourage C, Diop Gaye I (2007) Positionnement des sociologues dans la démarche de modélisation Domino. Nouvelles Perspectives en Sciences Sociales 2(2):103–126

Dart J (1999) A story approach for monitoring change in an agricultural extension project. In: Conference of the association for qualitative research, Melbourne, Australia

David D, Payet D, Botta A, Lajoie G, Manglou S, Courdier R (2007) Un couplage de dynamiques comportementales: le modèle DS pour l'aménagement du territoire. In: 15e Journées Francophones sur les Systèmes Multi-Agents, Carcassonne, France

Davies R (1998) An evolutionary approach to facilitating organisational learning: an experiment by the Christian Commission for Development in Bangladesh. Impact Assess Project Apprais 16(3):243–250

de Bandt J, Gadret J (1998) Relations de service, marchés de services. Collection Recherche et entreprise. CNRS Éditions, Paris 360 pp

De Koning K (1995) Participatory appraisal and education for empowerment? PLA Notes 24:34–37

Dessalles JL, Müller JP, Phan D (2007) Emergence in multi-agent systems: conceptual and methodological issues. In: Phan D, Amblard F (eds) Agent-based modelling and simulation in the social and human sciences. The Bardwell Press, Oxford, pp 327–355

Dias LC, Tsoukias A (2003) On the constructive and other approaches in decision aiding. In: Antunes CA, Figueira J, Climaco J (eds) Proceedings of the 56th meeting of the EURO MCDA Working Group Hengeller. Comissão de Coordenação e Desenvolvimento Regional do Centro, Coimbra, pp 13–28

Dinar A, Farolfi S, Patrone F, Rowntree K (2008) To negotiate or to game theorize: evaluating water allocation mechanisms in the Kat Basin, South Africa. In: Dinar A, Albiac J, Sánchez-Soriano J (eds) Game theory and policy making in natural resources and the environment. Routledge, London, pp 85–111

Diop Gaye I, Daré W, Fourage C (2007) Territoires, environnement et décentralisation: la participation en question. In: Les vocations actuelles de la sociologie francophone, 18–20 April 2007, Dakar, Senegal, 12 pp

Dray A, Perez P, Jones N, Le Page C, d'Aquino P, White I, Auatabu T (2006) The AtollGame experience: from knowledge engineering to a computer-assisted role playing game. J Artif Soc Soc Simul 9(1). http://jasss.soc.surrey.ac.uk/9/1/6.html

Dray A, Perez P, Page CL, d'Aquino P, White I (2007) Who wants to terminate the game? The role of vested interests and metaplayers in the ATOLLGAME experience. Simul Gaming 38:494–511

Drogoul A, Vanbergue D, Meurisse T (2003) Simulation orientée agent: où sont les agents? In: Le statut épistémologique de la simulation. 10e journées de Rochebrune: rencontres interdisciplinaires sur les systèmes complexes naturels et artificiels. École Nationale Supérieure des Télécommunications, Rochebrune, France, pp 65–84

Dubé L (1990) *Psychologie de l'apprentissage*. Presses de l'Université du Quebec, Quebec 381 p

Dubois JL, Mathieu FR (2002) La dimension sociale du développement durable: réduction de la pauvreté ou durabilité sociale? In: Développement durable? Doctrines, pratiques, évaluations. IRD Éditions, Paris, France, pp 73–94

Duchesne S, Haegel F (2004) L'enquête et ses méthodes: l'entretien collectif. Nathan, Paris, France 128 p

Ducrot R, Barban V (2008) From revindication to proposition: capacity building in negociation about water management in perirurban areas. The 'teraguas' approach in Sao Paulo, Brazil. In: IWRA. 13^{e}congrès mondial de l'eau. International Water Resources Association, Montpellier, France, p 14

Dupont C (1994) La négociation: conduite, théories et applications. Éditions Dalloz, Paris, France, 386 p

Echeverria J (2006) Modélisation d'accompagnement et gestion des problèmes de ruissellement érosif en Haute Normandie. Mémoire de fin d'études, Master Économie du développement durable, de l'environnement et de l'energie (EDDEE), Université Paris X Nanterre, France, 100 p

Edelenbos J (2005) Institutional implication of interactive governance: insights from Dutch practice. Governance 18(1):111–134

Edmonds B, Hales D (2003) Replication, replication and replication: some hard lessons from model alignment. J Artif Soc Soc Simul 6(4) http://jasss.soc.surrey.ac.uk/6/4/11.html

Elmqvist T, Folke C, Nyström M, Peterson G, Bengtsson J, Walker B, Norberg J (2003) Response diversity, ecosystem change, and resilience. Front Ecol Environ 1(9):488–494

Étienne M (2003) SYLVOPAST a multiple target role-playing game to assess negotiation processes in silvopastoral management planning. J Artif Soc Soc Simul 6(2) http://jasss.soc.surrey.ac.uk/6/2/5.html

Étienne M (2006) La modélisation d'accompagnement: un outil de dialogue et de concertation dans les réserves de biosphère. Biodiversité et acteurs, des itinéraires de concertation. *Réserves de Biosphère,* Notes techniques 1. UNESCO-MAB, Paris, France, pp 44–52

Étienne M (2008) Contexte. In: Barreteau O, Botta A, Étienne M, Gramaglia C, Souchère V (eds) Rapport ADD. Comparaison systématique des canevas des expériences ComMod. ADD, Montpellier, France

Étienne M (2009) Co-construction d'un modèle d'accompagnement selon la méthode ARDI: guide méthodologique. Cardère Éditions, Laudun 71 p

Étienne M, Barreteau O, Daré W, Mathevet R, Bousquet F, Le Page C (2008a) Usage des jeux de rôles en modélisation d'accompagnement. INRA/CIRAD/CEMAGREF, 225 p

Étienne M, Bioret F, Brua E, Courbet F, Fady B, Kerbiriou C, Lyet A, Reboul D Accepted in revision. Modelling biodiversity dynamics in French Biosphere Reserves: researchers and managers crossed viewpoints. Environ Model Softw

Étienne M, Bourgeois M, Souchère V (2008b) Participatory modelling on fire prevention and urbanisation in southern France: from co-constructing to playing with the model. In: Sànchez-Marrè M, Béjar J, Comas J, Rizzoli A, Guariso G (eds) iEMSs 2008: 4th Biennial meeting of the international congress on environmental modelling and software: integrating sciences and information technology for environmental assessment and decision making, 6–10 July 2008, Barcelona, Spain, pp 972–979

Étienne M, Cibien C, Génot J-C (2007) Un exemple de démarche d'accompagnement pouvant être utilisée lors de l'examen périodique. Réserves de Biosphère, Notes techniques 2. UNESCO-MAB, Paris, France, pp 53–59

Étienne M, Dobremez L, Guérin G, Rapey H, Simon C (2010) Une modélisation d'accompagnement pour la gestion combinée des systèmes d'élevage et des milieux boisés sur le Causse du Larzac. Cahiers Agric 19(2):84–89

Étienne M, Du Toit D, Pollard S (2008c) ARDI: a co-construction method for participatory modelling in natural resources management. In: Sànchez-Marrè M, Béjar J, Comas J, Rizzoli A, Guariso G (eds) iEMSs 2008: 4th Biennial meeting of the international congress on environmental modelling and software: integrating sciences and information technology for environmental assessment and decision making, 6–10 July 2008, Barcelona, Spain, pp 866–873

Étienne M, Le Page C (2004) Modéliser les dynamiques paysagères pour accompagner un projet d'aménagement du territoire: le cas du Causse Méjan. In: *Actes du colloque 'Gérer les paysages de montagne pour un développement concerté et durable'*. Parc national des Cévennes, Meyrueis, France, pp 1–15

Étienne M, Le Page C, Cohen M (2003) A step-by-step approach to building land management scenarios based on multiple viewpoints on multi-agent system simulations. J Artif Soc Soc Simul 6(2) http://jasss.soc.surrey.ac.uk/6/2/2.html

Étienne M, Monteil C, Legras M (2008d) Teaching companion modelling to agronomy students: an original way to integrate concept in agriculture and environment. In: Proceedings of the 8th European IFSA symposium, Clermont-Ferrand, France

Étienne M, Souchère V (2010) Didacticiel de la séquence pédagogique RuisselPois. Collectif ComMod, 7 p

Evans K, Velarde SJ, Prieto R, Rao SN, Sertzen S, Dávila K, Cronkleton P, de Jong W (2006) Field Guide to the Future: Four Ways for Communities to Think Ahead. Center for International Forestry Research/Alternatives to Slash-and-Burn Programme/World Agroforestry Centre, Nairobi 87 p

Eversol R (2003) Managing the pitfalls of participatory development: some insight from Australia. World Dev 31(5):781–795

Ewert F, Van Keulen H, Van Ittersum M, Gillera H, Leffelaara P, Roetter R (2006) Multi-scale analysis and modelling of natural resource management options. In: Voinov A, Jakeman AJ, Rizzoli AE (eds) Proceedings of the iEMSs third biennial meeting: summit on environmental modelling and software

Farolfi S, Gumede H, Rowntree K, Jones N (2008) Local water governance in South Africa: to which extent participatory approaches facilitate multi-stakeholder negotiations? The Kat River Valley experience. In: Proceedings of the XIII World Water Congress, 1–4 Sep, Montpellier, France

Farolfi S, Müller JP, Bonte B (2010) An iterative construction of multi-agent models to represent water supply and demand dynamics at the catchment level. Environ Model Softw 25(10):1130–1148

Farolfi S, Rowntree Kate (2007) Accompanying local stakeholders in negotiation processes related to water allocation through simulation models and role-playing games: an experience from South Africa. Empowers Insights 1(2):5–7

Faysse N (2006) Troubles on the way: an analysis of the challenges faced by multistakeholder platforms. Nat Res Forum 30:219–229

Ferber J (1995) Les systèmes multi-agents: vers une intelligence collective. InterÉditions, Paris, France 522 p

Ferber J, Guérin V (2003) Représentations et simulation: de la modélisation à la mise en situation. *In: Le statut épistémologique de la simulation.* 10e journées de Rochebrune: rencontres interdisciplinaires sur les systèmes complexes naturels et artificiels. École Nationale Supérieure des Télécommunications, Rochebrune, France, pp 85–98

Ferber J, Gutknecht O (1998) A meta-model for the analysis and design of organizations in multi-agents systems. In: Demazeau Y (ed) ICMAS'98. IEEE Press, New York, pp 128–135

Fishwick PA (1998) A taxonomy for simulation modeling based on programming language principles. IIE Trans 30(9):811–820

Fleury B, Fabre M (2005) Psychanalyse de la connaissance et problématisation des pratiques pédagogiques: la longue marche vers le processus 'apprendre'. Recherche et Formation 48:75–90

Flood RL, Romm NRA (1995) Enhancing the process of methodology choice in Total Systems Intervention (TSI) and improving chances of tackling coercion. Syst Pract 8(4):377–408

Folke C, Carpenter S (2002) Resilience and sustainable development: building adaptive capacity in a world of transformations. Ambio 31(5):437–440

Folke C, Holling CS (1996) Biological diversity, ecsosystems, and the human scale. Ecol Appl 6(4):1018–1024

Folke C, Pritchard L, Berkes F, Colding J, Svedin U (2007) The problem of fit between ecosystems and institutions: ten years later. Ecol Soc 12(1) http://www.ecologyandsociety.org/vol12/iss1/art30/

Follett MP (1940) Constructive conflict. In: Metcalf HC, Urwick L (eds) Dynamic administration: the collected papers of Mary Parker Follett. Harper, New York, pp 30–49

Forman RTT, Godron G (1986) Landscape ecology. Wiley, New York 640 p

Fox J (1998) Mapping the commons: the social context of spatial information technologies. Common Property Res Dig 45:1–4

Friedberg C (1992) Représentations, classifications: comment l'homme pense ses rapports au milieu naturel. In: Jollivet M (ed) Sciences de la nature, sciences de la société. Les passeurs de frontières. CNRS Éditions, Paris, France, pp 357–371

Froger G (2006) Significations et ambiguïtés de la gouvernance dans le champ du développement durable. Monde en Développement 4(136):11–28

Funtowicz SO, Ravetz JR (1993) Science for the post-normal age. Futures 25(7):739–755

Funtowicz SO, Ravetz JR (1994) The worth of a songbird: ecological economics as a post-normal science. Ecol Econ 10(3):197–207

Gibon A, Balent G, Alard D, Ladet S, Mottet A, Julien M-P (2004) L'usage de l'espace par les exploitations d'élevage de montagne et la gestion de la biodiversité. Fourrages 178:245–263

Gibon A, Sheeren D, Monteil C, Ladet S, Balent G (2010) Modelling and simulating change in reforesting mountain landscape using a social-ecological framework. Landsc Ecol 25(2):267–285

Giddens A (1984) The constitution of society: outline of the theory of structuration. Polity, Cambridge, 402 p

Gilbert N (2008) Agent-based models, quantitative applications in the social sciences. SAGE Publications, London 98 p

Gillet J-C (1995) Animation et animateurs: le sens de l'action. Technologie de l'action sociale. L'Harmattan, Paris, France, 326 p

Gillon Y, Chaboud C, Boutrais J, Mullon C (2000) Du bon usage des ressources renouvelables. IRD Éditions, Paris, France 471 p

Gilmour A, Walkerden G, Scandol J (1999) Adaptive management of the water cycle on the urban fringe: three Australian case studies. Conserv Ecol 3(1):11

Goffman E (1979) La mise en scène de la vie quotidienne - tome 2. Les relations en public. Le sens Commun, Éditions de Minuit, Paris, France

Gonzalez R (2000) Platforms and terraces: bridging participation and GIS in joint learning for watershed management with the Ifugaos of the Philippines. PhD, ITC-Wageningen University, Wageningen, 186 p

Goupil G, Lusignan G (1993) Apprentissage et enseignement en milieu scolaire. Gaëtan Morin, Montreal, Québec, 445 p

Gourmelon F, Étienne M, Rouan M, Kerbiriou C, Charles M, Bioret F, Chlous-Ducharme F, Guermeur Y, Levrel H (2008) Éléments de prospective environnementale dans une réserve de biosphère. Cybergeo. http://www.cybergeo.eu/index20343.html

Grimble R, Wellard K (1997) Stakeholder methodologies in natural resource management: a review of principles, contexts, experiences and opportunities. Agric Syst 55(2):173–193

Grimm V (1999) Ten years of individual-based modelling in ecology: what have we learned and what could we learn in the future? Ecol Model 115:129–148

Grimm V, Railsback SF (2005) Individual-based modeling and ecology. Princeton University Press, Princeton 480 p

Guermond Y (2005) La charte: une contribution à l'utilisation de la modélisation en sciences sociales. Natures Sciences Sociétés 13(2):181–183

Guesnerie R (2003) La recherche au service du développement durable. Rapport du groupe de travail sur la recherche au service du développement durable installé par Claudie Haigneré, ministre déléguée à la Recherche et aux nouvelles Technologies, et Tokia Saïfi, secrétaire d'État au développement durable, le 16 janvier 2003, et présidé par Roger Guesnerie, France, 43 p

Guijt I (1999) Participatory monitoring and impact assessment of sustainable agriculture initiatives. International Institute for Environment and Development, London 112 p

Guillaume-Hoffnung M (2005) La médiation. Que sais-je? Les Presses Universitaires de France (PUF), Paris, France, 127 p

Gumuchian H, Grasset E, Lajarge R, Roux E (2003) Les acteurs, ces oubliés du territoire. Anthropos, Paris, France, 192 p

Gurung TR, Bousquet F, Trébuil G (2006) Companion modeling, conflict resolution, and institution building: sharing irrigation water in the Lingmuteychu Watershed, Bhutan. Ecol Soc 11(2) http://www.ecologyandsociety.org/vol11/iss2/art36/

Habermas J (1984) The theory of communicative action, vol 1. Beacon Press, Boston 300 p

Habermas J (1987) Théorie de l'agir communicationnel. Éditions Fayard, Paris, France

Hämäläinen RP, Kettunen E, Ehtamo H, Marttunen M (2001) Evaluating a framework for multi-stakeholder decision support in water resources management. Group Decis Negotiat 10:331–353

Hare AP (1994) Types of roles in small groups: a bit of history and a current perspective. Small Group Res 25(3):433–448

Hare MP, Letcher RA, Jakeman AJ (2003) Participatory modelling in natural resource management: a comparison of four case studies. Integr Assess 4(2):62–72

Hare MP, Pahl-Wostl C (2002) Stakeholder categorisation in processes of participatory integrated assessment. Integr Assess 3:50–62

Hatchuel A (2000) Quel horizon pour les sciences de gestion? Vers une théorie de l'action collective. In: David A, Hatchuel A, Laufer R (eds) Les nouvelles fondations des sciences de gestion. Éléments d'épistémologie de la recherche en management. Éditions Vuibert, Paris, France

Hénocque Y, Denis J (2001) Des outils et des hommes pour une gestion intégrée des zones côtières. Guide méthodologique, vol II. UNESCO, Paris, France, 64 p

Hervé D (2005) La modélisation d'accompagnement: un débat nécessaire et salutaire. Natures Sciences Sociétés 13(2):174–176

Hisschemöller M, Tol RSJ, Vellinga P (2001) The relevance of participatory approaches in integrated environmental assessment. Integr Assess 2:57–72

Holling CS (1978) Adaptive environmental assessment and management. Wiley, Chicester 377 p

Holling CS (2001) Understanding the complexity of economic, ecological and social systems. Ecosystems 4(5):390–405

Holmes T, Scoones I (2000) Participatory environmental policy processes. Experiences from north and south. Institute of Development Studies, Brighton 68 p

Honey P, Mumford A (1992) The manual of learning styles, 3rd revision. Peter Honey Publications, Maidenhead 94 p

Höppner C, Frick J, Buchecker M (2007) Assessing psycho-social effects of participatory landscape planning. Landsc Urban Plan 83:196–207

Hutchins E (1995) Cognition in the wild. The MIT Press, Cambridge 381 p

Innes JE (2004) Consensus building: clarifications for the critics. Plan Theor 3(1):5–20

Innes JE, Booher DE (1999) Consensus building as role playing and bricolage: toward a theory of collaborative planning. J Am Plan Assoc 65(1):9–26

Ison R, Röling N, Watson D (2007) Challenges to science and society in the sustainable management and use of water: investigating the role of social learning. Environ Sci Policy 10:499–511

Jackson MC (2000) Systems Approaches to management, vol 302. Kluwer Academic, New York, pp 1915–1916

Jackson WJ, Ingles AW (1998) Participatory techniques for community forestry. International Union for Conservation of Nature, Gland, Schweitzerland, 124 p.

Jésus F (2001) P.A.C.T. A Pro-active conciliation tool analysing stakeholders inter-relations. Monograph 41. Centre for Research and Development of Coarse Grains, Pulses, Roots and Tuber Crops (CGPRT) Centre, Bogor, Indonesia, 70 p

Jiggins J, Roling N (2000) Adaptive management: potential and limitations for ecological governance. J Agric Res, Govern Ecol 1(1):28–42

Jollivet M (ed) (1992) Sciences de la nature, sciences de la société. Les passeurs de frontières. CNRS Éditions, Paris, France 589 p

Kalaora B (2003) La concertation, outil ou art de vivre ensemble? Montagnes méditerranéennes 18:121–126

Kareiva P, Wennergren U (1995) Connecting landscape patterns to ecosystem and population processes. Nature 373:299–302

Karstens SAM, Bots PWG, Slinger JH (2007) Spatial boundary choice and the views of different actors. Environ Impact Assess Rev 27:386–407

Kikker GA, Bridges TS, Varghese A, Seager TP, Linkov I (2005) Application of multicriteria decision analysis in environmental decision making. Integr Environ Assess Manage 1(2):95–108

Kim DH (1993) The link between individual and organizational learning. MIT Sloan Manage Rev 35(1):37–50

Kinzig A, Ryan P, Étienne M, Allyson H, Elmqvist T, Walker B (2006) Resilience and regime shifts: assessing cascading effects. Ecol Soc 11(1) www.ecologyandsociety.org/vol11/iss1/art20/

Knorr-Cetina K (1981) The manufacture of knowledge: an essay on the constructivist and contextual nature of science. Pergamon, Oxford 189 p

Knorr-Cetina K (1984) The fabrication of facts: towards a micro-sociology of scientific knowledge. In: Stehr N, Meja V (eds) Society and knowledge. Contemporary perspectives in the sociology of knowledge. Transaction Books, London, pp 223–246

Koeslter A (1967) The ghost in the machine. Hutchinson, London

Kolb DA (1984) Experiential learning: experiences as source of learning and development. Prentice-Hall, Englewood Cliffs 256 p

Korfmacher K (2001) The politics of participation in watershed modelling. Environ Manage 27(2):161–176

Laganier R, Villalba B, Zuindeau B (2002) Le développement durable face au territoire: éléments pour une recherche pluridisciplinaire. Développement Durable et Territoire. Dossier 1: Approches territoriales du développement durable: http://developpementdurable.revues.org/document774.html

Laloë F (2005) A propos de la charte concernant la modélisation comme outil d'accompagnement. Natures Sciences Sociétés 13:426–427

Landry M, Malouin J-L, Oral M (1983) Model validation in operations research. Eur J Operat Res 14(3):207–220

Lane DC (1992) Modelling as learning: a consultancy methodology for enhancing learning in management teams. Eur J Operat Res 59:64–84

Langlois Y (1995) Regarde comment tu me regardes. Techniques d'animation sociale en vidéo. Éditions Charles Léopold Mayer, Paris, France

Lankford B, Watson D (2007) Metaphor in natural resource gaming: insights from the River Basin Game. Simul Gaming 38(3):421–442

Lardon S (2005) Une charte pour la modélisation d'accompagnement: pour quoi faire? Natures Sciences Sociétés 13(2):177–179

Lascoumes P (2005) Expertise et action publique. Coll. Problèmes politiques et sociaux, 912. La Documentation française, Paris, 119 p

Lasseur J, Étienne M, Legeard J-P, Bataille J-F (2010) Modélisation des relations entre dynamiques des territoires et des systèmes d'élevage dans le massif du Luberon. Cahiers Agric 19(2):90–96

Latour B (1987) Science in Action. Harvard University Press, Cambridge 274 p

Latour B (1999) Politiques de la nature: comment faire entrer les sciences en démocratie. La Découverte, Paris, France 382 p

Lave J, Wenger E (1999) Situated learning: legitimate peripheral participation. Cambridge University Press, Cambridge 138 p

Lavigne Delville P, Selamna NE, Mathieu M (2000) Les enquêtes participatives en débat. Ambition, pratiques et enjeux. Karthala Éditions, International Centre for Development Oriented Research in Agriculture, Groupe de recherche de recherche et d'échanges technologiques, Paris, France, 543 p

Le Page C, Bommel P (2005) A methodology for building agent-based simulations of common-pool resources management: from a conceptual model designed with UML to its implementation in CORMAS. In: Bousquet F, Trébuil G, Hardy B (eds) Companion modeling and multi-agent systems for integrated natural resources management in Asia. International Rice Research Institute, Los Baños, pp 327–349

Le Page C, Étienne M, Bousquet F (2001) Using dynamics spatial entities in agent-based simulations. In: Ghassemi F, Post D, Sivapalan M, Vertessy R (eds) Integrating models for natural resources management across disciplines, issues and scales, 10–13 Dec 2001, Canberra (Australia). The Modelling and Simulation Society of Australia and New Zealand Inc, Canberra, pp 1129–1134

Lebel L, Garden P, Imamura M (2006) The politics of scale, position, and place in the governance of water resources in the Mekong region. Ecol Soc 10(2) www.ecologyandsociety.org/vol10/iss2/art18/

Lecocq P (2007). Apprentissage. Encyclopedia Universalis, DVDRom

Lederman LC (1992) Debriefing: toward a systematic assessment of theory and practice. Simul Gaming 23(2):145–160

Leeuwis C (2000) Reconceptualizing participation for sustainable rural development: towards a negotiation approach. Dev Change 31:931–959

Leeuwis C (2004) Communication for rural innovation. Rethinking agricultural extension. Blackwell Science, Oxford 412 p

Legay J-M (1997) L'expérience et le modèle. Un discours sur la méthode. Sciences en question. INRA Éditions, Versailles, France

Léopold A (1949) Almanach d'un comté des sables. 1995, Aubier, Paris, 289 p

Lesage M (2007) Une expérience de médiation foncière citoyenne dans une commune rurale réunionnaise, Ilet à Cordes. Apports à la théorie de la médiation. Université de Paris Sud 11, faculté Jean Monet, École doctorale des Sciences juridiques économiques et de gestion, 488 p

Levrel H, Étienne M, Kerbiriou C, Le Page C, Rouan M (2009) Comodeling process, negotiations and power relationships: some outputs from a MAB project in the island of Ouessant. Soc Nat Res

Levy J, Lussault M (2003) Dictionnaire de la géographie et de l'espace des sociétés. Édition Belin, Paris, France 1033 p

Lewin K (1946) Action research and minority problems. J Soc Issues 2:34–46

Lewin K (1959) Psychologie dynamique: les relations humaines. Les Presses Universitaires de France (PUF), Sciences humaines, section psychologie, Paris, France, 267 p

Lima DM (2002) Éthique et politique environnementale en Amazonie contemporaine. Lusotopie 1:13–23

Linton R (1977) Le fondement culturel de la personnalité. Dunod, Paris, France 138 p

List D (2006) Action research cycles for multiple futures perspectives. Futures 38(6):673–684

Lobry C (2005) La science et l'action. Natures Sciences Sociétés 13(2):180–180

Loeber A (2003) Practical wisdom in the risk society: methods and practice of interpretive analysis on questions of sustainable development. PhD thesis, University of Amsterdam, Amsterdam, The Netherlands

Long N (1990) From paradigm lost to paradigm regained: the case for an actor oriented sociology of development. Eur Rev Lat Am Caribb Stud 49:3–24

Long N, Van Der Ploeg JD (1989) Demythologizing planned intervention: an actor perspective. Sociologia Ruralis XXIX 3–4:226–249

Ludwig D (2001) Ecology, conservation and public policy. Ann Rev Ecol Syst 32:481–517

Lunch N, Lunch C (2006) Insight into participatory video: a handbook for the field. Insight Editions, San Rafael

Lussault M (2003) Agir communicationnel. In: Lévy J, Lussault M (eds) Dictionnaire de la géographie et de l'espace des sociétés. Édition Belin, Paris, France, pp 49–50

Lynam T, de Jong W, Sheil D, Kusumanto T, Evans K (2007) A review of tools for incorporating community knowledge, preferences and values into decision making in natural resources management. Ecol Soc 12(1) http://www.ecologyandsociety.org/vol12/iss1/art5/

MacArthur J (1997) Stakeholder analysis in project planning: origins, applications and refinement of the method. Proj Apprais 12:251–265

Maccio C (2002) Guide de l'animateur de groupes. Chroniques sociales, Lyon, France 127 p

Maples MF, Webster JM (1980) Thorndike's connectionism. In Gazda GM, Corsini RJ (eds) Theories of learning. A comparative approach. F.E. Peacock, Itasca

Margolouis R, Salafsky N (1998) Measures of success: designing, managing and monitoring conservation and development projects. Island Press, Washington DC 362 p

Marshall T (1984). Reparation, conciliation and mediation. Research and Planning Unit Paper No 27, Home Office, London, UK

Mathevet R, Le Page C, Étienne M, Lefebvre G, Poulin B, Gigot G, Proréol S, Mauchamp A (2007) ButorStar: a role-playing game for collective awareness of wise reedbed use. Simul Gaming 38(2):233–262

Mathevet R, Le Page C, Étienne M, Poulin B, Lefebvre G, Cazin F, Ruffray X (2008) Des roselières et des hommes, ButorStar: un jeu de rôles pour l'aide à la gestion collective. Revue Internationale de Géomatique (Eur J GIS Spat Anal) 18:375–395

Mathevet R, Poulin B (2006) De la biologie à la géographie de la conservation. Bulletin de l'Association de Géographes Français 3:341–354

Mathieu N, Jollivet M (1989) Du rural à l'environnement, la question de la nature aujourd'hui. ARF Éditions L'Harmattan, Paris, France 354 p

Maturana HR, Varela FJ (1992) The tree of knowledge: the biological roots of human understanding. Shambhala Publications, Boston 269 p

Maurel P, Craps M, Cernesson F, Raymond R, Valkering P, Ferrand N (2007) Concepts and methods for analysing the role of information and communication tools (IC-tools) in social learning processes for river basin management. Environ Model Softw 22(5):630–639

McCracken JA, Pretty JN, Conway GR (1998) An Introduction to rapid rural appraisal for agricultural development. International Institute for Environment and Development, London 96 p

Meadowcroft J (2002) Politics and scale: some implications for environmental governance. Landsc Urban Plan 61:169–179

Meadows DL (2001) Tools for understanding the limits to growth: comparing a simulation and a game. Simul Gaming 32(4):522–536

Meizoz J (2007) Postures littéraires. Mises en scène modernes de l'auteur. Slatkine Erudition, Geneva, Schwitzerland, 204 p

Miettinen R, Virkkunen J (2005) Epistemic objects, artefacts and organizational change. Organization 12(3):437–456

Mikkelsen B (1995) Methods for development work and research: a guide for practitioners. SAGE Publications, London

Millair L (2007) CauxOpération, un outil d'aide à la concertation locale pour la gestion du ruissellement érosif en bassin versant: quels modes de concertation pour la négociation et l'émergence d'actions? Illustration avec la construction d'un Jeu de rôles. Mémoire de fin d'études, Master 'Environnement, Milieux, Techniques, Sociétés', AgroParisTech, Paris, France, 131 p

Molle F (2008) Nirvana concepts, narratives and policy models: insight from the water sector. Water Altern 1(1):131–156

Monteil C, Simon C, Ladet S, Sheeren D, Étienne M, Gibon A (2008) Participatory modelling of social and ecological dynamics in mountain landscapes subjected to spontaneous ash reforestation. In: Paegelow M, Camacho-Olmedo MT (eds) Modelling environmental dynamics. Advances in geomatic solutions. Springer, Heidelberg, pp 199–222

Morand B (2000) Le diagramme: à la périphérie ou au cœur de la cognition? In: 9e Journées de Rochebrune: rencontres interdisciplinaires sur les représentations graphiques dans les systèmes complexes naturels et artificiels. École Nationale Supérieure des Télécommunications, Rochebrune, France

Morel F (1979) Un point de vue de physiologiste expérimentateur sur les modèles. In: Delattre P, Thellier M (eds) Élaboration et justification des modèles. Applications en biologie, 9–14 Oct 1978. ENS Ulm, Paris, France, pp 31–37

Morineau J (1998) L'esprit de la médiation. Éditions Erès, collection Trajets, Ramonville Saint-Agne, France, 176 p

Mostert E, Pahl-Wostl C, Rees Y, Searle B, Tàbara B, Tippet J (2007) Social learning in European river-basin management: barriers and fostering mechanisms from 10 river basins. Ecol Soc 12(1) http://www.ecologyandsociety.org/articles/1960.html

Mottet A, Julien M, Balent G, Gibon A (2007) Agricultural land-use change and ash (Fraxinus excelsior L.) colonization in Pyrenean landscapes: an interdisciplinary case study. Environ Model Assess 12(4):293–302
Mottet A, Ladet S, Coqué N, Gibon A (2006) Agricultural land-use change and its drivers in mountain landscapes: a case study in the Pyrenees. Agric, Ecosyst Environ 114(2–4):296–310
Moulaert F, Mehmood A (2008) Analysing regional development: from territorial innovation to path dependent geography. In: Davis J, Dolfsma W (eds) The Elgar handbook of socio-economics. Edward Elgar Publishing, Cheltenham, pp 607–631
Mullon C (2005) Vers un usage concerté de la modélisation et de la simulation dans les recherches interdisciplinaires. Natures Sciences Sociétés 13(2):172–173
Naivinit W, Le Page C, Trébuil G, Gajaseni N (2010) Participatory agent-based modeling and simulation of rice production and labor migrations in Northeast Thailand. Environ Model Softw 25:1345–1358
Napoléone M, Lasseur J, Étienne M (2008) Devices based on models to accompany stakeholders in enhancing collective learning and action on livestock farming systems. In: Proceedings of IFSA symposium 2008, 6–10 July 2008, Clermont-Ferrand, France, 4 p
Newell A (1982) The knowledge level. Artif Intell 18:87–127
Nguinguiri JC (2008) Gouvernance des aires protégées: l'importance des 'normes pratiques' de régulation de la gestion locale pour la faisabilité des réformes dans le Bassin du Congo. Le bulletin de l'APAD 26 http://apad.revues.org/document3563.html
Odum HT (1983) Systems ecology: an introduction. Wiley, New York
Olivier de Sardan J-P (1995) Anthropologie et développement: essai en socio-anthropologie du changement social. Karthala Éditions, Paris, France
Olivier de Sardan J-P (2001) Les trois approches en anthropologie du développement. Revue Tiers Monde 42(168):729–754
Olivier de Sardan J-P, Paquot E (1991) D'un savoir à l'autre. Les agents de développement comme médiateurs. Groupe de recherche de recherche et d'échanges technologiques (CTA), ministère français de la Coopération, La Documentation française, Paris, 204 p
Ollagnon H (1989) Une approche patrimoniale de la qualité du milieu naturel. In: Mathieu N, Jollivet M (eds) Du rural à l'environnement, la question de la nature aujourd'hui. L'Harmattan, Paris, pp 258–268
Ostrom E (1990) Governing the commons: the evolution of institutions for collective action. Cambridge University Press, Cambridge 280 p
Ostrom E (2005) Understanding institutional diversity. Princeton University Press, Princeton 376 p
Ostrom E, Gardner R, Walker J (1994) Rules, games & common-pool resources. University of Michigan Press, Ann Arbor
Pahl-Wostl C, Hare M (2004) Processes of social learning in integrated resources management. J Commun Appl Soc Psychol 14:193–206
Pahl-Wostl C, Tàbara D, Bouwen R, Craps M, Dewulf A, Mostert E, Ridder D, Taillieu T (2008) The importance of social learning and culture for sustainable water management. Ecol Econ 64(3):484–495
Pala Ö, Vennix JAM, van Mullekom T (2003) Validity in SSM: neglected areas. J Oper Res Soc 54:706–712
Patton MQ (1990) Qualitative evaluation and research methods. SAGE Publications, Thousand Oaks
Pavé A (2005) La modélisation et la simulation des objets et processus complexes. Questions scientifiques, méthodologiques et éthiques. Natures Sciences Sociétés 13(2):169–171
Peters V, Vissers GAN (2004) A simple classification model for debriefing simulation games. Simul Gaming 35(1):70–84
Peterson GD, Cumming GS, Carpenter SR (2003) Scenario planning: a tool for conservation in an uncertain world. Conserv Biol 17(2):358–366
Piaget J, Inhelder B (1984) Psychologie de l'enfant. Coll. Que sais-je?, 369. Les Presses Universitaires de France (PUF), Paris, France, 126 p

Pickett STA, Parker VT, Fielder P (1992) The new paradigm in ecology: implications for conservation biology above the species level. In: Fielder PL, Jain SK (eds) Conservation biology: the theory and practice of nature conservation, preservation and management. Chapman and Hall, New York, pp 65–88

Pimbert M (2004) Institutionalising participation and people-centered processes in natural resource management research and publications highlights. International Institute for Environment and Development, London, Institute of Development Studies, Brighton

Pimbert MP, Pretty JN (1997) Parks, people, and professionals: putting participation into protected area management. In: Ghimire K, Pimbert MP (eds) Social change and conservation. Earthscan, London, pp 297–332

Poujol G (1994) Éducation populaire. In: Champy P, Etève C (eds) Dictionnaire encyclopédique de l'éducation et de la formation. Nathan, Paris, France, pp 356–361

Pretty J (2003) Social capital and the collective management of resources. Science 302:1912–1914

Pretty JN (1995) Participatory learning for sustainable agriculture. World Dev 23(8):1247–1263

Primack RB (1993) Essentials of conservation biology. Sinauer Associates, Sunderland 564 p

Promburom P, Bousquet F (2008) Role-playing games and institutional engagement for modeling land and water management in a northern Thailand watershed. In: Second international forum on water and food, Addis Ababa, Ethiopia, 10–14 Nov 2008

Putnam RD (1993) The prosperous community: social capital and public life. Am Prosp 4(13)

Rambaldi G, Callosa-Tarr C 2000. Manual on participatory 3-D modeling for natural resource management. National Integrated Protected Areas Programme (NIPAP), Quezon City

Ramsden P (1992) Learning to teach in higher education. Routledge, London 290 p

Raynaud D (2006) Le contexte est-il un concept légitime de l'explication sociologique? L'Année Sociologique 56(1):309–329

Reed MS (2008) Stakeholder participation for environmental management: a literature review. Biol Conserv 141(10):2417–2431

Richard-Ferroudji A (2008) L'appropriation des dispositifs de gestion locale et participative de l'eau. Accueillir et composer une pluralité de valeurs, d'objectifs et d'attachements. PhD thesis, sociologie, l'École des hautes études en sciences sociales, Paris, France

Rist S, Chidambaranathan M, Escobar C, Wiesman U, Zimmermann A (2006) Moving from sustainable management to sustainable governance of natural resources: the role of social learning processes in rural India, Bolivia and Mali. J Rural Stud 23:23–27

Rodary E., Castellanet C., Rossi G., 2003. Conservation de la nature et développement: l'intégration impossible? Coll. Économie et développement. Karthala Éditions, Paris, France, 308 p

Röling N (2002) Beyond the aggregation of individual preferences; moving from multiple to distributed cognition in resources dilemmas. In: Leeuwis C, Pyburn R (eds) Wheelbarrows full of frogs; Social learning in rural resource management. Royal van Gorcum, Assen, pp 25–66

Röling NG, Wagemakers MA (1998) A new practice: facilitating sustainable agriculture. In: Röling NG, Wagemakers MA (eds) Facilitating sustainable agriculture: participatory learning and adaptive management in times of environmental uncertainty. Cambridge University Press, Cambridge, pp 3–22

Rosenau JN (1990) Turbulence in world politics. A theory of change and continuity. Princeton University Press, Princeton, 362 p

Rosenau JN (1992) Governance without government: order and change in world politics. Cambridge University Press, New York

Rouan M, Kerbiriou C, Levrel H, Étienne M (2010) Sheep, turf and bikes: a co-modelling process of social and natural dynamics on the isle of Ouessant. Environ Model Softw 25(11):1399–1412

Rouchier J (2003) Reimplementation of a multi-agent model aimed at sustaining experimental economic research: the case of simulations with emerging speculation. J Artif Soc Soc Simul 6(4): http://jasss.soc.surrey.ac.uk/6/4/7.html

Rouwette EAJA, Vennix JAM, van Mullekorn T (2002) Group model building effectiveness: a review of assessment studies. Syst Dyn Rev 18(1):5–45

Rowlands J (1995) Empowerment examined. Dev Pract 5(2):101–107

Ruankaew N, Le Page C, Dumrongrojwatthana P, Barnaud C, Gajaseni N, van Paassen A, Trébuil G (2010) Companion modeling for integrated renewable resource management: a new collaborative approach to create common values for sustainable development. Int J Sustain Dev World Ecol 17(1):15–23

Ryan T (2000) The role of simulation gaming in policy making. Syst Res Behav Sci 17:359–364

Sadler TD, Chambers FW, Zeidler DL (2004) Student conceptualisations of the nature of science in response to a socioscientific issue. Int J Sci Educ 26(4):387–410

Salafsky N, Margoluis R, Redford KH, Robinson JG (2002) Improving the practice of conservation: a conceptual framework and research agenda for conservation science. Conserv Biol 16(6):1469–1479

Salafsky N, Wollenberg E (2000) Linking livelihoods and conservation: a conceptual framework and scale for assessing the integration of human needs and biodiversity. World Dev 28(8):1421–1438

Salomon G (1993) Distributed cognitions. Cambridge University Press, New York 275 p

Santos M (1997) Nature de l'espace, technique et temps, raison et émotion. L'Harmattan, Paris 275 p

Sateesh PV (1999) An alternative to literacy. For, Trees People Newsl 40(41):9–13

Saunders-Newton D, Scott H (2001) 'But the computer said!' Credible uses of computational modeling in public sector decision making. Soc Sci Comput Rev 19(1):47–65

Schein EH (1985) Organizational culture and leadership. Jossey-Bass, San Francisco 358 p

Schein EH (1988) Process consultation. Addison-Wesley, New York 204 p

Scoones I, Thompson J (1994) Beyond farmer first: rural people's knowledge, agricultural research and extension practice. Intermediate Technology Publications, London, 288 p

Scoones I., Thompson J., 1999. *La reconnaissance du savoir rural. Savoir des populations, recherche agricole et vulgarisation.* Karthala Éditions, Paris, France, Technical Centre for Agricultural and Rural Cooperation (CTA), Wageningen, The Netherlands

Sen AK (1982) Equality of what? In: Sen AK (ed) Choice, welfare and measurement. Blackwell, Oxford (re-edited Harvard University Press, Cambridge, MA) pp 353–369

Shakun ME (1996) Modeling and supporting task-oriented group processes: purposeful complex adaptive systems and evolutionary systems design. Group Decis Negotiat 5:305–317

Shaw JC, Robertson C (1997) Participatory video: a practical approach to using video creatively in group development work. Routledge, London

Siebenhüner B, Barth V (2005) The role of computer modelling in participatory integrated assessments. Environ Impact Assess Rev 25:367–389

Simon C (2004) Approche multi-agents pour une gestion pastorale et forestière concertée: application aux espaces boisés de la Société Civile des Terres du Larzac. Mémoire FIF-ENGREF Nancy, 73 p

Simon C, Étienne M (2010) A companion modelling approach applied to forest management planning with the Société Civile des Terres du Larzac. Environ Model Softw 25(11):1371–1384

Six JF (1998) Dynamique de la médiation. Coll. Cultures de la paix. Desclée de Brouwer, Lonrai, France, 283 p

Smouts M-C (1998) Du bon usage de la gouvernance en relations internationales. Revue Internationale des Sciences Sociales 155:85–93

Snapp S, Heong KL (2003) Scaling up and out. In: Pound B, Snapp S, McDougall C, Braun A (eds) Managing natural resources for sustainable livelihoods: uniting science and participation. Earthscan, London, pp 67–83

Sneddon C (2002) Water conflicts and river basins: the contradictions of comanagement and scale in northeast Thailand. Soc Nat Res 15(8):725–741

Snowden D (1984) Eyes see, ears hear. Food and Agriculture Organization, Rome

Souchère V, Cerdan O, Dubreuil N, Le Bissonnais Y, King C (2005) Modelling the impact of agri-environmental scenarios on runoff in a cultivated catchment (Normandy, France). Catena 61(2–3):229–240

Souchère V, Millair L, Echeverria J, Bousquet F, Le Page C, Étienne M (2010) Co-constructing with stakeholders a role-playing game to initiate collective management of erosive runoff risks at the watershed scale. Environ Model Softw 25(11):1359–1370

Soulé ME (1986) Conservation biology. The science of scarcity and diversity. Sinauer Associates, Sunderland 584 p

Star SL, Griesemer JR (1989) Institutional ecology, 'translations' and boundary objects: amateurs and professionals in Berleley's Museum of Vertebrate Zoology, 1907–1939. Soc Stud Sci 19(3):387–420

Sterk B, Carberry P, Leeuwis C, Ittersum MKV, Howden M, Meinke H, Keulen HV, Rossing WAH (2008) The interface between land use systems research and policy: multiple arrangements and leverages. Land Use Policy 26:434–442

Steyaert S, Lisoir H (2005) Participatory methods toolkit. A practitioner's manual. King Baudouin Foundation and the Flemish Institute for Science and Technology Assessment, Brussels

Suchman L (1987) Plans and situated actions: the problem of human/machine communication. Cambridge University Press, New York

Suchman L, Trigg R, Blomberg J (2002) Working artefacts: ethnomethods of the prototype. Br J Sociol 53(2):163–179

Suphanchaimart N, Wongsamun C, Panthong P (2005) Role-playing games to understand farmers' land-use decisions in the context of cash-crop price reduction in upper northeast Thailand. In: Bousquet F, Trébuil G, Hardy B (eds) Companion modeling and multi-agent systems for integrated natural resource management in Asia. International Rice Research Institure, Los Baños, pp 121–140

Swyngedouw E (2004) Globalisation or 'glocalisation'? Networks, territories and rescaling. Camb Rev Int Aff 17(1):25–48

Theys J (2002) L'approche territoriale du 'développement durable' condition d'une prise en compte de sa dimension sociale. Développement Durable et Territoire. Dossier 1: Approches territoriales du développement durable. http://developpementdurable.revues.org/document 1475.html

Torres E (2002) Adapter localement la problématique du développement durable: rationalité procédurale et démarche-qualité. Développement Durable et Territoire. Dossier 1: Approches territoriales du développement durable. http://developpementdurable.revues.org/document 878.html

Trébuil G, Baron C, Bousquet F, Shinawatra Ekasingh B (2002a) Collective creation of artificial worlds helps the governance of concrete natural resources management problems. In: International symposium on sustaining food security and managing natural resources in Southeast Asia: challenges for the 21st century, Chiang Mai, Thailand

Trébuil G, Bousquet F, Ekasingh B, Baron C, Le Page C (2005) A multi-agent model linked to a GIS to explore the relationship between crop diversification and the risk of land degradation in northern Thailand highlands. In: Bousquet F, Trébuil G, Hardy B (eds) Companion modeling and multi-agent systems for integrated natural resource management in Asia. International Rice Research Institute, Los Baños, pp 167–190

Trébuil G, Shinawatra-Ekasingh B, Bousquet F, Thong-Ngam C (2002b) Multi-agent systems companion modelling for integrated watershed management: a northern Thailand experience. In: Jianchu X, Mikesell S (eds) Landcapes of diversity. Proceedings of 3rd international conference on Montane Mainland Southeast Asia. Yunnan Science and Technology Press, Yunnan, pp 349–358

Treuil J-P, Drogoul A, Zucker J-D (2008) Modélisation et simulation à base d'agents. Dunod, Paris, France 352 p

Trom D (1999) De la réfutation de l'effet Nimby considérée comme pratique militante. Revue Française de Science Politique 49(1):31–50

Tsoukias A (2007) On the concept of decision aiding process: an operational perspective. Ann Oper Res 154:3–27

Ulrich W (2003) Beyond methodology choice: critical systems thinking as critically systemic discourse. J Oper Res Soc 54:325–342

Urwin K, Jordan A (2008) Does public policy support or undermine climate change adaptation? Exploring policy interplay across different scales of governance. Global Environ Change 18:180–191

USAID (2006) Concepts and practices of 'Résilience': a compilation from various secondary sources. US Indian Ocean Tsunami warning system (IOTWS) program, working paper prepared for the Coastal Community Resilience Program. www.adpc.net/v2007/Programs/EWS/CCRConceptsandPracticesofResilence.pdf

Vachon B (1993) Le développement local. Théorie et pratique. Gaëtan Morin, Quebec 331 p

Van den Belt M (2004) Mediated modeling: a system dynamics approach to environmental consensus building. Island Press, Washington DC 339 p

van den Hove S (2001) Approches participatives pour la gouvernance en matière de développement durable: une analyse en termes d'effets. In: Froger G (ed) Gouvernance et développement durable. Helbing & Lichtenhahn, Bâle, pp 53–89

Vega D, Peñarrieta R, Faysse N (2006) Proceso multi-actor y uso de juego de roles: concertación para manejar los impactos de la urbanización sobre infraestructura de riego. In: Experiencias del proyecto Negowat en Bolivia: Facilitando Negociaciones sobre el Acceso al Agua y Uso de la tierra en Zonas Peri-Urbanas. Negowat, Cochabamba, Bolivia, pp 161–198

Vejpas C, Bousquet F, Naivinit W, Trébuil G (2004) Participatory modelling for managing rainfed lowland rice varieties and seed system in lower northeast Thailand. In: Mekong rice conference, 15–17 Oct 2004, Ho Chi Minh City. International Rice Research Institute, Los Baños, Philippines, 15 p

Vejpas C, Bousquet F, Naivinit W, Trébuil G, Srisombat N (2005) Participatory modeling for managing rainfed lowland rice variety and seed systems in lower northeast Thailand: methodology and preliminary findings. In: Bousquet F, Trébuil G, Hardy B (eds) Companion modeling and multi-agent systems for integrated natural resources management in Asia. International Rice Research Institute, Los Baños, pp 141–163

Veneklasen L, Miller V (2002) A new weave of power, people and politics: the action guide for advocacy and citizen participation. World Neighbors, Oklahoma City 346 p

Vennix JAM (1996) Group model building, facilitating team learning using system dynamics. Wiley, Chichester

Vermeulen S (2005) Power tools: handbook to tools and resources for policy influence in natural resource management. International Institute for Environment and Development, London 68 p

Vermunt JD (1998) The regulation of constructive learning processes. Br J Educ Psychol 68:149–171

Viard J (1994) La société d'archipel ou les territoires du village global. Éditions de l'Aube, La Tour d'Aïgues, France, 126 p

Vinck D (1999) Les objets intermédiaires dans les réseaux de coopération scientifique. Contribution à la prise en compte des objets dans les dynamiques sociales. Revue Française de Sociologie XL 2:385–414

Vira B, Dubois O, Daniels S, Walker G (1998) Le pluralisme institutionnel dans le secteur forestier: les outils analytiques et opérationnels. Unasylva 194. Concilier des intérêts multiples en foresterie. Revue Internationale des Forêts et des Industries Forestières 49

Voinov A, Brown EJ (2008) Lessons for successful participatory watershed modeling: a perspective from modeling practitioners. Ecol Model 216:197–207

Water-Bayer A, Bayer W (1995) Planification avec des pasteurs: MARP et au-delà, un compte-rendu de méthodes centrées sur l'Afrique. Deutsche Gesellschaft für Technische Zusammenarbeit (GTZ), Göttingen, Germany, 186 p

Weber J (1995) Gestion des ressources renouvelables: fondements théoriques. CIRAD, Montpellier, France 21 p

Weber J (1998) Perspectives de gestion patrimoniale des ressources renouvelables. In: Lavigne Delville P (ed) Quelle politique foncière en Afrique rurale?. Karthala Éditions, ministère français de la Coopération, pp 534–552

Weber J, Reverêt J-P (1993) Biens communs, les leurres de la privatisation. Le Monde Diplomatique, Savoirs, pp 71–73

Weber M (1968) Economy and Society: an outline of interpretive sociology. Bedminster Press, New York

Weber M (1995) Économie et société. Presses Pocket, Paris, France

Webler T (1999) The craft and theory of public participation. J Risk Res 2(1):55–71

Weible C, Sabatier PA, Lubell M (2004) A comparison of a collaborative and top-down approach to the use of science in policy: establishing marine protected areas in California. Policy Stud J 32(2):187–207

Weick KE (1995) Sensemaking in organizations. SAGE Publications, London 231 p

Weick KE (2001) Making sense of the organization. Blackwell, Oxford 483 p

Wenger E (1998) Communities of practice: learning, meaning and identity. Cambridge University Press, Cambridge 318 p

Wollenberg E, Anderson J, Edmunds D (2001) Pluralism and the less powerful: accommodating multiple interests in local forest management. J Agric Resour Govern Ecol 1(3–4):199–222

World Bank Operations Evaluation Department (2004) Monitoring and evaluation: some tools, methods and approaches. World Bank, Washington DC

Worrapimphong K (2005) Companion modelling for razor clam Solen regularis conservation at Don Hoi Lord, Samut Songkhram Province Thailand. Chulalongkorn University, Bangkok

Worrapimphong K (2010) Integrated and collaborative ecological and socio-economic modelling for sustainable razor clam management at Don Hoi Lord Ramsar site, Thailand. PhD in Agricultural Technology, Chulalongkorn University, Bangkok, Thailand, 201 p

Worrapimphong K, Gajaseni N, Le Page C, Bousquet F (2010) A companion modeling approach applied to fishery management. Environ Model Softw 25(11):1334–1344

Young OR (1995) The problem of scale in human environment relationships. In: Ostrom E, Kehoane R (eds) Local commons and global interdependance: heterogeneity and cooperation in two domains. The Harvard University Center for International Affairs/SAGE Publications, Cambridge

Young O (2006) Vertical interplay among scale-dependent environmental and resource regimes. Ecol Soc 11(1):27. http://www.ecologyandsociety.org/vol11/iss1/art27/

MIX
Papier aus verantwortungsvollen Quellen
Paper from responsible sources
FSC® C105338

If you have any concerns about our products,
you can contact us on
ProductSafety@springernature.com

In case Publisher is established outside the EU,
the EU authorized representative is:
Springer Nature Customer Service Center GmbH
Europaplatz 3, 69115 Heidelberg, Germany

Printed by Libri Plureos GmbH
in Hamburg, Germany